CHICAGO PUBLIC LIBRARY
BUSINESS / SCIENCE / TECHNOLOGY
400 S. STATE ST.   60605

R01049 98002

```
QE          Korth, William W.
882
.R6         The Tertiary record
K68          of rodents in North
1994         America.
```

$85.00

The Tertiary Record of Rodents in North America

TOPICS IN GEOBIOLOGY

Series Editors: F. G. Stehli, DOSECC, Inc., Gainesville, Florida
D. S. Jones, University of Florida, Gainesville, Florida

Volume 1   SKELETAL GROWTH OF AQUATIC ORGANISMS
Biological Records of Environmental Change
Edited by Donald C. Rhoads and Richard A. Lutz

Volume 2   ANIMAL–SEDIMENT RELATIONS
The Biogenic Alteration of Sediments
Edited by Peter L. McCall and Michael J. S. Tevesz

Volume 3   BIOTIC INTERACTIONS IN RECENT AND FOSSIL BENTHIC COMMUNITIES
Edited by Michael J. S. Tevesz and Peter L. McCall

Volume 4   THE GREAT AMERICAN BIOTIC INTERCHANGE
Edited by Francis G. Stehli and S. David Webb

Volume 5   MAGNETITE BIOMINERALIZATION AND MAGNETORECEPTION IN ORGANISMS
A New Biomagnetism
Edited by Joseph L. Kirschvink, Douglas S. Jones, and Bruce J. MacFadden

Volume 6   *NAUTILUS*
The Biology and Paleobiology of a Living Fossil
Edited by W. Bruce Saunders and Neil H. Landman

Volume 7   HETEROCHRONY IN EVOLUTION
A Multidisciplinary Approach
Edited by Michael L. McKinney

Volume 8   GALÁPAGOS MARINE INVERTEBRATES
Taxonomy, Biogeography, and Evolution in Darwin's Islands
Edited by Matthew J. James

Volume 9   TAPHONOMY
Releasing the Data Locked in the Fossil Record
Edited by Peter A. Allison and Derek E. G. Briggs

Volume 10   ORIGIN AND EARLY EVOLUTION OF THE METAZOA
Edited by Jere H. Lipps and Philip W. Signor

Volume 11   ORGANIC GEOCHEMISTRY
Principles and Applications
Edited by Michael H. Engel and Stephen A. Macko

Volume 12   THE TERTIARY RECORD OF RODENTS IN NORTH AMERICA
William W. Korth

The Tertiary Record of Rodents in North America

William W. Korth
Rochester Institute of Vertebrate Paleontology
Penfield, New York

Plenum Press • New York and London

Library of Congress Cataloging-in-Publication Data

Korth, William W.
  The Tertiary record of rodents in North America / William W. Korth.
     p.   cm. -- (Topics in geobiology ; v. 12)
  Includes bibliographical references and index.
  ISBN 0-306-44696-0
  1. Rodents, Fossil--North America. 2. Paleontology--Tertiary.
3. Paleontology--North America.   I. Title. II. Series.
QE882.R6K68 1994
569'.323--dc20   94-15242
                                                                CIP

ISBN 0-306-44696-0

©1994 Plenum Press, New York
A Division of Plenum Publishing Corporation
233 Spring Street, New York, N.Y. 10013

All rights reserved

No part of this book may be reproduced, stored in a retrieval system, or transmitted in any form or by any means, electronic, mechanical, photocopying, microfilming, recording, or otherwise, without written permission from the Publisher

Printed in the United States of America

Acknowledgments

The author would like to thank the following individuals for their helpful discussions and suggestions, which became integral parts of this book: John H. Wahlert, Charles A. Repenning, Mary R. Dawson, Michael R. Voorhies, Emily Oaks, Volker Fahlbusch, and Malcolm C. McKenna. J. H. Wahlert provided the micrographs of incisor enamel. Special thanks to Mary R. Dawson, who introduced me to the wonders of rodent paleontology and continues to be my strongest supporter and critic.

Contents

I • Rodents—What, Where, and When

Chapter 1 • Introduction 3

Chapter 2 • Specialized Cranial and Dental Anatomy of Rodents 7
1. Introduction 7
2. Skull 7
3. Dentition 16
4. Postcranial Skeleton 21

Chapter 3 • Origin of Rodents 23

Chapter 4 • Classification of Rodents 27
1. Superordinal Classification 27
2. Subordinal Classification 29

II • Systematic Description of Rodent Families

Chapter 5 • Ischyromyidae 37
1. Characteristic Morphology 37
2. Evolutionary Changes in the Family 41
3. Fossil Record 44
4. Phylogeny 46
5. Problematical Taxa 49
6. Classification 51

vii

Chapter 6 • Sciuravidae ... 55

1. Characteristic Morphology ... 55
2. Evolutionary Changes in the Family ... 58
3. Fossil Record ... 59
4. Phylogeny ... 61
5. Problematical Taxa ... 64
6. Classification ... 65

Chapter 7 • Cylindrodontidae ... 67

1. Characteristic Morphology ... 67
2. Evolutionary Changes in the Family ... 70
3. Fossil Record ... 70
4. Phylogeny ... 71
5. Problematical Taxa ... 75
6. Classification ... 76

Chapter 8 • Protoptychidae ... 77

1. Characteristic Morphology ... 77
2. Evolutionary Changes in the Family ... 80
3. Fossil Record ... 81
4. Phylogeny ... 81
5. Problematical Taxon: *Presbymus* ... 83
6. Classification ... 83

Chapter 9 • Aplodontidae ... 85

1. Characteristic Morphology ... 85
2. Evolutionary Changes in the Family ... 89
3. Fossil Record ... 89
4. Phylogeny ... 91
5. Problematical Taxa ... 94
6. Classification ... 96

Chapter 10 • Mylagaulidae ... 99

1. Characteristic Morphology ... 99
2. Evolutionary Changes in the Family ... 104
3. Fossil Record ... 105
4. Phylogeny ... 105
5. Problematical Taxon: "*Mesogaulus*" *novellus* ... 108
6. Classification ... 108

Contents ix

Chapter 11 • Sciuridae 111

1. Characteristic Morphology 111
2. Evolutionary Changes in the Family 116
3. Fossil Record 117
4. Phylogeny 119
5. Problematical Taxon: ?*Protosciurus jeffersoni* 122
6. Classification 122

Chapter 12 • Eutypomyidae 125

1. Characteristic Morphology 125
2. Evolutionary Changes in the Family 127
3. Fossil Record 129
4. Phylogeny 129
5. Problematical Taxa 132
6. Classification 132

Chapter 13 • Castoridae 135

1. Characteristic Morphology 135
2. Evolutionary Changes in the Family 141
3. Fossil Record 141
4. Phylogeny 142
5. Problematical Taxa 146
6. Classification 147

Chapter 14 • Eomyidae 149

1. Characteristic Morphology 149
2. Evolutionary Changes in the Family 151
3. Fossil Record 154
4. Phylogeny 155
5. Problematical Taxa 158
6. Classification 161

Chapter 15 • Heliscomyidae 163

1. Characteristic Morphology 163
2. Evolutionary Changes in the Family 167
3. Fossil Record 167
4. Phylogeny 167
5. Problematical Taxon: *Akmaiomys incohatus* 170
6. Classification 171

Chapter 16 • Heteromyidae ... 173

1. Characteristic Morphology ... 173
2. Evolutionary Changes in the Family ... 177
3. Fossil Record ... 178
4. Phylogeny ... 179
5. Problematical Taxa ... 182
6. Classification ... 186

Chapter 17 • Florentiamyidae ... 189

1. Characteristic Morphology ... 189
2. Evolutionary Changes in the Family ... 192
3. Fossil Record ... 193
4. Phylogeny ... 193
5. Problematical Taxon: *Florentiamys agnewi* ... 196
6. Classification ... 197

Chapter 18 • Geomyidae ... 199

1. Characteristic Morphology ... 199
2. Evolutionary Changes in the Family ... 203
3. Fossil Record ... 204
4. Phylogeny ... 204
5. Problematical Taxa ... 208
6. Classification ... 210

Chapter 19 • Zapodidae ... 213

1. Characteristic Morphology ... 213
2. Evolutionary Changes in the Family ... 216
3. Fossil Record ... 216
4. Phylogeny ... 217
5. Problematical Taxa: *Plesiosminthus* and *Schaubeumys* ... 220
6. Classification ... 221

Chapter 20 • Cricetidae—Part 1 ... 223

1. Characteristic Morphology ... 223
2. Evolutionary Changes in the Family ... 227
3. Fossil Record ... 228
4. Phylogeny ... 228
5. Problematical Taxa ... 233
6. Classification ... 234

Contents

Chapter 21 • Cricetidae—Part 2 (Microtidae) 239
1. Characteristic Morphology 239
2. Evolutionary Changes in the Family 241
3. Fossil Record 241
4. Phylogeny 243
5. Problematical Taxon: *Propliophenacomys* 245
6. Classification 245

Chapter 22 • Monotypic Families 249
1. Simimyidae 249
2. Armintomyidae 252
3. Laredomyidae 253

Chapter 23 • Rodents Not Referable to Recognized Families 255
1. *Diplolophus* 256
2. *Griphomys* 256
3. *Guanajuatomys* 258
4. *Jimomys* 260
5. *Marfilomys* 262
6. *Nonomys* 263
7. *Pipestoneomys* 264
8. *Texomys* 265
9. *Zetamys* 266

III • Dynamics of the North American Rodent Fauna through Time

Chapter 24 • Changes in the Rodent Fauna 271
1. Tertiary 271
2. Pleistocene 277

Chapter 25 • Tertiary Rodent Faunas of Other Continents 279
1. Europe 279
2. Asia 282
3. Africa 283
4. South America 284

References 287

Taxonomic Index 307

The Tertiary Record of Rodents in North America

I
Rodents—What, Where, and When

Chapter 1
Introduction

Nearly half of the known species of mammals alive today (more than 1600) are rodents or "gnawing mammals" (Nowak and Paradiso, 1983). The diversity of rodents is greater than that of any other order of mammals. Thus, it is not surprising that the fossil record of this order is extensive and fossil material of rodents from the Tertiary is known from all continents except Antarctica and Australia. The purpose of this book is to compile the published knowledge on fossil rodents from North America and present it in a way that is accessible to paleontologists and mammalogists interested in evolutionary studies of rodents. The literature on fossil rodents is widely scattered between journals on paleontology and mammalogy and in-house publications of museums and universities. Currently, there is no single source that offers ready access to the literature on a specific family of rodents and its fossil history. This work is presented as a reference text that can be useful to specialists in rodents (fossil or recent) as well as mammalian paleontologists working on whole faunas.

Because the diversity of rodents in the world is essentially limitless, any monograph that included all fossil rodents would similarly be limitless. Hence, this book is limited to the record of Tertiary rodents of North America. The several species of South American (caviomorph) rodents that invaded North America near the end of the Tertiary are also not included in this text.

The first description of Tertiary rodents from North America was by Leidy (1856) from the fossiliferous "White River" badlands of Nebraska. He recognized three species: *Stenofiber nebrascensis* (a beaver), *Ischyromys typus* (primitive squirrellike rodent), and *Eumys elegans* (a field mouse). By the end of the 19th century, some 60 species of Tertiary rodents had been described from North America (Hay, 1902). However, rodents, being generally small and easily overlooked in the field, were not a major part of the fossil record in the early history of North American vertebrate paleontology simply because of collecting bias. With the advent of screen washing techniques for the collection of vertebrate microfossils beginning in the 1940s (Hibbard, 1949; McKenna, 1962), the representation of rodents in the fossil record has greatly increased and rodents are now, more often than not, the most common mammals recovered from any fossil-collecting locality. Since the time of Leidy, the described species of Tertiary rodents from North America number in the hundreds and that number is continually increasing every year as the fossil record continues to grow.

The first and only previous attempt at a review of all of the Tertiary rodents of North America was by Robert W. Wilson (1949a). Wilson's review was thorough and provided insights into the phylogeny of rodents that have been followed by virtually all later workers. Since the time of Wilson's paper, the record of fossil rodents has increased exponentially; thus, this book is an updating of Wilson's work.

As classically defined, the Tertiary begins with the Paleocene (about 66 million years ago) and ends with the Pliocene (about 1.67 million years ago).

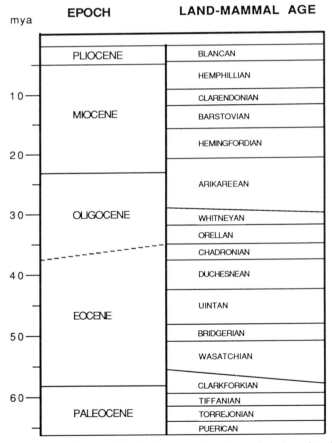

FIGURE 1.1. Land-mammal ages of North America. Boundaries based on data from Berggren et al. (1985) with some modifications suggested by Swisher and Prothero (1990).

Introduction

Because the first rodent appeared in North America in the latest Paleocene (around 58 mya), the majority of the Paleocene will not be of concern here. Since the epochs of the Tertiary were defined in Europe and based on marine macrofossils, there has always been difficulty in correlating the boundaries of these epochs with the North American continental stratigraphic sequence. Because of this, a committee of vertebrate paleontologists headed by Horace Wood, II defined provincial ages for the continental rocks of the North American Tertiary based on well-known vertebrate faunas (Wood et al., 1941). Since the time of the Wood Committee, the fossil record of Tertiary vertebrates has increased manyfold, and a revision of the original report has recently been published (Woodburne, 1987). Throughout the text presented here, the terms for the North American provincial ages (land-mammal ages) will be used in place of the traditional Tertiary epochs. The boundaries of these ages, as used here, follow the work of Berggren et al. (1985) with some minor modifications by Swisher and Prothero (1990), and the definitions of these ages follow the work presented in Woodburne (1987) by several different committees (Fig. 1.1). The major fossil-collecting localities of the North American Tertiary are also summarized in Woodburne (1987).

Chapter 2
Specialized Cranial and Dental Anatomy of Rodents

1. Introduction	7
2. Skull	7
2.1. Rostrum	7
2.2. Zygomasseteric Structure	9
2.3. Glenoid Fossa	13
2.4. Auditory Bulla	14
2.5. Mandible	14
3. Dentition	16
3.1. Dental Formula	16
3.2. Incisors	16
3.3. Primitive Cheek Tooth Morphology	18
3.4. Diastema	21
4. Postcranial Skeleton	21

1. Introduction

Rodents can be separated from all other eutherian mammals on the basis of a number of characteristics. Because the focus of this book is fossil rodents, the characters cited here are only those that are commonly preserved in fossils, that is, the skeleton and dentition. The skull bones and cranial foramina of rodents that are discussed throughout this work are diagrammed in Fig. 2.1. In the systematic chapters, reference to size of rodents as small, medium, and large is defined as follows: small, similar in size to modern field mice (*Peromyscus*) or chipmunk (*Tamias*) or smaller; medium, no larger than a Recent tree squirrel (*Sciurus*) or muskrat (*Ondatra*); and large, the largest rodents near the size of modern beaver (*Castor*) or larger.

2. Skull

2.1. Rostrum

The rostrum in rodents is elongated by enlargement of the premaxillary bones. The incisive foramina are also enlarged and are more posteriorly placed. The nasals are elongated and flared anteriorly. The infraorbital canal

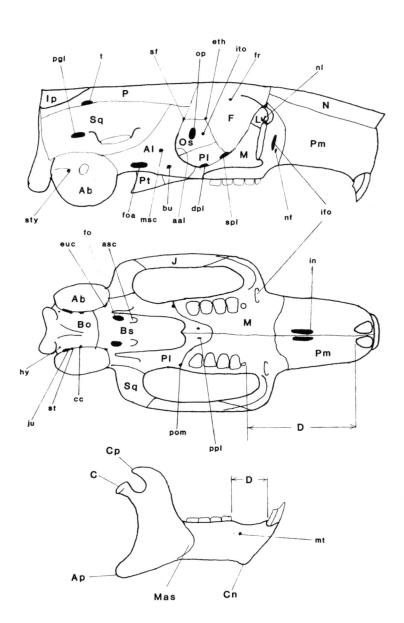

is greatly shortened, and the infraorbital foramen is primitively situated ventral to the orbit.

2.2. Zygomasseteric Structure

2.2.1. Musculature

In order for rodents to enhance their peculiar form of mastication, the masseter muscle is enlarged, differentiated, and extended anteriorly onto the rostrum to increase the mechanical advantage. Even in the most primitive rodents the anterior root of the zygomatic arch is more anterior than in primitive mammals where the zygoma originates posterior to the tooth row. The masseter muscle, which primitively originates on the ventral surface of the zygoma and inserts on the lateral side of the ascending ramus of the mandible, is differentiated into three major parts in rodents (Fig. 2.2). The masseter superficialis, the most superficial part, has its origin limited to the anterior root of the zygoma and inserts along the base of the ascending ramus, at the angle of the mandible. Deep to this is the masseter lateralis, which has an origin that extends much more posteriorly on the jugal but has an insertion similar to that of the m. superficialis. The deepest division is the masseter medialis, primitively the smallest part of the masseter. The m. medialis originates just posterior to the origin of the m. superficialis as does the m. lateralis but does not extend posteriorly to more than the middle of the zygoma.

Expansion onto the rostrum of the various parts of the masseter muscle is accomplished by several different methods. Four arrangements of zygomasseteric structure are recognized: (1) protrogomorphy, (2) sciuromorphy, (3) hystricomorphy, and (4) myomorphy.

FIGURE 2.1. Schematic drawing of generalized rodent skull and mandible, showing bones, cranial foramina, and landmarks of mandible. (After Wahlert, 1974, Fig. 1.) (A) Lateral view of skull (zygoma removed); (B) ventral view of skull; (C) lateral view of mandible.

Abbreviations of bones and landmarks: Ab, auditory bulla; Al, alisphenoid; Ap, angular process; Bo, basioccipital; Bs, basisphenoid; C, mandibular condyle; Cn, chin process; Cp, coronoid process; D, diastema; F, frontal; J, jugal; Ip, interparietal; L, lacrimal; M, maxilla; Mas, masseteric scar (mandible); N, nasal; Os, orbitosphenoid; P, parietal; Pl, palatine; Pm, premaxilla; Pt, pterygoid; Sq, squamosal.

Abbreviations of foramina: aal, anterior alar fissure; asc, alisphenoid canal; bu, buccinator; cc, carotid canal; dpl, dorsal palatine; euc, eustachian canal; eth, ethmoid; fo, foramen ovale; foa, foramen ovale accessorius; fr, frontal; hy, hypoglossal; ifo, infraorbital; in, incisive; ito, interorbital; ju, jugal; msc, masticatory; mt, mental; nf, nutritive; nl, nasolachrymal; op, optic; pgl, postglenoid; pom, posterior maxillary; ppl, posterior palatine; sf, sphenofrontal; spl, sphenopalatine; st, stapedial; sty, stylomastoid; t, temporal.

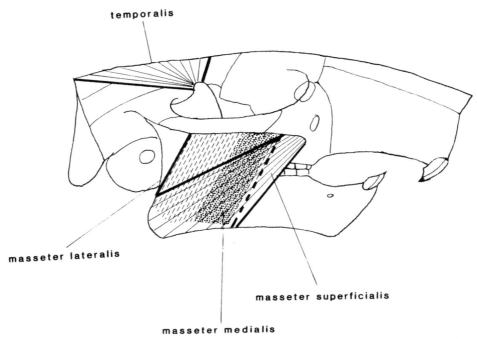

FIGURE 2.2. Schematic diagram of the primitive conditions of musculature of rodents used in mastication. Masseter superficialis (lines), most superficial; masseter medialis (stippled), deepest.

2.2.2. Protrogomorphy

In protrogomorphy (Fig. 2.3A), the primitive zygomasseteric condition for rodents (described above), the masseter is limited to the ventral surface of the zygoma, and the rostrum is unmodified.

2.2.3. Sciuromorphy

Sciuromorphy (Fig. 2.3B,C) is attained by extension of the m. lateralis onto the rostrum, and tilting and broadening of the ventral margin of the zygoma forming a zygomatic plate. The m. medialis is limited to the center of the zygoma, much the same as in protrogomorphy. The m. superficialis extends anteriorly onto the ventral side of the rostrum below the infraorbital foramen. In sciurids (squirrels) and castorids (beavers), the infraorbital foramen is smaller than in protrogomorphous rodents and has migrated ventrally and is laterally compressed by the m. lateralis. The origin of the m. superficialis is marked by a small process or swelling of bone lateral and ventral to the infraorbital foramen. In heteromyids (pocket mice), which are also fully sciuromorphous, the infraorbital foramen is at middepth of the rostrum, is

Specialized Cranial and Dental Anatomy

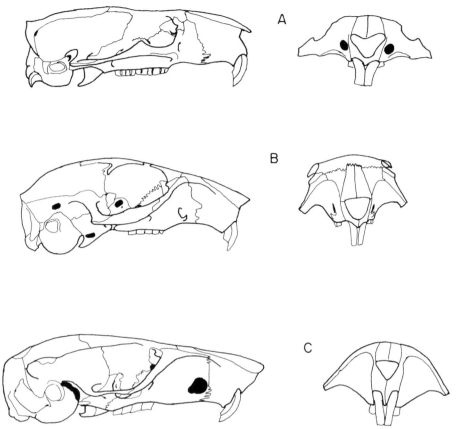

FIGURE 2.3. Sciuromorphy and protrogomorphy. Anterior and lateral views of skulls. (A) Protrogomorphous "mountain beaver" *Aplodontia rufa*; (B) sciuromorphous squirrel *Sciurus niger*; (C) sciuromorphous pocket mouse *Heteromys*. Skulls not drawn to same scale.

laterally compressed, and has migrated anteriorly onto the rostrum. The infraorbital foramen is not reduced as in sciurids or castorids, and there is no bony process for the origin of the m. superficialis.

2.2.4. Hystricomorphy

Hystricomorphy (Fig. 2.4) is characterized by enlargement of the m. medialis which invades the infraorbital foramen (also causing its enlargement) as it extends onto the rostrum. The origin of the m. superficialis is limited anteriorly on the zygoma and the m. lateralis extends nearly the entire length of the zygoma. Not all hystricomorphous skulls are identical. In dipodoid rodents (Fig. 2.4A), the infraorbital foramen is enlarged as in all

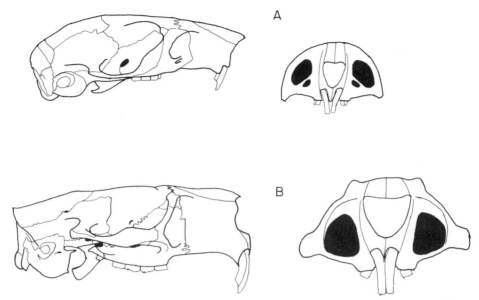

FIGURE 2.4. Hystricomorphy. Anterior and lateral views of skulls. (A) Meadow mouse *Zapus hudsonius*; (B) New World porcupine *Erethizon dorsatum*. Skulls not drawn to same scale.

other hystricomorphs, but a smaller accessory foramen is medial to the infraorbital foramen through which the blood vessels and nerves are allowed to pass, separate from the masseter. In other hystricomorphs, these nerves and vessels are contained within the infraorbital foramen along with the muscle.

2.2.5. Myomorphy

Myomorphy (Fig. 2.5) is a combination of the hystricomorphous and the sciuromorphous modifications. The m. medialis invades the infraorbital foramen and extends onto the rostrum as in hystricomorphs, the m. lateralis extends dorsally onto the rostrum, and the m. superficialis is limited to a single point ventral to the infraorbital foramen as in sciuromorphs. The infraorbital foramen is never as large as in hystricomorphs, and is usually limited to the top half of the rostrum. The expansion of the zygomatic plate to accommodate the m. lateralis differs in different genera of myomorphous rodents and laterally compresses the infraorbital foramen to differing degrees. In some rodents, the m. lateralis does not extend as far dorsally as the infraorbital foramen, thus causing no change in its shape. In others the m. lateralis has expanded dorsally to almost reach the top of the rostrum. In these cases the infraorbital foramen is laterally compressed and becomes slitlike.

Specialized Cranial and Dental Anatomy

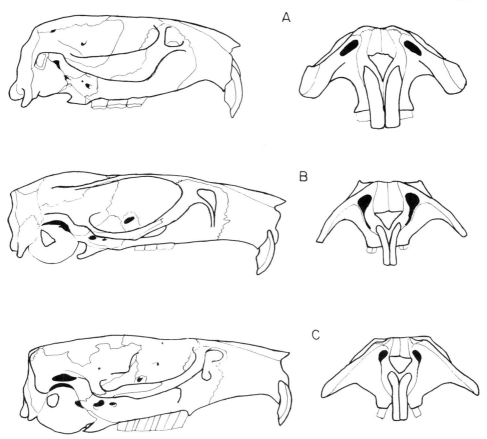

FIGURE 2.5. Myomorphy. Lateral and anterior views of skulls. (A) Bamboo rat *Rhizomys pruinosis*; (B) common rat *Rattus rattus*; (C) muskrat *Ondatra zibethica*. A and lateral view of B from Ellerman (1940). Skulls not drawn to same scale.

And, of course, there are intermediates in which only the ventral half of the infraorbital foramen is compressed by the zygomatic plate.

2.3. Glenoid Fossa

Primitively in mammals, the articulation of the mandible with the skull is a hinge structure with the fossa and condyle laterally elongate, allowing for a vertical opening and closing of the mouth. In order to isolate the use of the incisors and disengage the cheek teeth, or occlude the cheek teeth and disengage the incisors, the fossa and condyle in rodents are anteroposteriorly elongate, allowing the forward movement of the entire mandible.

2.4. Auditory Bulla

The bullae in all rodents that possess them are made entirely of the ectotympanic bone. This condition is known for only a few other groups of mammals and is used as a character for determining superordinal relationships (Chapter 4, Section 1).

2.5. Mandible

2.5.1. General Morphology

The condyle of the mandible in rodents primitively rises above the level of the tooth row. The coronoid process is small and posteriorly curving (Fig. 2.1C). The small size of the coronoid process is directly related to the general reduction in the temporalis muscle which is used mainly in vertical movement of the jaw. In association with the modification of the zygoma, the attachment of the masseter on the lateral side of the mandible (masseteric fossa) is also modified. In primitive protrogomorphous rodents, the fossa for the masseter muscle extends anteriorly to below the last two molars. The fossa for the masseter in primitive eutherian mammals does not even reach the most posterior molar. In rodents in which the masseter has extended onto the rostrum, the masseteric fossa on the mandible extends even farther anteriorly, and in many cases extends to a point anterior to the entire cheek tooth row.

2.5.2. Angle of the Jaw

2.5.2a. Displacement of the Angular Process. The orientation of the angle of the mandible relative to the horizontal ramus is commonly used in higher-level taxonomy of rodents. Two different orientations of the angular process of the mandible were first recognized by Tullberg (1899): sciurognathy and hystricognathy (Fig. 2.6). If the angle of the mandible is in the same plane as the horizontal ramus, it is referred to as sciurognathous. The other named condition occurs when the angle is no longer in the same plane as the horizontal ramus. This is called hystricognathy.

2.2.2b. Hystricognathy. Hystricognathy is often difficult to determine and has been the subject of much rodent literature because of a number of factors. The major cause of confusion (see Wilson, 1986) is that a number of authors beginning with Matthew (1910) interpreted hystricognathy as the separation of the angle of the jaw from the plane of the incisor. It is evident,

Specialized Cranial and Dental Anatomy 15

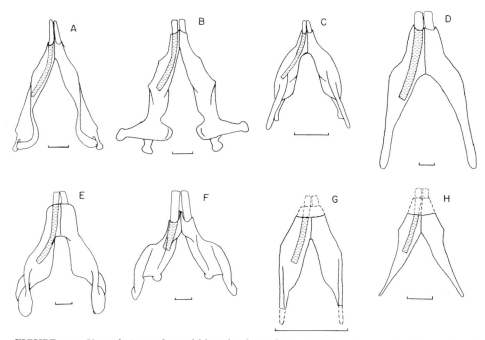

FIGURE 2.6. Ventral views of mandibles of rodents demonstrating sciurognathy (top row) and hystricognathy or "subhystricoagnathy" (bottom row). Incisor trace on right side of mandible. Dashed lines in fossil species represent reconstructed areas. Left side of mandible in G and H reflected from right side, but not preserved in fossil. Specimens not drawn to same scale. All bars = 1 cm.

(A) Recent squirrel *Sciurus niger*. (B) Recent "mountain beaver" *Aplodontia rufa*. (C) Recent North African *Ctenodactylus gundi*. (D) Uintan paramyine ischyromyid (Manitshini) *Pseudotomus petersoni*. (E) Recent New World porcupine *Erethizon dorsatum*. (F) Recent mole rat *Bathyergus suillus*. (G) Uintan "subhystricognathous" sciuravid *Prolapsus sibilatoris*. (H) Clarkfordian "subhystricoagnathous" paramyine ischyromyid (Pseudoparamyini) *Franimys amherstensis*. C and F from Ellerman (1940), G from Wood (1973), H from Wood (1962).

however, that the lower incisor in rodents forms as a spiral (or coil) and does not exist in a single plane. There seems to be a great deal of argument over the degree of separation of the angle of the mandible, and how much separation is necessary for the condition to be called hystricognathy. A number of early Tertiary rodents have been described as "incipiently hystricognathous" or "subhystricognathous" by some authors (Wood, 1962, 1975, 1984) but not by others (Dawson, 1977; Korth, 1984; Wilson, 1986). The compromise solution is that in these early, primitive rodents, the mandible was neither sciurognathous nor hystricognathous (Fig. 2.6D,G,H) but shows some ancestral condition that lent itself to modification in both directions (Landry, 1957; Korth, 1984; Wilson, 1986).

3. Dentition

3.1. Dental Formula

The primitive dental formula for eutherian mammals consists of three incisors, one canine, four or five premolars (see Novacek, 1986a, for explanation and historical review), and three molars in each quadrant of the mouth, denoted as $\frac{3143}{3143}$ (or $\frac{3153}{3153}$). The maximum number of teeth in rodents is 22. This includes one incisor, no canines, two upper premolars and one lower, and three molars ($\frac{1023}{1013}$). The two upper premolars are numbered P^3 and P^4, and the lower premolar is P_4, though their true homology may be P^4 and $P_{\bar{3}}^5$.

3.2. Incisors

3.2.1. Morphology and Number

The incisors of rodents are ever-growing, and only a single incisor occupies each of the quadrants of the skull and mandible. Traditionally (and throughout this text), these teeth are referred to as I^1 and I_1, respectively, though it has been demonstrated that the true homologies of these teeth are $I_{\bar{2}}^2$ or their deciduous equivalent (Luckett, 1985). The possession of ever-growing incisors is not unique to rodents. Other groups of modern mammals such as marsupials, lagomorphs, primates, artiodactyls, and proboscidians include species that develop ever-growing incisors. Similarly, many fossil mammals are known to have developed them as well (e.g., multituberculates, tillodonts). The incisors of rodents are unique because: (1) the enamel on the incisors is limited to the anterior surface and perhaps slightly onto the lateral and medial sides; (2) the base of these incisors extends posteriorly well into the posterior part of the mandible or into the maxilla; and (3) the enamel has a two-layered microstructure.

3.2.2. Microstructure

The microstructure of the enamel of rodents has been studied in depth for more than a century (Tomes, 1850; Korvenkontio, 1934; Wahlert, 1968, 1989; Sahni, 1985; von Koenigswald, 1985). The prisms of the outer layer of enamel (portio externa) are radially oriented in all rodents. The internal layer (portio interna) appears in three basic patterns (= Schmelzmuster): pauciserial, multiserial, and uniserial (Fig. 2.7). The recognition of these three enamel types depends on the number and orientation of the enamel prisms called Hunter–Schreger bands. In pauciserial enamel the Hunter–Schreger bands are normal to or slightly inclined with respect to the dentine–enamel boundary and vary from two to six prisms thick, often branching irregularly (Fig. 2.7A). In multiserial enamel the thickness of the bands is four to nine prisms, and they are inclined toward the apex of the tooth at a greater angle than in

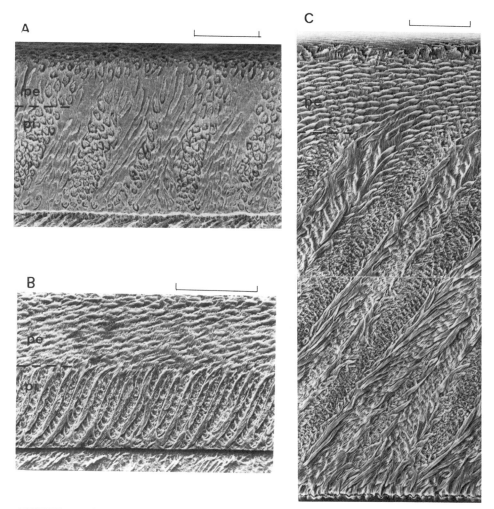

FIGURE 2.7. Scanning electron micrographs of rodent incisor enamel. (A) Pauciserial enamel, *Paramys delicatus* (Bridgerian). (B) Uniserial enamel, *Eumys* sp. (Orellan). (C) Multiserial, *Erethizon dorsatum* (Recent). Abbreviations: pe, portio externa; pi, portio interna. In all panels, anterior is to the right, the external surface is toward the top of the page. Bars = 40 μm. All micrographs from Wahlert (1989).

pauciserial enamel (Fig. 2.7C). Uniserial enamel is characterized by single-prism Hunter–Schreger bands either normal or inclined (Fig. 2.7B).

It is generally accepted that the pauciserial pattern of enamel is primitive for rodents (Wilson, 1986) because it is the most random in pattern and occurs in all early Eocene rodents. Transitions from pauciserial to uniserial enamel have been recorded in many families or superfamilies of rodents such as the

Theridomyidae, Ischyromyidae, Cylindrodontidae (see Wahlert, 1968, Table 5, for genera tested), and Dipodoidea (Dawson et al., 1990). Similarly, the earliest ctenodactyloids from Asia have pauciserial enamel, later to develop multiserial (Dawson et al., 1984). A few recent authors have proposed that the multiserial pattern of enamel is primitive because a similar pattern occurs in other mammals that develop ever-growing incisors and that the pauciserial enamel is derived from it (von Koenigswald, 1980, 1985; Wahlert, 1984a, 1989).

3.3. Primitive Cheek Tooth Morphology

Basic dental terminology for rodents is presented in Fig. 2.8. Any modifications of this terminology are presented separately in the systematic chapters for those families that require them. Interpretation of what is the primitive rodent (or protorodent) occlusal morphology of the cheek teeth is clearly dependent on the interpretation of the ancestry of the order (see Chapter 3). Following the arguments of several authors (Hartenberger, 1980; Dawson et al., 1984; Korth, 1984; Li and Ting, 1985; Luckett and Hartenberger, 1985; Flynn et al., 1986), it is accepted here that rodents were derived from the Paleocene and Eocene Asian mammals, the Eurymylidae, through a primitive ctenodactyloid-like ancestor. Based on the shared dental characters of the eurymylids and primitive ctenodactyloids, a generalized primitive dental morphotype for rodents can be determined (Fig. 2.9).

Primitively, rodent cheek teeth were brachydont and bunodont. The third upper premolar is a single-rooted tooth with a simple conical crown, much smaller than any of the other cheek teeth. The upper fourth premolar primitively consisted of one central major buccal cusp (paracone) and one lingual cusp (protocone). In all but the most primitive ctenodactyloids, a second buccal cusp (metacone) is developed, and is equal in size to the paracone. A small hypocone and conules are present. The upper molars are transversely elongate and have the basic tribosphenic pattern. They also possess at least a moderately sized hypocone and conules. The metaloph and protoloph converge on the protocone.

FIGURE 2.8. Dental terminology for cheek teeth of rodents: RP^4–M^1 (above); RP_4–M_1 (below).
Abbreviations for upper cheek teeth: ac, anterior cingulum; acn, anterocone; el, ectoloph; hcn, hypocone; mcl, metaconule; mcn, metacone; ml, metaloph; mst, mesostyle; pan, paracone; pc, posterior cingulum; pcl, protoconule; pcr, protocone crest; pl, protoloph; prn, protocone; pst, parastyle.
Abbreviations for lower cheek teeth: ac, anterior cingulum (metlophulid I); acd, anteroconid; ecd, entoconid; eld, ectolophid; hcd, hypoconid; hcld, hypoconulid; hld, hypolophid; mcd, metaconid; mld, mesolophid; msd, mesoconid; mstc, metastylid crest; mstd, mesostylid (= metastylid); mtld, metlophlid II (posterior arm of the protoconid); pc, posterior cingulum; pcd, protoconid; tal, talonid basin; tri, trigonid.

Specialized Cranial and Dental Anatomy

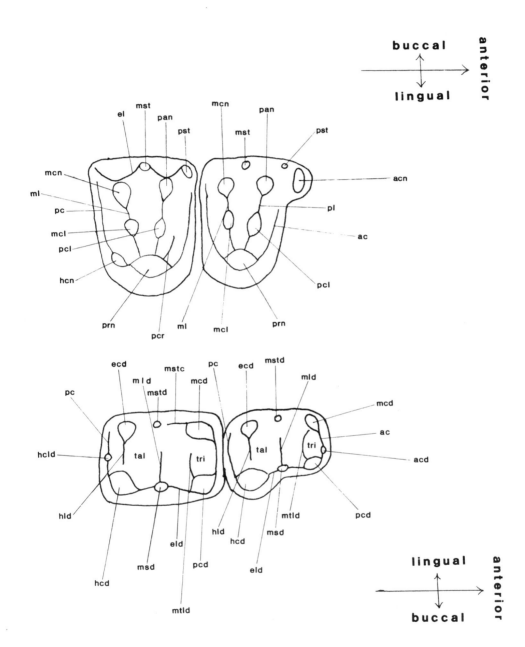

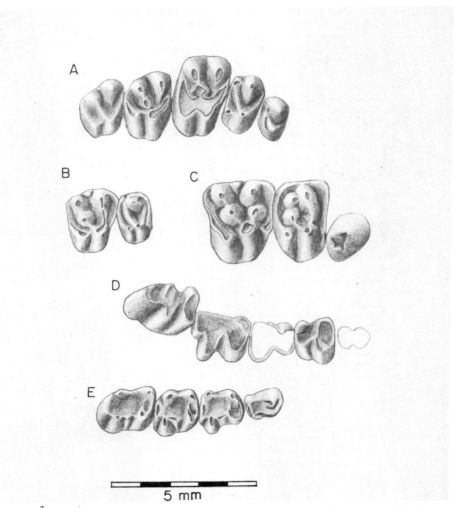

FIGURE 2.9. Comparisons between dentitions of eurymylids and early ctenodactyloids from Asia. (A, D) Late Paleocene eurymlid *Heomys orientalis*. (A) RP^3–M^3; (D) RP$_4$–M$_3$ with roots for P$_3$. (B, C, E) Early Eocene ctenodactyloids. (B) *Bumbanomys edestus*, RP^4–M^1; (C) *Chkhikvadzomys elpisma*, RP^3–M^1; (E) *Cocomys lingchaensis*, RP$_4$–M$_3$. B from Shevyreva (1989), C from Shevyreva (1984), E from Dawson et al. (1984).

The lower fourth premolar is not molariform. The trigonid is wider than the talonid and has two main cusps. The talonid is reduced with cusps not distinguishable. The lower molars lack a paraconid on the trigonid, maintaining only the metaconid and protoconid. The talonid, which is wider than the trigonid, is dominated by three subequal cusps, hypoconid, entoconid, and hypoconulid. M_3 is equal to or larger than the anterior molars in size.

The dental patterns of the cheek teeth of all other rodents discussed here are believed to have been derived ultimately from a eurymylid/primitive ctenodactyloid-like morphology.

3.4. Diastema

A long diastema between the incisor and the cheek teeth exists in all rodents. Often the diastema is longer than the tooth row. The upper diastema is always longer than the lower. This allows rodents to disengage the cheek teeth while gnawing (incisor occlusion), and likewise, to isolate the incisors during chewing (cheek tooth occlusion). This elongation of the rostrum is clearly functionally associated with the lengthening of the diastemata.

4. Postcranial Skeleton

Because of the great adaptability of rodents, the range of their postcranial skeletal specializations is remarkable. Many rodents have specialized skeletons for burrowing (fossorial) habits (e.g., palaeocastorine beavers, mylagaulids, gophers, prairie dogs, and several Old World families) while other groups are specialized for aquatic habitats (e.g., castorine beavers, muskrats), gliding (e.g., "flying" squirrels), arboreal habitats (e.g., tree squirrels, Old World dormice), increased jumping ability (e.g., kangaroo rats, Old World jerboas), and various others. The postcranial specializations are rarely limited to specific families or superfamilial groups of rodents; in nearly every case, very distantly related families have achieved the same adaptations convergently or in parallel.

Little work has been done to determine a suite of postcranial skeletal characters that defines the order Rodentia. The most recent work on primitive rodent postcranials as a tool for determining superordinal relationships was done by Szalay (1977, 1985). He noted a few characters of the ankle joint that characterized what he called the protorodent condition and used these to relate rodents to some primitive "insectivore" groups (see Chapters 3 and 4). Because the tarsal elements of rodents are poorly represented in the fossil record, and Szalay's conclusions have not been fully verified (see Luckett and Hartenberger, 1985, for discussion), little can be said about a primitive morphotype for the rodents other than that it follows the general scheme of eutherian mammals.

Chapter 3
Origin of Rodents

Osborn (1902) viewed the origin of rodents from the primitive Paleocene insectivore family the Mixodectidae for which he proposed the term Proglires. Mixodectids have since been shown to be most closely related to other insectivores (notably the leptictids and tupaiids) and have nothing to do with rodents (Szalay, 1969, pp. 237–239). Other early workers suggested the Mesozoic and early Tertiary Multituberculata as the ancestors of rodents, noting the enlarged lower incisor and diastema in many families of this Mesozoic and early Tertiary group (Forsyth Major, 1983; Hinton, 1926; Friant, 1932). The multituberculates are, however, specially derived prototherians and cannot be ancestral to any eutherian group. The condition of the incisor and diastema in the multituberculates is only convergent with that of rodents.

Wood (1962) suggested that the primitive dentition and skull of rodents most closely resembled that of the Paleocene primate *Plesiadapis*, hence, the ancestor of the rodents could be found within the order Primates. This view was briefly noted by some other authors, but never discussed in depth (McKenna, 1961; Van Valen, 1966; Lillegraven, 1969). Wood (1962) suggested that the reduction in the number of premolars, enlarged central incisors, long diastema, primitive bunodont tribosphenic upper molars, and reduction of the paraconid and trigonid on the lower molars in *Plesiadapis* could easily be modified into the morphology of the earliest known rodents. *Plesiadapis* (or a *Plesiadapis*-like primate) was also ideal temporally as an ancestor for the rodents, *Plesiadapis* being from the middle and late Paleocene of North America and the earliest true rodent known from the very latest Paleocene of North America. Wood (1962, Fig. 91) reconstructed his hypothetical ancestral rodent based on this presumption and visualized this animal as having upper molars with reduced conules and no hypocone, and lower molars lacking a hypoconulid with an elevated trigonid. The last lower premolar had a trigonid narrower than the talonid, the former consisting of a single conical cusp. The upper fourth premolar was two-cusped with one buccal and one lingual cusp. All but the last of these characters are no longer considered primitive for rodents (see Chapter 2, Section 1).

In his review of Paleocene plesiadapids, Gingerich (1976, p. 86) considered the similarities between *Plesiadapis* and early rodents as convergence owing to similar feeding habits. Few authors have accepted Wood's hypothesis of the primate origin of rodents (see Hartenberger, 1985, for literature review).

Szalay (1985) suggested that rodents were directly derived from the fossil insectivore family Leptictidae based on features of the ankle bones. However, the similarity of the postcranials of rodents and leptictids has been viewed as convergence because of similarity in the method of locomotion (Novacek, 1977, 1980). Also, Szalay's conclusions have been questioned because only one small part of the skeleton was used for study (Luckett and Hartenberger, 1985).

More recently than Wood's work, numerous authors have suggested that the early Tertiary Eurymyloidea of Asia are more likely ancestral to the rodents (Hartenberger, 1980; Dawson et al., 1984; Korth, 1984; Li and Ting, 1985; Luckett and Hartenberger, 1985; Li et al., 1987; Dashzeveg and Russell, 1988). This conclusion has resulted from the recovery of early Tertiary eurymyloids from Asia, not known at the time of Wood's work (Sych, 1971; Li, 1977; Zhai, 1978; Dashzeveg and Russell, 1988). These animals have many of the gliriform adaptations of rodents: (1) continuously growing incisors with enamel limited to the anterior and lateral surfaces; (2) reduced number of premolars; (3) elongate premaxillae; (4) upper and lower diastemata of different lengths; (5) auditory bulla entirely made of the ectotympanic bone; and (6) anterior position of the anterior root of the zygomatic arch (see also Li et al., 1987). Sych (1971) introduced the order Mixodontia to include the eurymyloids. Recently, two families have been recognized in this order (or superfamily): the Eurymylidae and the Mimotonidae (Li and Ting, 1985; Dashzeveg and Russell, 1988). The Eurymylidae contain eight genera from the late Paleocene to the late Eocene of Asia. Their dental formula is $\frac{1023}{1023}$, possessing only one additional lower premolar than is primitive for rodents. The upper third premolar is bicuspid rather than the single conical cusp found in rodents. Among the eurymylids, two genera *Eurymylus* and *Hoemys* have at least the beginnings of the two-layered enamel of the incisors characteristic of rodents (Flynn et al., 1987, p. 153).

The best evidence for relationship of the eurymylids with rodents comes from the great similarity of the early Eocene ctenodactyloids, the Cocomyidae (or Tamquammyidae; Li et al., 1979; Dawson et al., 1984), and the late Paleocene eurymyloid *Heomys* (Li, 1977). Both have molars with large conules and hypocone, and lower molars with large hypoconulids and isolated entoconids (Fig. 2.9). Hartenberger (1980) argued that *Heomys* was not a rodent and was too derived to be considered the true ancestor of all later rodents but noted its striking similarity to the early rodents from Asia. Along with this similarity in tooth morphology, the reduced dental formula of eurymylids is also a convincing argument for an ancestral relationship of a *Heomys*-like eurymylid with rodents.

However, Dashzeveg (1990) recognized a new family of rodents, the Alagomyidae from the early Eocene of Asia, that had similar lower cheek teeth and upper premolar to that of the earliest ctenodactyloids, but lacked the well-developed hypocone of the upper molars of eurymyloids and ctenodactyloids. This led him to conclude that the most primitive rodent lacked hypocones on the upper molars (as previously proposed by Wood, 1962) and that the ctenodactyloid-like molars were derived from this simpler morphology beginning sometime in the Paleocene. He argued that this made it impossible for the eurymyloids to be ancestral to the rodents.

The other family of eurymyloids, the Mimotonidae, while sharing many of the gliriform characters of eurymylids and rodents, is distinctive in the possession of an even greater number of teeth (two upper incisors, and possibly two lower incisors in some cases; and three premolars), unilateral hypsodonty of molars, and a single layer of enamel on the incisors. This family is generally believed to be ancestral to the lagomorphs rather than the rodents (Li and Ting, 1985; Dashzeveg and Russell, 1988) and has been included in the order Lagomorpha (Bleedfeld and McKenna, 1985; Li and Ting, 1993).

The origin of the Eurymyloidea appears to be within the "insectivore" families Zalambdalestidae (McKenna, 1975) or Pseudictopidae (Van Valen, 1966) of the late Cretaceous or earliest Tertiary of Asia, both included in the superordinal grouping the Anagalida (*sensu* McKenna, 1975). The strongest arguments for such an origin are based on dental features (Van Valen, 1966) and cranial features studied by Novacek (1986b). Li and Ting (1985) observed that a specimen referred to a late Cretaceous species of zalambdalestid from Mongolia, *Barunlestes butleri* (Kielan-Jaworowska and Trofimov, 1980), possessed an elongate lower incisor that extended posteriorly to below the molars and it also had a reduced trigonid of the lower molars, reminiscent of eurymyloids and rodents (see Li et al., 1987). Li et al. (1987) interpreted this specimen as representing a species that could have been a possible precursor to the gliriform condition in rodents and eurymylids, thus tracing the first occurrence of the Glires (rodentlike mammals) into the Cretaceous.

Chapter 4
Classification of Rodents

1. Superordinal Classification	27
1.1. History	27
1.2. Classification	29
2. Subordinal Classification	29
2.1. History	29
2.2. Classification	33

1. Superordinal Classification

1.1. History

Since the earliest classifications of mammals, rabbits and hares were considered very closely related to the rodents. Linnaeus (1758) included these groups in the same order, Glires. This arrangement was followed by nearly all of the early workers (e.g., Cuvier, 1817; Brandt, 1855). Alston (1876) recognized Linnaeus's order with two suborders, Simplicidentata (rodents) and Duplicidentata (lagomorphs). Gidley (1912) first named the order Lagomorpha separate from the Rodentia and considered the two orders as having had independent origins, with no special relationship. Simpson (1945) recognized the Rodentia and Lagomorpha as separate orders but maintained them in the cohort Glires. However, Simpson considered this association quite tentative because he, like Gidley, believed that there were no real characters that united these two groups and that the gliriform adaptations were merely convergences. Simpson's view was followed by several workers (e.g., Wilson, 1949a; Dawson, 1967) and was generally accepted for many years.

    In an attempt at the classification of all mammals, McKenna (1975) employed cladistic analysis and left the Rodentia as *incertae sedis* in his cohort Epitheria which contained all of the eutherian mammals except the edentates. He did, however, include the lagomorphs in the grandorder Anagalida (Szalay and McKenna, 1971) along with the order Macroscelidea (including macroscelidids, anagalids, pseudictopids, and zalambdalestids) based on dental homologies as separate from the rodents.

    Szalay (1977) used the tarsal elements of mammals as the basis for his analysis of mammalian classification and concluded that both rodents and lagomorphs were contained in the cohort Glires along with a new order, Leptictimorpha, that included the early Tertiary "insectivore" families Leptictidae, Pantolestidae, and Palaeoryctidae; a family of archaic mammals, the Taeniodontidae; and a probable primate family, the Microsyopidae. Szalay also included a number of orders of "insectivores" as suborders of the Lagomorpha (Anagalida, Mixodontia, Macroscelida) essentially equivalent to McKenna's (1975) magnorder Ernotheria.

Li and Yan (1979) included the Eurymylidae in the suborder Mixodontia (Sych, 1971) within the Rodentia based on the great similarity of the then recently reported late Paleocene eurymyloid *Heomys* and early rodents. Previously, the Eurymylidae had been included within Glires, but not with the rodents or lagomorphs specifically (Matthew et al., 1929) or placed in the Lagomorpha as the most primitive family (Wood, 1942). Shortly after this proposal by Li and Yan, Hartenberger (1980) argued that *Heomys* was not a rodent and therefore should not be included in the order.

A number of studies on recent mammals included in a symposium provided no definite classifications, but concluded that the concept of Glires (rodents closely related to lagomorphs) was either endorsed or not disproved (Luckett and Hartenberger, 1985). These studies included amino acid sequencing of hemoglobin (Shoshani et al., 1985), eye lens protein (de Jong, 1985), and embryological studies of the dentition and placentation (Luckett, 1985).

The most recent analysis of higher categories of mammals was reported by Novacek (1985, 1986b). Novacek studied the cranial and dental morphology of eutherian mammals. He recognized Glires as a grandorder that contained the Rodentia and Lagomorpha, emphasizing the gliriform adaptations of the dentition and skull as the strongest evidence of relationship. He also included Glires within the superorder Anagalida (sensu McKenna, 1975). Though no formal classification was proposed, Bleedfeld and McKenna (1985) parenthetically included the Eurymylidae in the Rodentia and the Mimotonidae in the Lagomorpha along with a number of earlier Asian species that lacked the gliriform adaptations of lagomorphs but had similar molar morphology (see also McKenna, 1982).

Dashzeveg and Russell (1988), in their systematic survey of Paleocene and Eocene eurymyloids, formally included the Mixodontia (including both the Eurymylidae and Mimotonidae) in the cohort Glires as a separate order along with the Rodentia and Lagomorpha. They recognized that the concept of the Mixodontia was, in part, one of grade (primitive gliriform mammals in which the ancestors for both rodents and lagomorphs could be found) but felt that ultimately this was the best taxonomic arrangement.

Li et al. (1987) proposed a classification of gliriform mammals. They recognized the cohort Glires which contained two superorders, Duplicidentata (orders Mimotonida and Lagomorpha) and Simplicidentata (orders Mixodontia and Rodentia). The mixodonts included the families Eurymylidae and Rhombomylidae (a more specialized family of eurymyloids), and the Mimotonida included the mimotonids and the late Cretaceous specimen cited by Li and Ting (1985) as having gliriform characteristics of the lower dentition. Most recently, Li and Ting (1993) demonstrated that the Mimotonidae were referable to the Lagomorpha based on new and more complete cranial material.

Currently, it is generally accepted that the mammals most closely related to the Glires appear to be the elephant shrews (Macroscelidea) from the Oligocene to Recent of Africa, and the Anagalida from the early Tertiary of Asia (see Carroll, 1987). Thus, the hierarchical arrangement should be similar to that of Novacek (1986b) with the modifications proposed by Li *et al.* (1987) and Li and Ting (1993) for Glires. The rank of each of the taxa above order is, in a sense, arbitrary and is a compromise between the rankings proposed by the previous authors (Fig. 4.1).

1.2. Classification

Cohort Anagalida Szalay and McKenna, 1971
    Superorder Glires Linnaeus, 1758
        Grandorder Duplicidentata Alston, 1876
            Order ? (unnamed Cretaceous specimen)
            Order Lagomorpha Brandt, 1855
        Grandorder Siplicidentata Alston, 1876
            Order Mixodontia Sych, 1971
            Order Rodentia Bowdich, 1821

2. Subordinal Classification

2.1. History

The recognition of rodent suborders or infraorders has been a source of controversy and discussion for the better part of the last two centuries. In his classification of rodents, Wood (1955a) presented a thorough review of all of the previous attempts at the subordinal classifications of rodents. These earlier classifications will be dealt with in less detail here.

The first to assign subordinal names to groups of rodent families was de Blainville (1816) who identified three groups of rodents: Grimpeurs (sciurids), Fouisseurs (muroids), and Marcheurs (hystricoids). Like many of the early workers, de Blainville included the lagomorphs in the Rodentia and named a fourth suborder to contain them. Since lagomorphs are not the focus of this book, this last suborder (and those of other workers who included the rabbits and hares with the rodents) will not be discussed. In the year after de Blainville's first division of the rodents, Cuvier (1817) used the criterion of the presence or absence of a clavicle to divide the rodents into two groups. He did not assign formal taxonomic names to these two groups, but those with

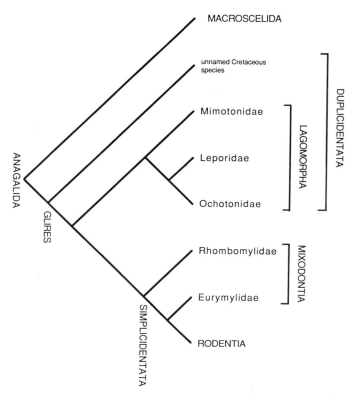

FIGURE 4.1. Cladogram of relationships of gliriform mammals. Based on that of Li et al. (1987).

clavicles were equivalent to de Blainville's Grimpeurs and Fouisseurs, and those without were the Marcheurs.

Brandt (1855), like de Blainville, recognized a tripartite division of the rodents and proposed the names Sciuromorpha, Hystricomorpha, and Myomorpha, all of which have survived into the current literature. As is evidenced by the chosen names of the suborders, the major character used to divide these groups was the zygomasseteric structure of the skull. Not surprisingly, Brandt's divisions are nearly identical to those of de Blainville with the Grimpeurs equivalent to the Sciuromorpha, the Marcheurs equal to the Hystricomorpha, and the Fouisseurs being the same as the Myomorpha.

All of these previous classifications had included only Recent rodents with little or no consideration of the fossil record which was admittedly quite sparse at that time. Alston (1876) and Schlosser (1884) were the first to include fossil rodents in a classification. They both used Brandt's three-suborder classification and simply included the fossil families in them. A fourth suborder was proposed by Zittel (1893) to include the primitive fossil forms: Protrogomorpha. This suborder was ignored by a number of later authors in the early 20th century (most notably Miller and Gidley, 1918) until it was resurrected by Wood (1937).

In his renown anatomical study of rodents, Tullberg (1899) incorporated the alignment of the mandible into his classification, along with the zygomasseteric structure. He divided the rodents into two suborders: Sciurognathi and Hystricognathi. In the former, he included the Sciuromorphi and Myomorphi (equivalent to Brandt's suborders of similar name) and subdivided the Hystricognathi into the Hystricomorphi (all known hystricomorphous rodents) and the Bathyergomorphi (African family of rodents with a protrogomorphous skull). Clearly, Tullberg felt that the alignment of the angle of the mandible was a stronger character on which to base his classification than Brandt's criteria of the zygomasseteric structure. Nearly all subsequent classifications of rodents have followed either Brandt's threefold division or Tullberg's twofold division of the order with only minor modifications. Winge (1924), the exception, recognized eight subordinal groupings of rodents but his conclusions have not been followed by any later workers.

Lavocat (1951) utilized Tullberg's classification but added a third suborder, the Atypognathes. This group was essentially the same as the Protrogomorpha, as employed by Wood (1937). Schaub (1953) abandoned the use of the zygomasseteric structure because of the probable derivation of the different types in more than one lineage (Stehlin and Schaub, 1951). Schaub believed that the dentition was the best criterion for classification. He did not reclassify the families previously referred to the Myomorpha and Sciuromorpha, but concentrated on the hystricomorph rodents. He believed that the primitive condition in these rodents was five-crested cheek teeth, from which later forms could be derived. Schaub proposed two suborders for the hystricomorphous rodents: Pentalophodonta (including the European early Tertiary family Theridomyidae, hystricoids, and castorids) and Nototrogomorpha (South American hystricomorphs).

The next major classification of rodents was by Wood (1955a). He generally followed Brandt's three subdivisions, recognizing the Sciuromorpha, Myomorpha, and Hystricomorpha, but introduced a number of new suborders and changed the content of some of the previously named suborders. Though Wood (1937) had earlier recognized the Protrogomorpha, he did not in his 1955a classification, but later did reintroduce it (Wood, 1958). He included all of the protrogomorph families in the Sciuromorpha in his 1955a work. He also included the Geomyoidea (geomyids, heteromyids, and eomyids), a sciuromorphous group of rodents, in the suborder Myomorpha mainly based on the arguments of Wilson (1949a). There were four new suborders introduced by Wood (1955a): Theridomyomorpha, Castorimorpha, Caviomorpha, and Bathyergomorpha. The Theridomyomorpha included only the theridomyoids which had previously been included in the Protrogomorpha or their own separate suborder (Schaub, 1953). The Castorimorpha included the sciuromorphous castoroids (Castoridae and Eutypomyidae) which had previously been classified as sciuromorphs. The Caviomorpha included all of the hystricomorphous–hystricognathous rodents of South America which Wood believed evolved separately from the Old World hystricoids (Wood and Patterson, 1959). The last suborder, Bathyergomorpha, was a resurrection of Tullberg's infraorder Bathyergomorphi which included only the African bathyergoids that had the unique combination of a protrogomorphous skull and a hystricognathous mandible. Wood (1955a) also left a number of families as *incertae sedis*: Ctenodactylidae, Pedetidae, and Gliridae.

Ten years later, in his last revision, Wood (1965) employed a gradal concept to the problem and totally rearranged his earlier classification. Still roughly following Brandt's subdivisions, Wood (1965) recognized only three suborders, the Caviomorpha, the Myomorpha (again containing the geomyoids), and the Protrogomorpha (containing the primitive protrogomorphous rodents and the aplodontoids). The remaining 15 families were not assigned to a suborder.

Another suborder, the Geomorpha, was introduced by Thaler (1966) to accommodate the geomyoids (Geomyidae, Heteromyidae) and the fossil family Eomyidae, which removed this group from the Myomorpha. Lavocat (1969) proposed a new suborder, the Phiomorpha, which included the Oligocene and Miocene hystricomorphous–hystricognathous rodents of Africa, again believing in a separate evolution of this group from other hystricomorphs. Later, the concept of Phiomorpha was expanded to include the remainder of the Old World hystricomorphous–hystricognathous rodents (Chaline and Mein, 1979).

Bugge (1974) studied the cranial arterial circulation of Recent rodents and proposed a classification that did not include any fossil families. His classification recognized all of Brandt's suborders as well as Wood's Castorimorpha, Protrogomorpha (restricted to Aplodontidae), and Caviomorpha. The only change in these groupings was the inclusion of the European myomorphous Gliridae (dormice) in the Sciuromorpha, whereas previously they had been included in the Myomorpha. Bugge introduced two new suborders: Erethizontomorpha and Anomaluromorpha. He believed that the New World porcupines (Erethizontidae) were not closely related to the other South American hystricomorphs and represented yet another suborder. The Anomaluromorpha included two problematical families of African rodents that are hystricomorphous and sciurognathous, the Anomaluridae and Pedetidae. Both of these families had previously been included in either the Hystricomorpha or *incertae sedis*. Bugge suggested that these families may have been derived from the theridomyids.

More recently, Wood (1975, 1980, 1981, 1984, 1985) has rethought his earlier classifications and adopted Tullberg's two suborders, Sciurognathi and Hystricognathi. The only major change in Tullberg's arrangement was the introduction of a new infraorder of hystricognath, Franimorpha, which consisted of a number of Eocene and Oligocene North American rodents that supposedly possessed "incipient hystricognathy." While some have accepted Wood's concept of the Franimorpha (Chaline and Mein, 1979), a number of other authors have questioned the existence of "incipient hystricognathy," and considered the Franimorpha a group of primitive but unrelated rodents (Dawson, 1977; Korth, 1984; Wilson, 1986). Currently, it is generally accepted that the hystricomorphous–hystricognathous rodents of both the Old World and South America should be included in the same suborder (summarized in Luckett and Hartenberger, 1985) and that the condition of having both hystricomorphy and hystricognathy has evolved only once.

The only other previously proposed group of rodents above the level of superfamily was the infraorder Ctenodactylomorpha by Chaline and Mein (1979) for the hystricomorphous–sciurognathous ctenodactyloid rodents of Asia and northern Africa.

The subordinal classification presented below is that which will be followed in this book and includes only the families of rodents from the North American Tertiary.

2.2. Classification

Order Rodentia Bowdich, 1821
  Suborder Sciuromorpha Brandt, 1855
    Superfamily Ischyromyoidea Alston, 1876
      Family Ischyromyidae Alston, 1876 (including Paramyidae)
    Superfamily Aplodontoidea Trouessart, 1897
      Family Aplodontidae Trouessart, 1897
      Family Mylagaulidae Cope, 1881
    Superfamily Sciuroidea Gray, 1821
      Family Sciuridae Gray, 1821
  Suborder Sciuromorpha *incertae sedis*
      Family Cylindrodontidae Miller and Gidley, 1918
    Superfamily Castoroidea Gray, 1821
      Family Castoridae Gray, 1821
      Family Eutypomyidae Miller and Gidley, 1918
  Suborder Myomorpha Brandt, 1855
  Infraorder Myodonta Schaub, 1958
    Superfamily Muroidea Miller and Gidley, 1918
      Family Cricetidae Rocheburne, 1883 (including Microtidae)
    Superfamily Dipodoidea Weber, 1904
      Family Zapodidae Coues, 1875
      Family Simimyidae Wood, 1980
      Family Armintomyidae Dawson, Krishtalka, and Stucky, 1990
  Infraorder Geomorpha Thaler, 1966
    Superfamily Geomyoidea Bonaparte, 1845
      Family Geomyidae Bonaparte, 1845
      Family Heteromyidae Gray, 1868
      Family Florentiamyidae Wood, 1936

      Family Heliscomyidae Korth, Wahlert, and Emry, 1991
   Superfamily Eomyoidea Depéret and Douxami, 1902
      Family Eomyidae Depéret and Douxami, 1902
Suborder ?Myomorpha *incertae sedis*
      Family Sciuravidae Miller and Gidley, 1918
Suborder uncertain
      Family Protoptychidae Schlosser, 1911
      Family Laredomyidae Wilson and Westgate, 1991

II

Systematic Description of Rodent Families

Chapter 5
Ischyromyidae

1. Characteristic Morphology	37
1.1. Skull	37
1.2. Dentition	39
1.3. Skeleton	40
2. Evolutionary Changes in the Family	41
2.1. Paramyinae	42
2.2. Reithroparamyinae	42
2.3. Ailuravinae	43
2.4. Ischyromyinae	43
3. Fossil Record	44
4. Phylogeny	46
4.1. Origin	46
4.2. Intrafamilial Relationships	46
4.3. Extrafamilial Relationships	48
5. Problematical Taxa	49
5.1. *Ischyromys* and *Titanotheriomys*	49
5.2. *Apatosciuravus*	50
5.3. "*Paramys*" *simpsoni*	51
6. Classification	51

1. Characteristic Morphology

1.1. Skull

Where known, the skulls of ischyromyids are robust with a heavy rostrum and small incisive foramina (Fig. 5.1). Commonly there is a well-defined central sagittal crest on the cranium. The postorbital constriction is pronounced. The attachment for the masseter on the skull is limited to the ventral base of the zygoma (protrogomorphous) and the infraorbital foramen is small, opens anteriorly, and is not enlarged by muscle invasion or compressed by any modification of a zygomatic plate.

    The auditory bullae are believed not to have been ossified and attached to the skull, though Wood (1962) showed that in larger manitshines the bulla was ossified but not attached. The bulla in *Reithroparamys* and the ischyromyines is fully ossified to the skull. In the middle ear region, the presence of a stapedial artery is variable (Wahlert, 1974). The attachment of the stapedial

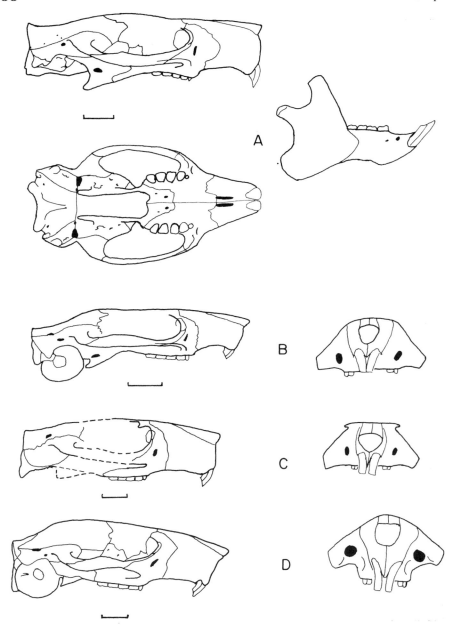

FIGURE 5.1. Skulls of Ischyromyidae. (A) Wasatchian *Paramys copei* (Paramyini), lateral (above) and ventral (below) views of skull, and lateral view of mandible (right). (B) Bridgerian *Reithroparamys delicatissimus* (Reithroparamyinae), lateral and anterior views. (C) Clarkforkian *Franimys amherstensis* (Pseudoparamyini). (D) Orellan *Ischyromys typus* (Ischyromyinae). All skulls not to same scale, Bars = 1 cm. Lateral views of A and D from Wahlert (1974).

muscle was limited to inside the bulla and there is a large segment of the mastoid bone separating the bulla from the posterior margin of the skull. These latter two features are believed to be unique and derived for ischyromyids by Lavocat and Parent (1985, p. 336).

The mandible is also usually robust. There is a sharp ridge on the dorsal margin of the mandibular diastema. The angle of the jaw is not deflected either laterally or medially from the horizontal ramus (sciurognathy). The masseteric scar is a shallow depression bordered dorsally and ventrally by distinct, rounded ridges that unite anteriorly below the posterior molars.

1.2. Dentition

In all ischyromyids (with the exception of European pseudoparamyines) a complete rodent dental formula is maintained with two upper premolars and one lower (Fig. 5.2). The upper third premolar is a simple peg. In the European pseudoparamyines the P^3 is lost, although in at least one of the two European genera, *Plesiarctomys*, a deciduous P^3 is known to have existed (Wood, 1970). The incisor enamel microstructure in ischyromyids is pauciserial or uniserial. Only in the youngest subfamily, Ischyromyinae, is the enamel uniserial; in all other species of the remainder of the family, the enamel is pauciserial.

The cheek teeth of the Ischyromyidae follow the same general plan with only slight variations. All of the cheek teeth are brachydont and cuspate, developing moderate lophodonty only in the ischyromyines. The development of irregularities of the enamel in the basins of the cheek teeth is variable in ischyromyids at the level of genus.

The upper molars are generally squared in occlusal outline. The conules are generally large but not as large as the major cusps. The protoloph and metaloph converge lingually at the protocone. The hypocone is usually slightly smaller than the protocone and always present, connecting with the posterior slope of the protocone after wear. Stylar cusps are commonly present but never enlarged to modify the shape of the tooth. M^3 is always reduced posteriorly, the metacone, hypocone, and metaloph being reduced and occasionally absent. P^4 is perhaps the most variable tooth, being molariform but not as large as the molars.

The lower molars are rectangular in occlusal outline with a shallow talonid basin having marginal cusps. The trigonid basin is always small. The development of a hypolophid and hypoconulid is variable among genera and subfamilies. M_3 is usually longer than the anterior molars with a posteriorly extended posterolophid and the hypoconid and entoconid commonly reduced. As with P^4, the lower premolar is the most variable lower cheek tooth. It is always smaller than the molars and is narrower anteriorly (trigonid) than posteriorly (talonid).

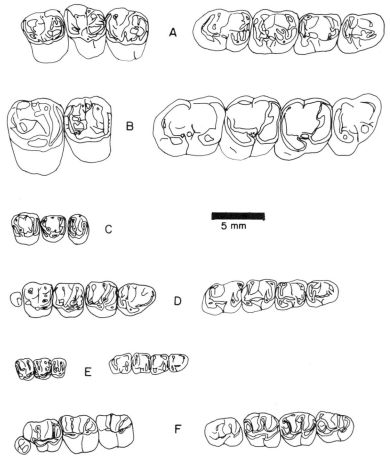

FIGURE 5.2. Dentitions of Ischyromyidae. (A) Bridgerian *Paramys delicatus* (Paramyini), LP^4–M^2 and RP$_4$–M$_3$. (B) Bridgerian *Pseudotomus robustus* (Manitshini) RP^4–M^1 and RP$_4$–M$_3$. (C) Wasatchian *Franimys buccatus* (Pseudoparamyini) P^4–M^2 (lowers unknown). (D) Bridgerian *Reithroparamys delicatissimus* (Reithroparamyini), LP^3–M^3 and RP$_4$–M$_3$. (E) Uintan *Microparamys tricus* (Microparamyini) RP^4–M^2 and RP$_4$–M$_3$. (F) Chadronian *Ischyromys veterior* (Ischyroparamyinae), LdP^3–M^2 and LP$_4$–M$_3$. A–E from Wood (1962), F from Wood (1937).

1.3. Skeleton

The skeleton of ischyromyids, as with the skull and mandible, is robust and apparently adapted for terrestrial quadrupedal locomotion (Fig. 5.3). The tibia and fibula are never fused. The scaphoid and lunate bones of the wrist become fused in *Ischyromys* and possibly *Reithroparamys* but remain separate in all other genera. There is an epicondylar foramen on the distal

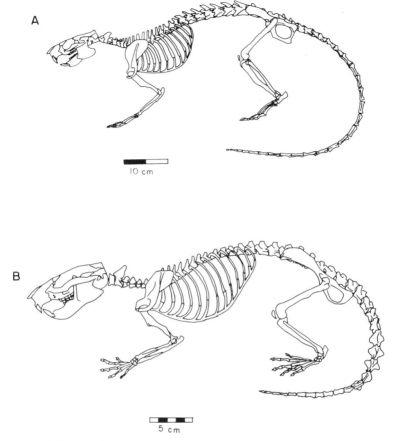

FIGURE 5.3. Skeletons of Ischyromyidae. (A) Bridgerian *Paramys delicatus*; (B) Orellan *Ischyromys typus*. A from Wood (1962), B from Wood (1937).

humerus, and the femur retains a third trochanter, both primitive conditions for eutherian mammals.

2. Evolutionary Changes in the Family

Once the ischyromyids were differentiated into the recognized subfamilies, little evolutionary change occurred. This separation of the subfamilies occurred early. In the late Clarkforkian of North America (almost the earliest occurrence of the family) already three subfamilies are known: Paramyinae, Reithroparamyinae, and Pseudoparamyinae (Korth, 1984). At roughly the same time in Europe, the Ailuravinae had also appeared (Godinot, 1981). The evolutionary changes within the subfamilies will be dealt with separately.

2.1. Paramyinae

The Paramyinae are medium to large rodents that appear to have simplified their dentitions through time. The early Wasatchian species of *Paramys* develop crenulations of the enamel that were enhanced in later Bridgerian species. The skull and dentition show an increase in size through time. The Bridgerian–Chadronian tribe of paramyines, the Manitshini, are the largest of the ischyromyids and developed slightly higher crowned cheek teeth having a much simplified occlusal pattern with shallow basins. In the manitshine *Pseudotomus*, the rostrum of the skull is elongated more than in other paramyines. In the largest species (and genus) *Manitsha tanka* from the Chadronian, the zygoma is modified. It arises from the maxilla slightly more anteriorly than in the earlier species and is slightly tilted dorsally in a lateral direction.

The Pseudoparamyini (here considered a tribe of Paramyinae) is chiefly a European subfamily and is represented by a single genus, *Franimys*, in North America from the Clarkforkian and Wasatchian. This tribe is distinguished from other paramyines by its relatively shorter rostrum of the skull and reduction of the upper premolars. The only known skull of *Franimys* also possesses a postorbital process that is unknown in any other ischyromyid, including the European pseudoparamyines. In the European species of this subfamily, P^3 is completely lost (Michaux, 1968; Wood, 1970). In *Franimys*, P^3 is retained. In all known specimens of *Franimys*, only the alveolus for P^3 is preserved and appears to be of equivalent size to that of other ischyromyids. However, P^4 is greatly reduced in all species of *Franimys*. The P^4 in *Franimys* in its first occurrence (*F. amherstensis*) completely lacks a paracone and protoloph. The paracone and protoloph became better developed in the later species, but the tooth retained a small oval occlusal outline and was never quite as molariform as it is in other ischyromyids.

2.2. Reithroparamyinae

The Reithroparamyinae, here separated into the tribes Reithroparamyini and Microparamyini (= Microparamyinae of Wood, 1962), consists of species that are smaller than other ischyromyids. The origin of the zygoma on the maxilla is more anterior than in most other ischyromyids. The first two upper molars are more nearly square and equal in size (in others M^2 is generally slightly larger), the hypocones are larger, a partial hypolophid is commonly developed on the lower molars, and the posterolophid ends lingually before fusing with the entoconid. The skeleton of *Reithroparamys* is also more slightly built than that of paramyines, and in the known skulls (two of *Reithroparamys*) there is no central sagittal crest, but two lower, more lateral parasagittal crests that nearly merge posteriorly. The rostrum tapers more gently anteriorly and is not as heavy as in other ischyromyids. The auditory bulla in reithroparamyines is made of bone and is ossified to the skull. It appears that the reithroparamyines increased in size through time. A hypolophid appeared on the lower cheek teeth and lophodonty of the cheek teeth increased from the earliest species to those of the late Eocene in both *Microparamys* and the *Acritoparmys–Reithroparamys* lineage.

2.3. Ailuravinae

The Ailuravinae, similar to pseudoparamyines, are mainly a European subfamily (Wood, 1976a). Two genera, *Mytonomys* and possibly *Eohaplomys*, are known from North America and are common only in the Uintan, though some species are reported from the Wasatchian, Duchesnean, and Chadronian (Korth, 1988a). These genera are distinguished from other primitive ischyromyids by their generally large size (but not as large as manitshines), crenulate molars, less reduced M^3, shortened lower incisor, and large hypocone on the upper molars (see also Michaux, 1968; Wood, 1976a). In *Mytonomys* the only change through time appears to be an increase in size and complexity of the irregularity of the enamel in the basins of the cheek teeth. *Eohaplomys* from California closely approaches the European species of the subfamily in morphology (Korth, 1988a), and only shows an increase in size and complexity of the molars.

2.4. Ischyromyinae

The Ischyromyinae are the youngest and most advanced subfamily. These rodents developed lophodonty of the cheek teeth, uniserial microstructure of the incisor enamel (pauciserial in all other subfamilies; see Wahlert, 1968). The auditory bulla is ossified and attached to the skull, and many of the cranial foramina of the medial orbital wall are more anterior than in paramyines (Wahlert, 1974). It has been argued that the infraorbital foramen in some species was invaded by the medial masseter, therefore altering the zygomasseteric structure into a modified myomorphous condition (Wood, 1976b). Because of these differences in the skull and dentition, many authors have placed the ischyromyines in their own family (Ischyromyidae) and placed all of the remaining subfamilies in the family Paramyidae (Wahlert, 1974; Wood, 1976b). Other authors have insisted that the similarity to the Eocene paramyines (or reithroparamyines) is so great that the ischyromyines are just the most derived subfamily of a single family, the Ischyromyidae

(Black, 1968, 1971; Korth, 1984). Because the zygomasseteric structure is variable within species of ischyromyines, it is difficult to use this as a familial character. The difference in the enamel microstructure may also be questionable because a change from pauciserial to uniserial is also seen within the Eocene European rodent family Theridomyidae, in fact, within a single genus of theridomyid (Wahlert, 1968, pp. 15–16).

Within the Ischyromyinae, again little change other than size increase is seen. In the cheek teeth the frequency of accessory cuspules appears to be greater in the youngest specimens of *Ischyromys* (Heaton, 1993).

3. Fossil Record

The reithroparamyine ischyromyid *Acritoparamys atavus* from the early Clarkforkian (latest Paleocene) of Montana is the earliest known rodent in the world. In North America, ischyromyids rapidly diversified in the Clarkforkian and Wasatchian (Fig. 5.4), reaching their greatest diversity in Uintan times (23 species and 11 genera). During the Oligocene, the diversity of the ischyromyids diminished gradually until a single species remained in the Whitneyan (Howe, 1966; Heaton, 1993). In Europe, ischyromyids are nearly as diverse as in North America and are known from the earliest Eocene through the late Eocene (Dawson, 1977). In Asia, ischyromyids were rare but are known from the early Eocene of Mongolia (Shevyreva, 1984) and the middle Eocene to early Oligocene of China (see Li and Ting, 1983).

Among the subfamilies, the time of greatest diversity differed greatly. The reithroparamyines had their greatest diversity in the Wasatchian (5 genera, 11 species) and gradually reduced in number of species until the Chadronian where only a single species persisted, *Microparamys perfossus*. The larger paramyines (including the Manitshini) are the most diverse among the subfamilies of ischyromyids. They had their greatest diversity in the Uintan (7 genera, 16 species) but were more diverse than any other subfamily as well in the Bridgerian (5 genera, 13 species), and just slightly less diverse than the reithroparamyines in the Wasatchian (4 genera, 10 species). In the Duchesnean and Chadronian, the paramyines (dominated by manitshines) are considerably less diverse, but remain represented by more species than any other subfamily. The Chadronian *Manitsha tanka* is the youngest, largest, and most derived paramyine.

The earliest record of the Ischyromyinae is from the Duchesnean of Wyoming (Black, 1971). They reach their greatest diversity in the Chadronian and Orellan where as many as three or four species may be known. *Ischyromys* is the last known ischyromyid ranging into the early Whitneyan.

Both the Ailuravinae and Pseudoparamyinae are rare relative to other ischyromyines. The Pseudoparamyinae, restricted to *Franimys* in North America, is known only from the early Eocene Clarkforkian and Wasatchian and from only three species. The Ailuravinae, represented in North America

Ischyromyidae

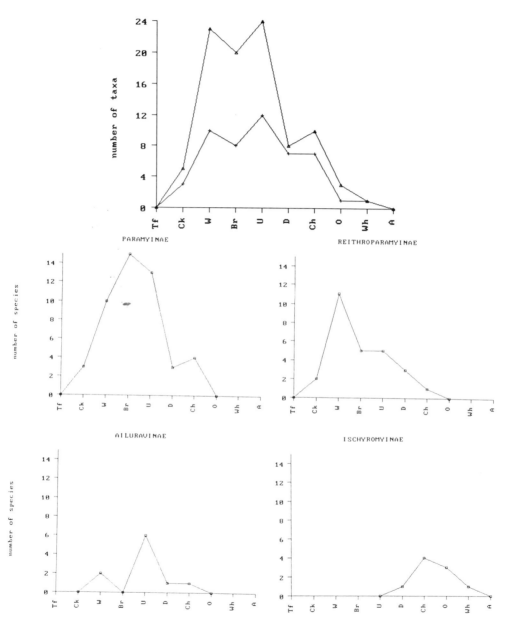

FIGURE 5.4. Occurrence of ischyromyids through the Tertiary of North America by subfamily (below) and total numbers of species (△) and genera (+) (above). Abbreviations for North American land-mammal ages used on horizontal axis (and those used in later chapters): Tf, Tiffanian; Ck, Clarkforkian; W, Wasatchian; Br, Bridgerian; U, Uintan; D, Duchesnean; Ch, Chadronian; O, Orellan; Wh, Whitneyan; A, Arikareean; Hf, Hemingfordian; B, Barstovian; Hp, Hemphillian; Bl, Blancan.

by *Mytonomys* and *Eohaplomys*, are most abundant in the Uintan but do persist into the Chadronian of Texas and Mexico.

4. Phylogeny (Fig. 5.5)

4.1. Origin

Because of the extensive Eocene record of ischyromyids in North America, and because the earliest known rodent is an ischyromyid, many early authors believed that the primitive ischyromyids, namely the paramyines, represented the basal rodent morphotype from which all later families could be derived (Matthew, 1910; Wilson, 1949a; Wood, 1962). However, with a more extensive collection and knowledge of the earliest Tertiary record of rodents from Asia, others have removed the Ischyromyidae from this ancestral position and derived only a few later families (e.g., sciurids, castorids) from the ischyromyid stock (Hartenberger, 1980; Dawson et al., 1984; Korth, 1984; Flynn et al., 1986). If indeed all rodents are derived from a *Heomys*-like eurymyloid ancestor, the ischyromyids developed a simplified dental pattern reducing lophodonty, conule, and hypocone size on the upper molars; reducing the hypoconulid on the lower molars; and developing a more molariform last premolar. Korth (1984) named a Wasatchian species *Reithroparamys ctenodactylops* that most closely approached the ancestral type of morphology. It is most likely that the earliest rodents migrated to North America in the late Paleocene (?Tiffanian) and rapidly radiated throughout the Eocene.

No rodent known from either Asia or North America is intermediate in morphology between a eurymyloid ancestor and the characteristic ischyromyid in dental pattern. However, at present it seems best to view ischyromyids as being derived directly from the eurymyloid (?ctenodactyloid) ancestor.

4.2. Intrafamilial Relationships

Dentally, perhaps the most primitive subfamily of ischyromyids, based on similarity to the hypothetical ancestral morphotype, is the Reithroparamyinae. The remainder of the subfamilies of ischyromyids can be derived from them (see Korth, 1984). Recently, Meng (1990) challenged the ancestral position of the reithroparamyines based on the basicranial morphology of the Bridgerian *Reithroparamys delicatissimus*. He referred this subfamily to its own family (Reithroparamyidae) and viewed it as most closely related to the aplodontids and sciurids, being more derived than the paramyines rather than being more primitive. However, reithroparamyines remain the most primitive ischyromyids dentally, and the basicranium is not known for the earliest and most primitive genus of the subfamily, *Acritoparamys*. Charac-

Ischyromyidae 47

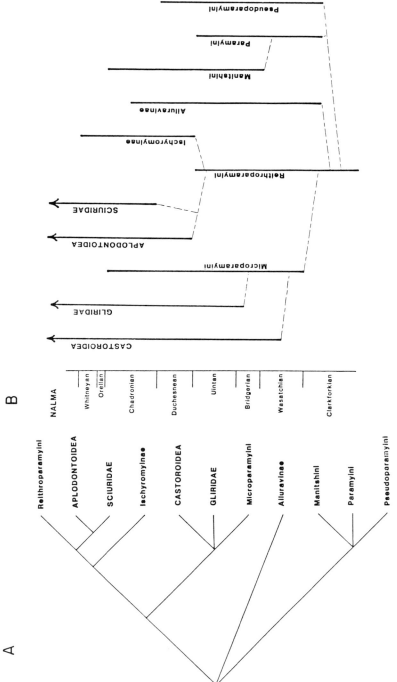

FIGURE 5.5. Phylogeny of the Ischyromyidae. (A) Cladogram of ischyromyid relationships. (B) Dendrogram of the Ischyromyidae with age of occurrence. Width of each North American land mammal age (NALMA) based on data from Berggren et al. (1985). Solid vertical lines indicate period of occurrence of taxon; arrows at top indicate continued occurrence beyond limit of vertical axis. Dashed lines indicate proposed relationships between taxa.

ters of the mandible, dentition, and middle ear have already been pointed out as allying *Reithroparamys* with the earliest aplodontids (Emry and Korth, 1989). That reithroparamyines were ancestral to later sciuromorphs is not surprising. However, it does not seem likely that they are quite so distant from the other subfamilies of ischyromyids.

Based on the dentition, the paramyines and ailuravines can be derived from reithroparamyines by simplifying the cheek tooth pattern and generally increasing in size. However, it is possible, because of the almost instantaneous appearance of four of the ischyromyid subfamilies in the earliest Eocene in both North America and Europe, that all of the subfamilies were derived from a more primitive form that radiated quite rapidly into the recognizable primitive subfamilies rather than through a reithroparamyine.

The Ischyromyinae may also be directly derived from the Reithroparamyinae based on shared dental as well as skeletal characters (Korth, 1984, p. 66) sometime in the latest Eocene.

In a series of papers on the origin of South American rodents ("Caviomorpha"), Wood (1975, 1981) put the reithroparamyines, the paramyine *Tapomys*, a sciuravid, and some cylindrodontids in an infraorder Franimorpha (within the suborder Hystricognathi) based on the orientation of the angle of the mandible. He believed that from this group all Oligocene and later South American rodents were derived. Other authors have questioned this interpretation and considered that the "incipient hystricognathy" of these rodents was either nonexistent or a condition paralleling that in true hystricognaths, and that the "Franimorpha" is a mixture of many unrelated taxa that are better retained within their originally designated families (Dawson, 1977; Korth, 1984; Luckett and Hartenberger, 1985; Wilson, 1986).

4.3. Extrafamilial Relationships

The later Tertiary families of rodents that appear to be directly derived from the ischyromyids are: the Sciuridae, Aplodontidae, Old World Gliridae, and possibly Eutypomyidae (Castoroidea). Wood (1962) named a reithroparamyine *Uriscus* from the Uintan and noted that the general occlusal outline of the lower molars greatly resembled that of sciurids. Black (1963) in his review of the Tertiary sciurids agreed with Wood but noted that the masseteric scar on the mandible of *Uriscus* terminated anteriorly below the posterior molars (typical for ischyromyids) and not below the premolar as in sciurids, therefore barring it from being a sciurid. Later, Black and Sutton (1984) referred *U. californicus* to *Reithroparamys* but still viewed the ischyromyids as ancestral to the sciurids. More recent authors have viewed the derivation of sciurids from the paramyine ischyromyids by a continuation of the simplification of the dental pattern and the addition of cranial specializations (Wahlert, 1972; Korth, 1984). Meng (1990) suggested that the sciurids as well as the aplodontids may have arisen from reithroparamyines based on the morphology of the middle ear.

Primitive aplodontids (Prosciurinae) have long been recognized as dentally similar to ischyromyids and have been included as a subfamily of the Ischyromyidae or Paramyidae (Wilson, 1949a; Wood, 1962; Black, 1971). Recently, with the description of the upper dentition of the Bridgerian *Reithroparamys huerfanensis*, it has been suggested that this species is at least morphologically ancestral to alter prosciurines (Emry and Korth, 1989). This relationship has been supported by the shared characters of the basicranium of *Reithroparamys* and aplodontids (Meng, 1990).

Hartenberger (1971) demonstrated convincingly, on the basis of dentition, that the Old World Gliridae were directly derived from European Eocene species of *Microparamys*. Some authors, however, maintain that the glirids are more closely related to the Myomorpha (Wahlert, 1978, 1983).

It has been suggested that dentally the Eutypomyidae can be derived from the Uintan genus *Janimus* which in turn can be derived from the Wasatchian and possibly Bridgerian genus *Mattimys*. Both of these genera have been questionably included in the Microparamyinae (= Reithroparamyinae) of the Ischyromyidae (Black, 1971). The eutypomyids are most closely related to the castorids among later Tertiary rodents (Wahlert, 1977). Because of this series of relationships, it has been proposed that the Castoroidea are derived from the reithroparamyines. A great number of characters of the castorids and eutypomyids are not shared with any other group of rodents (Bugge, 1974; Luckett and Hartenberger, 1985) which obscures their ancestry. Wahlert (1977) viewed all of the characters of the skull shared by eutypomyids and ischyromyids as primitive, making for an indefinite ancestry of the Eutypomyidae. At present it seems best to view the castoroids (particularly the eutypomyids) as being split from a reithroparamyine stock near *Microparamys* sometime in the Wasatchian or late Clarkforkian and subsequently developed independently of all other rodents.

The Cylindrodontidae have long been classified near the ischyromyids (Wilson, 1949a; Wood, 1965; Wahlert, 1974; Black and Sutton, 1984). However, they are quite distinct dentally and cranially from the ischyromyids (Wahlert, 1974; Korth, 1984). The earliest reported cylindrodont, *Dawsonomys*, is clearly dentally distinct from contemporary ischyromyids, and therefore it is unlikely that the cylindrodontids were derived from any ischyromyid.

5. Problematical Taxa

5.1. *Ischyromys* and *Titanotheriomys*

Ischyromys was first named by Leidy (1856) and is the most common rodent in the Orellan of the White River Group of the Great Plains. Matthew

(1910) referred a Chadronian species originally referred to *Ischyromys* to a new subgenus *Titanotheriomys*. Later, Miller and Gidley (1920) recognized the latter as a separate genus distinct from *Ischyromys*. Since that time, several workers have argued for both separate genera (notably Wood, 1976b) and for synonymy of these genera (Black, 1968).

There are no consistent dental criteria to separate these two genera. This fact has been accepted by virtually all authors who have addressed this problem. The characters used to differentiate the two genera are cranial. Wood (1976b) contends that in *Titanotheriomys* the masseter muscle has extended onto the rostrum of the skull from the ventral margin of the zygoma and possibly through the infraorbital foramen, thus modifying the zygomasseteric structure to a myomorphous-like complex. In conjunction with this, the central sagittal crest typical of *Ischyromys* has been reduced and two lower parasagittal crests are present on the skull. However, examples of skulls with the *Titanotheriomys*-like zygoma and an *Ischyromys*-like sagittal crest exist from both the Chadronian and Orellan. And, as argued by Black and Sutton (1984, p. 78), there are specimens with the *Titanotheriomys*-like zygoma that are transitional between the extreme examples pointed out by Wood and the typical *Ischyromys* protrogomorphous zygoma. There has also been no research into the ontogenesis of this zygomasseteric structure which may explain the reason for the wide range of variation.

Clearly, until large populations including juvenile individuals of Oligocene ischyromyines can be studied, it is most economical to include all ischyromyines in a single genus *Ischyromys*. Because most of the fossil remains are dental elements, it is also very difficult to identify different genera distinguished entirely by cranial differences.

5.2. *Apatosciuravus*

Korth (1984) first described *Apatosciuravus* from the Clarkforkian and early Wasatchian of Wyoming and referred it questionably to the Reithroparamyinae. He noted that a number of features of the upper molars (such as a large hypocone) approached the condition in the earliest sciuravids. However, in the lower molars there was no indication of the beginnings of lophs on the molars or the isolated mesoconid typical of sciuravids. P^4 had a poorly developed metaloph and metaconid, apparently a transitional morphology between the supposed ancestral type (no metaloph or metacone) and typical ischyromyids and sciuravids. Later, Flanagan (1986) described a second species of this genus from the Wasatchian of Arizona that had better development of lophs on the lower molars and suggested that this new species *A. jacobsi* could be easily considered ancestral to the first true sciuravid *Knightomys*.

Most recently, Ivy (1990) referred all of the lower dentitions allocated by Korth (1984) to *Apatosciuravus bifax* from the Clarkforkian to *Acritoparamys atavus*. This means that the lower dentition of the earliest species of *Apatosciuravus* is unknown and may have been much less ischyromyid-like.

If *Apatosciuravus* was indeed ancestral to the sciuravids, it can be interpreted as either the earliest sciuravid that is transitional between the primitive "ctenodactyloid–eurymyloid" ancestor of rodents or it can be considered as a reithroparamyine. If the latter is accepted, then the Sciuravidae and all of their proposed descendants were ultimately derived from the Ischyromyidae.

5.3. "*Paramys*" *simpsoni*

This species is based on a single skull that differs from all other ischyromyids in the position of the maxillary–premaxillary suture on the lateral side of the rostrum, size of the infraorbital foramen (large), and presence of a small central groove in the center of the upper incisors (Korth, 1984). The cheek teeth of this species vary only slightly from those of typical Wasatchian ischyromyids. Korth (1988b) excluded this species from the genus *Paramys* but did not assign it to any other known genus. Clearly, this species should be included in a new genus.

6. Classification

For this and all following classifications, the abbreviations in parentheses after each taxon indicate age of occurrence (see Fig. 5.4 for definitions). An asterisk follows the type species of each genus. Only North American species are listed.

Ischyromyidae Alston, 1876
  Paramyinae Haeckel, 1895
    Paramyini Haeckel, 1895
      Paramys Leidy, 1871 (Ck–U)
        P. delicatus Leidy, 1871* (Br)
        P. delicatior Leidy, 1871 (Br)
        P. copei Loomis, 1907 (?Ck–Br)
        P. excavatus Loomis, 1907 (W)
        P. compressidens (Peterson, 1919) (U)
        P. taurus Wood, 1962 (Ck–W)
        P. nini (Wood, 1962) (W)
        ?*P. simpsoni* Korth, 1984 (W)
        P. pycnus Ivy, 1990 (W)
      Leptotomus Matthew, 1910 (Br–Ch)
        L. leptodus (Cope, 1873)* (Br–U)
        ?*L. caryophilus* Wilson, 1940 (U)
        ?*L. parvus* Wood, 1959 (Br)

    L. bridgerensis Wood, 1962 (Br)
    L. guildayi Black, 1971 (U–Ch)
    ?*L. coelumensis* Wilson and Stevens, 1986 (D)
  Thisbemys Wood, 1959 (W–U)
    T. uintensis (Osborn, 1895) (U)
    T. medius (Peterson, 1919) (U)
    T. corrugatus Wood, 1959* (Br)
    T. perditus Wood, 1962 (W)
    T. plicatus Wood, 1962 (Br)
    T. elachistos Korth, 1984 (Br)
  Notoparamys Korth, 1984 (W).
    N. costilloi (Wood, 1962)* (W)
    N. arctios Korth, 1984 (W)
  Quadratomus Korth, 1984 (Br, ?Ch)
    Q. grandis (Wood, 1962)* (Br)
    ?*Q. gigans* (Wood, 1974) (Ch)
    Q. sundelli (Eaton, 1982) (Br)
    Q. grossus Korth, 1985 (Br)
  Rapamys Wilson, 1940 (U)
    R. fricki Wilson, 1940* (U)
    R. wilsoni Black, 1971 (U)
  Tapomys Wood, 1962 (U)
    T. tapensis (Wilson, 1940) (U)
Manitshini Simpson, 1941
  Manitsha Simpson, 1941 (?Ch)
    M. tanka Simpson, 1941* (?Ch)
  Pseudotomus Cope, 1872 (Br–Ch)
    P. hians Cope, 1872* (Br)
    P. robustus (Marsh, 1872) (Br)
    P. petersoni (Matthew, 1910) (U)
    P. eugenei (Burke, 1935) (U)
    P. californicus (Wilson, 1940) (U)
    P. littoralis (Wilson, 1940) (U)
    P. horribilis (Wood, 1962) (Br)
    P. johanniculi (Wood, 1974) (Ch)
    P. timmys Storer, 1988 (D)
Pseudoparamyini Michaux, 1964
  Franimys Wood, 1962 (Ck–W)
    F. buccatus (Cope, 1877) (W)
    F. amherstensis Wood, 1962* (Ck)
    F. ambos Korth, 1984 (W)
Reithroparamyinae Wood, 1962
Reithroparamyini Wood, 1962
  Reithroparamys Matthew, 1920 (W–U)
    R. delicatissimus (Leidy, 1871)* (Br)

   R. *sciuroides* (Scott and Osborn, 1887) (U)
   R. *debequensis* Wood, 1962 (W)
   R. *huerfanensis* Wood, 1962 (Br)
   R. *ctenodactylops* Korth, 1984 (W)
  Acritoparamys Korth, 1984 (Ck–Br)
   A. *atwateri* (Loomis, 1907) (W)
   A. *atavus* (Jepsen, 1937) (Ck)
   A. *wyomingensis* (Wood, 1959) (Br)
   A. *francesi* (Wood, 1962)* (W)
   A. *pattersoni* (Wood, 1962) (W)
  Uriscus Wood, 1962 (U)
   U. *californicus* (Wilson, 1940)* (U)
 Microparamyini Wood, 1962
  Microparamys Wood, 1959 (Ck–Ch)
   M. *minutus* (Wilson, 1937)* (Br)
   M. *tricus* (Wilson, 1940) (U)
   M. *dubius* (Wood, 1949) U–D
   M. *perfossus* Wood, 1974 (Ch)
   M. *solidus* Storer, 1984 (U)
   M. *nimius* Storer, 1988 (D)
   M. *sambucus* Emry and Korth, 1989 (Br)
   M. *cheradius* Ivy, 1990 (Ck–W)
   M. *hunterae* Ivy, 1990 (W)
  Lophiparamys Wood, 1962 (W)
   L. *murinus* (Matthew, 1918)* (W)
   L. *debequensis* Wood, 1962 (W)
   L. *woodi* Guthrie, 1971 (W)
  Anonymus Storer, 1988 (D)
   A. *baroni* Storer, 1988* (D)
 Possible reithroparamyines
  Apatosciuravus Korth, 1984 (Ck–W)
   A. *bifax* Korth, 1984* (Ck)
   A. *jacobsi* Flanagan, 1986 (W)
Ailuravinae Michaux, 1968
  Mytonomys Wood, 1959 (W–Ch)
   M. *robustus* (Peterson, 1919)* (U–D)
   M. *burkei* (Wilson, 1940) (U)
   M. *wortmani* (Wood, 1962) (W)
   M. *mytonensis* (Wood, 1962) (U)
   ?M. *coloradensis* (Wood, 1962) (W)
   M. *gaitania* Ferrusquia and Wood, 1969 (Ch)
  Eohaplomys Stock, 1935 (U)
   E. *serus* Stock, 1935 (U)
   E. *matutinus* Stock, 1935 (U)
   E. *tradux* Stock, 1935 (U)

Ischyromyinae Alston, 1876
>> *Ischyromys* Leidy, 1856 (D–Wh)
>> *I. typus* Leidy, 1856* (O)
>> *I. veterior* Matthew, 1903 (Ch)
>> *I. parvidens* Miller and Gidley, 1920 (O)
>> *I. pliacus* Troxell, 1922 (O–Wh)
>> *I. douglassi* Black, 1968 (Ch)
>> *I. junctus* Russell, 1972 (Ch)
>> *I. blacki* Wood, 1974 (Ch)

Chapter 6
Sciuravidae

1. Characteristic Morphology	55
1.1. Skull	55
1.2. Dentition	56
1.3. Skeleton	58
2. Evolutionary Changes in the Family	58
3. Fossil Record	59
4. Phylogeny	61
4.1. Origin	61
4.2. Intrafamilial Relationships	62
4.3. Extrafamilial Relationships	62
5. Problematical Taxa	64
5.1. *Tillomys* and *Taxymys*	64
5.2. *Prolapsus*	64
5.3. AMNH 12118	65
6. Classification	65

1. Characteristic Morphology

1.1. Skull

Sciuravids are generally small rodents. The skull is generally smaller and more gracile than that in ischyromyids with a less pronounced postorbital constriction (Fig. 6.1). A weak, low sagittal crest is present in the early Eocene *Knightomys* (Korth, 1984) but lost in the known middle Eocene specimens of *Sciuravus* (Dawson, 1961). The anterior root of the zygoma originates anterior to the tooth row. A large, distinct interparietal bone is present. The infraorbital foramen is small, circular, and unmodified; attachment for the masseter muscle is limited to the ventral side of the zygoma (protrogomorphous). Wood (1981) stated that the skull of *Prolapsus* was hystricomorphous, but other authors have refuted this and recognized the genus as protrogomorphous (Korth, 1984; Wilson, 1986; Wilson and Runkel, 1991). The auditory bullae are ossified but only weakly attached to the skull. There is no carotid canal.

The mandible is similarly less robust than in ischyromyids. There is a sharp ridge on the dorsal margin of the mandibular diastema as in ischyromyids. The mental foramen is multiple. A second, smaller mental foramen is present posterior to the main foramen. The masseteric scar terminates

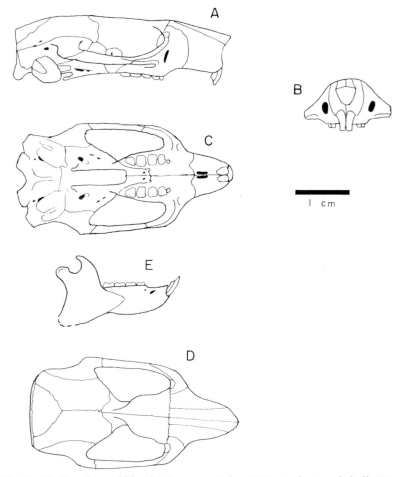

FIGURE 6.1. Skull and mandible of *Sciuravus nitidus*. (A) Lateral view of skull; (B) anterior view; (C) ventral view; (D) dorsal view; (E) lateral view of mandible. A and C modified from Wahlert (1974)

anteriorly below the boundary between M_1 and M_2 in a distinct V-shape, farther anterior than is common in ischyromyids. The angle of the mandible is generally in the plane of the horizontal ramus (sciurognathous) with perhaps some slight lateral deflection in *Prolapsus* (Wood, 1972, 1973).

1.2. Dentition

The cheek teeth are brachydont and cuspate with only the beginnings of lophodonty developing (Fig. 6.2). There is a complete rodent dental formula

Sciuravidae

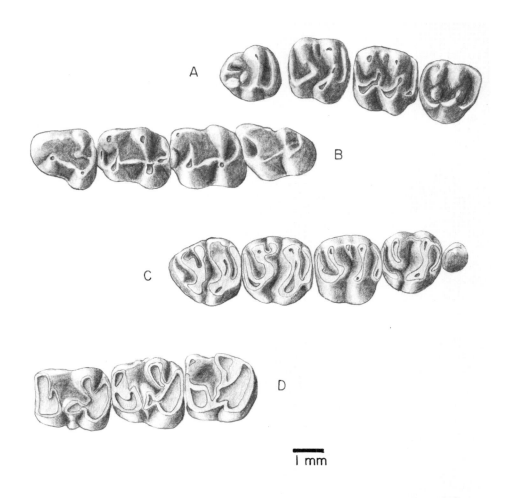

FIGURE 6.2. Cheek teeth of Bridgerian Sciuravidae. (A) *Sciuravus* sp., RP^4–M^3. (B) *Sciuravus* sp., LP$_4$–M$_3$. (C) *Taxymys lucaris*, RP^3–M^3. (D) *Tillomys senex*, LM$_1$–M$_3$. (E) *Pauromys perditus*, LP$_4$–M$_3$. (F) *Pauromys* sp., RM^1–M^2. C from Wilson (1938c), D from Wilson (1938b), F from Dawson (1968a).

with P³ small and peglike. Upper molars are approximately squared in outline. Hypocones are large, conules are reduced or absent, protoloph is complete, metaloph (if complete) joins the hypocone, and cingula are well-developed lophs. Lower molars are rectangular in occlusal outline. The entoconid is isolated from the posterolophid. At least a partial hypolophid is developed. The mesoconid is isolated or weakly attached to the hypoconid and often buccolingually elongate, extending into the talonid basin (= mesolophid). The posterior premolars are smaller than the first molars and submolariform with a distinct hypocone on P⁴. Premolars are greatly reduced in some species. The incisor enamel has a pauciserial microstructure (Wahlert, 1968).

1.3. Skeleton

The postcranial skeleton of sciuravids has never been fully described. The only author to mention it was Matthew (1910, p. 60) who noted that the skeleton of *Sciuravus* differed from that of *Paramys* only in smaller size and narrower width of the distal humerus. This implies that there is no fusion of the tibia and fibula nor any other specializations of the skeleton in *Sciuravus*. No skeletal material of any other genus of sciuravid has been reported.

2. Evolutionary Changes in the Family

Only three genera (and possibly a fourth) of sciuravids are definitely known from more than a single age, so little can be said about the evolutionary changes at the level of genus. Two of the longer-lived genera, *Knightomys* and *Sciuravus*, show a distinct size increase over time. In *Knightomys*, the youngest species, *K. huerfanensis*, is larger and has more bulbous, robust cheek teeth than the earlier species of the genus. Also in *Sciuravus*, the known Wasatchian species, *S. wilsoni*, is the smallest (Gazin, 1962), the Bridgerian species are larger, and the Uintan species are the largest (see Wilson, 1938a; Dawson, 1966, 1968a).

The third long-lived sciuravid genus is *Pauromys*. The overall size of the specimens from the Wasatchian is very near that of the type species and some specimens from the Bridgerian (Dawson, 1968a; Korth, 1984), but an early Bridgerian species, *P. exallos*, is the largest species presently described (Emry and Korth, 1989). The only recognizable trend in the morphology of the dentition of *Pauromys* is the reduction of P_4. The referred lower premolars from the Wasatchian and early Bridgerian are larger relative to the molars than in the type species.

The earliest genus, *Knightomys*, is morphologically ancestral to some later sciuravids (*Tillomys*, *Taxymys*, *Sciuravus*, *Prolapsus*, *Floresomys*). The

Sciuravidae

lower dentition of Knightomys need only become slightly more lophate and slightly higher crowned with a stronger development of the hypolophid to develop the morphology found in Tillomys. The same is true for the upper molars of Knightomys with respect to those of the Bridgerian Taxymys. In Sciuravus, Floresomys, and Prolapsus, the upper molars lose the connection of the metacone and the hypocone present in the upper molars of Knightomys and develop a stronger anterior cingulum. The lower molars of Sciuravus generally reduce the size of the mesoconid and it often becomes united with the hypoconid, while in Floresomys the mesoconid is lost completely (see Black and Stephens, 1973).

3. Fossil Record

The Sciuravidae are known exclusively from North America. A number of species from Asia were referred to the Sciuravidae (Li, 1963; Shevyreva, 1971, 1972). However, these rodents have since been shown to belong to a family of primitive ctenodactyloids (Wood, 1977; Dawson, 1977). Most recently, a new genus, Zelomys, from the late Eocene of China has been referred to the Sciuravidae (Wang and Li, 1990). This genus is protrogomorphous as in sciuravids, but the morphology of the cheek teeth most closely resembles that of the namatomyine eomyids. The age of Zelomys approximately coincides with the first appearance of eomyids in North America. This genus is considered here as a primitive eomyid rather than a sciuravid (see Chapter 11).

The earliest definite sciuravid is Knightomys from the early Wasatchian. The Sciuravidae attain their highest diversity in the Bridgerian where there are 6 known genera and 15 species (Fig. 6.3). In the Uintan, the number of species is reduced to 3. Dawson (in Black and Sutton, 1984) reported that the record of sciuravids extended into the Duchesnean, but to date no species have been formally described from this time.

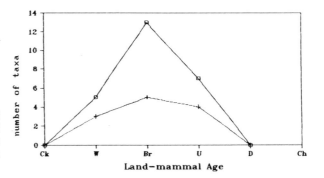

FIGURE 6.3. Occurrence of Sciuravidae in the Tertiary of North America (□, species; +, genera). Abbreviations as in Fig. 5.4.

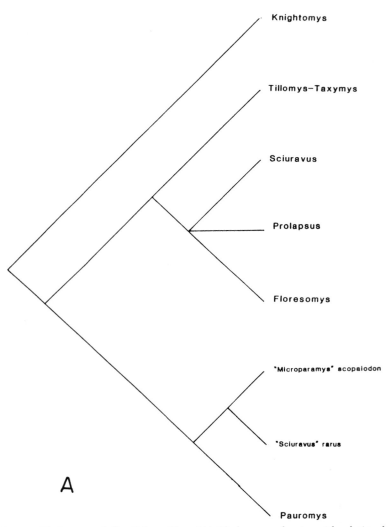

FIGURE 6.4. Phylogeny of the Sciuravidae. (A) Cladogram of sciuravid relationships. (B) Dendrogram of the Sciuravidae with age of occurrence (see Fig. 5.5 for explanation).

Sciuravidae   61

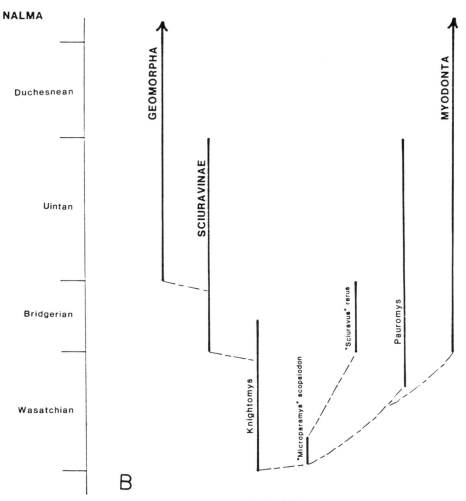

FIGURE 6.4. (Continued)

4. Phylogeny (Fig. 6.4)

4.1. Origin

The origin of the Sciuravidae appears to be through an *Apatosciuravus*-like ancestor (enlarged hypocone and reduction of the protoconule on the upper molars; isolation of the entoconid on the lower molars). The occurrence of *Apatosciuravus* (Clarkforkian) predates the earliest record of *Knightomys*, therefore supporting this ancestral interpretation. As discussed in the chapter on ischyromyids, the question remains whether *Apatosciuravus* is a reithroparamyine ischyromyid or a sciuravid that is transitional between the Sciuravidae and the ctenodactyloid–eurymyloid ancestor.

4.2. Intrafamilial Relationships

It appears that Bridgerian and younger species of *Sciuravus*, *Prolapsus*, *Tillomys*, *Taxymys*, and *Floresomys* can be derived from Wasatchian *Knightomys*. This appears likely both morphologically and chronologically. The Wasatchian and Bridgerian *Pauromys* differs greatly from the *Knightomys* type of cheek tooth pattern and cannot be directly derived from it. The most distinctive character of the dentition of *Pauromys* is the reduced lower (and possibly upper) fourth premolar. This feature is shared with two poorly known species, "*Microparamys*" *scopaiodon* from the early Wasatchian (Korth, 1984) and "*Sciuravus*" *rarus* from the Bridgerian (Wilson, 1938a). In the original description of "*M.*" *scopaiodon*, much was made of the similarity with *Pauromys* (Korth, 1984). However, in dental morphology (degree of lophodonty, reduction of cusps, and number of roots on P_4) the former more closely approaches "*S.*" *rarus*. It is quite possible that "*M.*" *scopaiodon* and "*S.*" *rarus* represent the same genus.

4.3. Extrafamilial Relationships

Sciuravids have long been considered the ancestral stock for geomyoid, muroid, and dipodoid rodents (Matthew, 1910; Wilson, 1949a; Wood, 1959; Black, 1965; Fahlbusch, 1979). This has been based on dental similarity. In the dipodoids and muroids, the cheek teeth are primitively developed with marginal cusps and a series of cross-lophs (buccolingual) along with a central anteroposteriorly directed loph (= mure) uniting the cross-lophs. Also, in both muroids and dipodoids, there is a reduction in the dental formula to no premolars in the muroids and only a small peglike P^4 and no longer premolars in the dipodoids. The beginnings of the characteristic cross-lophs of muroids and dipodoids are present in a number of sciuravid genera beginning with *Knightomys*. The central mure is beginning to develop in a number of species of Bridgerian *Sciuravus* (see especially Wilson, 1938a). The reduction in the dental formula is not completed in any known sciuravid, but *Pauromys* shows a marked reduction of the last premolar (particularly P_4).

Recently, Flynn *et al.* (1986) suggested that these myodont rodents (= muroids and dipodoids) were derived from Eocene ctenodactyloids. This was proposed because the first known true muroid is from the middle Eocene of China (Tong, 1992) and because the myomorphous zygomasseteric structure of the muroids was primitively hystricomorphous (Lindsay, 1977) as in the dipodoids and ctenodactyloids. Luckett and Hartenberger (1985) noted that the lack of the internal carotid artery (carotid canal) in the skull of *Sciuravus* bars the Sciuravidae from the ancestry of the geomyoids and myodont rodents because they retain this artery. However, the presence or absence of the internal carotid is known only for the Bridgerian *Sciuravus* and not for *Knightomys* and the most probable ancestral morphotype for the myodonts, *Pauromys*. Thus, the dental evidence for relationships of these later groups of rodents to the Sciuravidae appears the strongest case for ancestral–descendant relationships.

The earliest known dipodoids are from the very earliest Bridgerian of North America: *Elymys*, zapodid (Emry and Korth, 1989), and *Armintomys*, a unique genus put in its own family (see Chapter 22, Section 2). Whereas the slightly younger *Elymys* had the dental formula of zapodids (P^3 and P_4 lost, P^4 peglike), *Armintomys* retained the primitive rodent upper dental formula (lower dentition not known). In both of these genera, the dentition lacks some of the characteristic features of dipodoids [mesolophs (-ids), central anteroposteriorly oriented loph] but the morphological step from a sciuravid ancestor to either of these genera is very simple.

Among the geomyoid rodents (= Geomorpha of Wahlert, 1978), the Eomyidae are most easily derived from a sciuravid ancestor. The earliest eomyid, "*Namatomys*" from the Uintan of California and Saskatchewan (Lindsay, 1968; Storer, 1984), has developed the "omega" pattern of the molars typical of all eomyids, but the teeth are only slightly more lophate than those of typical sciuravids. The lophs on the molars of eomyids are complete and a mesoloph (-id) is developed, however, unlike the morphology in sciuravids. The loss of P^3 in eomyids is also a very easy step to take because this tooth is always minute at best in all rodents. All known eomyids (and geomyoids) are fully sciuromorphous with a more anteriorly extended masseteric scar on the mandible, whereas all known sciuravids are protrogomorphous with a more posteriorly ending masseteric scar. The sciuromorphy in geomyoids most likely was achieved in the very earliest eomyids, uniting the latter with the remainder of the geomyoids beyond the sciuravid ancestor.

The remainder of the geomyoid rodents (Geomyidae, Heteromyidae, Florentiamyidae, Heliscomyidae) are characterized by having cheek teeth that consist of two rows of three cusps that develop into two distinct buccolingual lophs of varying degrees. The closest sciuravid to develop this type of dental morphology is the Bridgerian *Taxymys*, particularly ?*T. progressus*, in which the protoloph and metaloph are complete, separate, and the cusps are less strongly developed than the connecting lophs (Wilson, 1938b). This would imply derivation of both the geomyid-like dentition and the eomyid-like dentition separately from a sciuravid ancestor which seems unlikely because of the great number of shared morphologies between the geomyoids and the eomyids (Wilson, 1949b). Thus, it is more likely that the eomyids were the ancestral stock for the remainder of the geomyoids, though there is currently no known transitional form of eomyid that bridges the morphological gap to geomyoids.

5. Problematical Taxa

5.1. *Tillomys* and *Taxymys*

In 1872, Marsh named both *Tillomys* and *Taxymys* from the Bridgerian of Wyoming. *Tillomys* was represented by only mandibles and lower dentitions, and *Taxymys* was only known from maxillary elements. Because of this, Troxell (1923a) synonymized the two. However, in his study of the Bridgerian sciuravids, Wilson (1938b,c) noted that there were significant differences in the upper and lower dentitions referred to these genera and that a synonymy was not definite. He recognized both genera pending the recovery of associated upper and lower dental elements.

Bown (1982) named a new species from the Bridgerian, *Taxymys cuspidatus*, to which he referred a number of upper and lower dentitions. However, this species is much more cuspate than other *Taxymys* and *Tillomys* and is identical to *Knightomys*. In size, *T. cuspidatus* is equal to *K. depressus* (Bown, 1982, Table 8; Korth, 1984, Table 10). It is most likely that *T. cuspidatus* is synonymous with *K. depressus* which has been previously recorded from the earliest Bridgerian (= Gardnerbuttean). Even if "*T.*" *cuspidatus* is a distinct species of *Knightomys*, it cannot be used to settle the question of synonymy of *Taxymys* and *Tillomys*.

No associated upper and lower dental elements of *Tillomys* or *Taxymys* have been recorded since Wilson's (1938b,c) review, so his comments remain valid.

5.2. *Prolapsus*

In his description of *Prolapsus*, Wood (1972, 1973) noted the great similarity of its dentition with that of *Sciuravus* but failed to put this genus in the Sciuravidae because of the "hystricognathous" mandible. Whereas he viewed the mandible as "fully hystricognathous" (Wood, 1973, p. 22), most later authors have noted that it is not the same as that of true hystricognaths (see Wilson, 1986) and may constitute only a slight variation in the typical sciuravid morphology. Similarly, Wood (1981) identified the skull of *Prolapsus* as being hystricomorphous, a condition not recognized by others (Korth, 1984; Wilson, 1986). Recently, Wilson and Runkel (1991, Fig. 2A) have investigated the skull of *Prolapsus* and noted that it is clearly protrogomorphous. Hence, *Prolapsus* is here viewed as a sciuravid (based on dental and skull morphology) with the beginnings of hystricognathy in the mandible.

5.3. AMNH 12118

An unusual specimen from the Bridgerian was described by Dawson (1962). This specimen differed from other sciuravids mainly in the proportions of the premolars (Fig. 8.2A). She compared this specimen (maxilla with P^3–M^1) extensively with other sciuravids and with paramyines, as well as with primitive ctenodactyloids from Asia and the earliest hystricomorph from South America. Since its original description, no specimens (either upper or lower dentitions) have been recovered from the Bridgerian that can be referred to the same species as AMNH 12118, so little can be added to Dawson's (1962) discussion. The attachment of the metaloph to the protoloph in AMNH 12118 is otherwise only known in the Clarkforkian *Apatosciuravus*. However, the larger size of the hypocone, and several other characters of the premolars and molars of AMNH 12118 clearly separate it from *A. bifax*.

6. Classification

Sciuravidae Miller and Gidley, 1918
    Sciuravus Marsh, 1871 (W–U)
      S. nitidus Marsh, 1871* (Br)
      S. bridgeri Wilson, 1938 (Br)
      S. altidens Peterson, 1919 (U)
      S. eucristadens Burke, 1937 (Br)
      S. wilsoni Gazin, 1961 (W)
      S. popi Dawson, 1966 (U)
    Tillomys Marsh, 1872 (Br)
      T. senex Marsh, 1872* (Br)
      ?*T. parvidens* (Marsh, 1872) (Br)
    Taxymys Marsh, 1872 (Br)
      T. lucaris Marsh, 1872* (Br)
      ?*T. progressus* Wilson, 1938 (Br)
    Knightomys Gazin, 1961 (W–Br)
      K. depressus (Loomis, 1907) (W–Br)
      K. senior (Gazin, 1952)* (W)
      K. huerfanensis (Wood, 1962) (W–Br)
      K. minor (Wood, 1965) (W)
      ?*K. cuspidatus* (Bown, 1982) (Br)
      K. reginensis (Korth, 1984) (W)
      K. cremneus Ivy, 1990 (W)
    Prolapsus Wood, 1973 (U)
      P. sibilatoris Wood, 1973* (U)
      P. junctionis Wood, 1973 (U)
    Floresomys Fries, Hibbard, and Dunkle, 1955 (?U)
      F. guanajuatoensis

Fries, Hibbard, and Dunkle, 1955* (?U)
Pauromys Troxell, 1923 (W–Br)
  P. perditus Troxell, 1923* (Br)
  P. schaubi Wood, 1959 (Br)
  P. sp. Korth, 1984 (W)
  P. exallos Emry and Korth, 1989 (Br)
  P. texensis Walton, 1993 (U)
  P. simplex Walton, 1993 (U)
Unnamed genus
  "*Sciuravus*" *rarus* Wilson, 1938 (Br)
  "*Microparamys*" *scopaiodon* Korth, 1984 (W)

Chapter 7
Cylindrodontidae

1. Characteristic Morphology	67
1.1. Skull	67
1.2. Dentition	68
1.3. Skeleton	70
2. Evolutionary Changes in the Family	70
3. Fossil Record	70
4. Phylogeny	71
4.1. Origin	71
4.2. Intrafamilial Relationships	74
4.3. Extrafamilial Relationships	74
5. Problematical Taxa	75
5.1. *Sespemys*	75
5.2. Tsaganomyinae	75
5.3. ?*Anomeomys lewisi*	75
6. Classification	76

1. Characteristic Morphology

1.1. Skull

The skull of pre-Chadronian cylindrodonts is unknown, so this description is based on the known skulls of derived Chadronian species. In general shape, the skull is low and broad with a short, dorsoventrally deep rostrum (Fig. 7.1). The infraorbital foramen is small and unmodified by attachment of the masseter muscle (protrogomorphous). A low sagittal crest runs the length of the cranium. The orbital foramina are situated more anteriorly than in primitive ischyromyids. The bullae are inflated, ossified, and strongly attached to the skull. The mandible is deep with a strongly convex ventral margin. The scar for the masseter extends anteriorly only to below the posterior molars. The angle of the mandible is broad with a robust ventral ridge of the masseteric scar which gives the appearance that there is some lateral extension (considered "subhystricognathous" by Wood, 1980, 1984).

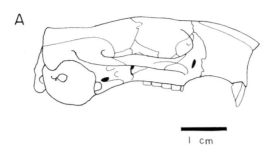
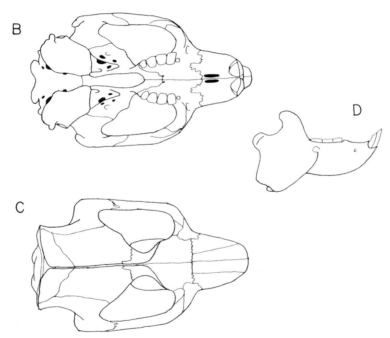

FIGURE 7.1. Skull and mandible of Cylindrodontidae. (A–C) Skull of *Ardynomys occidentalis*. (A) Lateral view; (B) ventral view; (C) dorsal view. (D) Lateral view of the mandible of *Cylindrodon fontis*. A and B from Wahlert (1974), C from Wood (1974), D from Wood (1984).

1.2. Dentition

The incisors are broad and thick with a rounded anterior surface. The enamel microstructure is pauciserial or uniserial (Wahlert, 1968). P³ is retained in all cylindrodonts except some species of the Chadronian *Cylindrodon*. Crown height of cheek teeth ranges from brachydont to hypsodont with unilateral hypsodonty on the lingual side of the upper molars in some intermediate species. Molars are weakly lophate to lophate with reduced or absent conules on the upper cheek teeth (Fig. 7.2). Mesolophs (-ids) are lacking on all cheek teeth. Enamel of the cheek teeth is thicker than in paramyines. Cheek teeth are circular (cylindrodontines) to ovate (jaywilsonomyines) in occlusal outline. The number of lophs varies from three (jaywilsonomyines) to four (cylindrodontines).

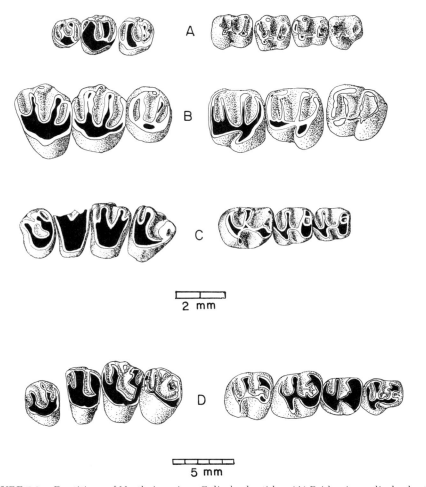

FIGURE 7.2. Dentitions of North American Cylindrodontidae. (A) Bridgerian cylindrodontine *Mysops parvus*, LP^4–M^2 and RP_4–M_3. (B) Chadronian cylindrodontine *Ardynomys occidentalis*, LM^1–M^3 and LM_1–M_3. (C) Uintan jaywilsonomyine *Pareumys grangeri*, LP^4–M^3 and RM_1–M_3. (D) Chadronian jaywilsonomyine *Jaywilsonomys ojinagaensis*, L^4–M^3 and RP_4–M_3. D to different scale (below). A from Wilson (1938b), B and D from Wood (1974).

1.3. Skeleton

No part of the postcranial skeleton of a North American cylindrodont has been described. Wood (1980), however, assumed that the skeleton of these rodents paralleled that of other fossorial rodents (short, broad limb bones with wide articular surfaces) because of his proposed relationship between cylindrodonts and the Asian tsaganomyines for which the skeleton is known to be fossorially adapted. The relationship between the Tsaganomyinae and cylindrodonts is highly questionable (see discussion below) and therefore this inference is not wholly valid. However, the skull of Oligocene cylindrodonts is broad and heavy, as in some other rodents that are known to be fossorial (such as the palaeocastorine beavers, Mylagaulidae, and Old World rhizomyids and bathyergids). The correlation of the skull shape to the development of fossorial adaptations of the postcranium is not definite, so any suggestion that the postcranial skeleton of cylindrodonts is anything more than the primitive condition found in paramyine ischyromyids is clearly conjecture.

2. Evolutionary Changes in the Family

The dental changes from weakly lophate Wasatchian and Bridgerian species occurred quickly in both subfamilies. Within the Cylindrodontinae (*Cylindrodon*), total hypsodonty was attained by the Chadronian without any known morphologically intermediate species from the Uintan or Duchesnean. In the Jaywilsonomyinae, unilateral hypsodonty (or its beginnings) is present even in the earliest species of *Pareumys*.

Typically, size also appears to increase from the Bridgerian and Uintan to the Chadronian species. The Chadronian species have also developed the more robust mandible than the earlier species, and the cheek teeth of the Chadronian cylindrodonts are higher crowned even in the genera that did not attain complete hypsodonty.

3. Fossil Record

The first record of the Cylindrodontidae in North America is from the Wasatchian of Wyoming (Korth, 1984). The last reported record is in the Chadronian (at least middle Chadronian and possibly late). Interestingly, the cylindrodonts attained their greatest diversity in the Chadronian (4 genera, 10 species) just before their extinction (Fig. 7.3). The only record of cylindrodontids outside of North America are the genera *Ardynomys*, *Morosomys*, and *Anomeomys* from the early and middle Oligocene of Asia (Matthew and Granger, 1925; Shevyreva, 1972; Wang, 1986). Flynn *et al.* (1986) suggested that the early Oligocene genus *Hulgana* Dawson (1968b) from China was referable

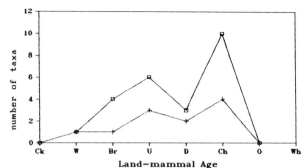

FIGURE 7.3. Occurrence of Cylindrodontidae in the Tertiary of North America (□, species; +, genera). Abbreviations as in Fig. 5.4.

to the Cylindrodontidae. However, the simplified tooth pattern of *Hulgana* and its lack of lophodonty (attained even in the earliest cylindrodonts in North America) make this allocation less than likely.

During Bridgerian through Duchesnean times, the diversity of cylindrodonts remained nearly the same, but there was a dominant genus during each time that contained nearly all of the known species. In the Bridgerian the dominant genus was *Mysops*, known from four species, all from the Bridger Basin of Wyoming. During the Uintan and Duchesnean, the dominant genus was *Pareumys*, known from four (or possibly five) species. It is also unusual that these two genera represent the two recognized subfamilies (*Mysops* = Cylindrodontinae; *Pareumys* = Jaywilsonomyinae). There were species of *Pseudocylindrodon* (a cylindrodontine) in the Uintan and Duchesnean, but these were found only in Saskatchewan and central Wyoming, generally farther north than the typical occurrence of Uintan *Pareumys*. This may indicate a regional separation which allowed the separate evolution of cylindrodontines in the north and jaywilsonomyines in the south.

4. Phylogeny (Fig. 7.4)

4.1. Origin

The origin of the family is difficult to trace. The earliest proposed cylindrodontid, *Dawsonomys woodi* from the Wasatchian, already had many of the familial characters developed (lophodonty of cheek teeth, reduced mesoconid, elevated trigonid, deeper mandible, broad I_1). The morphology of the cheek teeth of *Dawsonomys* is not easily derivable from any primitive ischyromyid or sciuravid. Since *D. woodi* is only known from a single lower jaw, the possibilities for comparison are clearly limited.

Mysops and *Dawsonomys* have been previously considered sciuravids because of the degree of lophodonty of the cheek teeth and relative small size of the species (Wilson, 1949a; Gazin, 1961). However, the sciuravids com-

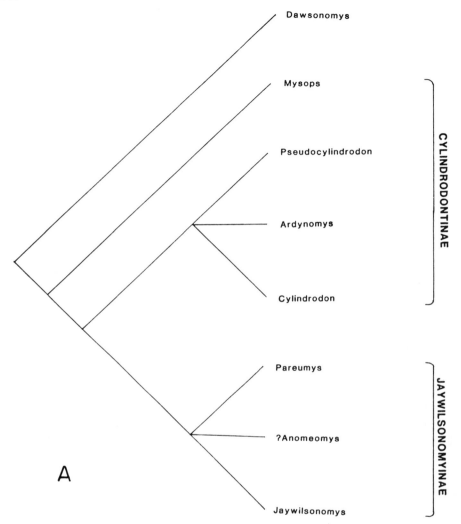

FIGURE 7.4. Phylogeny of the Cylindrodontidae. (A) Cladogram of cylindrodontid relationships. (B) Dendrogram of the Cylindrodontidae with age of occurrence (see Fig. 5.5 for explanation).

monly have transversely elongate mesoconids (= mesolophid) on the lower molars unlike *Mysops* and the cylindrodonts, and the upper molars of sciuravids are characterized by an enlarged hypocone, whereas in cylindrodonts the hypocone is reduced or absent.

It is possible that the cylindrodontid dental pattern was derived directly from the primitive ctenodactyloid pattern. In the lower molars there would

Cylindrodontidae

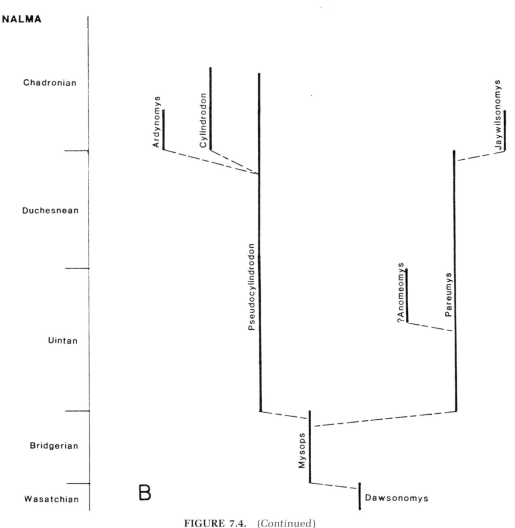

FIGURE 7.4. (Continued)

have to be a reduction in the hypoconulid and increased lophodonty, but the general morphology of the cheek teeth would remain little changed with the small mesoconid, directly anteriorly oriented ectolophid, and elevated trigonid with no anterior cingulum. The upper molars, however, are more difficult to derive. The hypocone and conules would have to be greatly reduced. It is possible that when the upper cheek teeth of *Dawsonomys* are known, they may bridge the morphological gap between the earliest Eocene ctenodactyloid dental pattern and that of the Bridgerian cylindrodonts.

4.2. Intrafamilial Relationships

Among the Cylindrodontinae, it is evident that the cheek teeth of *Pseudocylindrodon* can be easily derived from that of *Mysops* with an increase in lophodonty and hypsodonty. With even greater crown height of the cheek teeth, the *Pseudocylindrodon* dentition could give rise to that of *Cylindrodon*. It is quite possible that one of the known Uintan or Duchesnean species of *Pseudocylindrodon* was ancestral to *Cylindrodon* from the Chadronian, but no known species of cylindrodontine has a crown height intermediate between these two genera.

Among the Jaywilsonomyinae, there seems to be a similar pattern. The upper molars are ovate in occlusal outline rather than circular as in cylindrodontines and the upper cheek teeth are unilaterally hypsodont even in the most primitive species (Wood, 1973). Instead of having four transverse lophs on the cheek teeth, the jaywilsonomyines have three lophs and a partial fourth (metalophulid II on the lower molars, metaloph on the upper molars). The Uintan and Duchesnean *Pareumys* can be easily considered ancestral to the Chadronian *Jaywilsonomys* with only an increase in the size, robusticity, and crown height of the cheek teeth.

The most primitive *Pareumys* based on dental morphology is *P. boskeyi* from Texas. This species was originally referred to *Mysops* by Wood (1973) because it retained many primitive dental features of the latter. However, the oval shape and unilateral hypsodonty of the upper molars and the incompleteness of the fourth loph (-id) make it clearly a jaywilsonomyie. The retention of the primitive features of the molars may indicate that *Pareumys* (and in turn the Jaywilsonomyinae) were originally derived from a *Mysops*-like ancestor.

As discussed by Korth (1984), the dental pattern of *Mysops* is easily derivable from the Wasatchian *Dawsonomys*, thus implying that *Dawsonomys* was ancestral to all later cylindrodonts. Because of the poor record of *Dawsonomys*, it cannot be placed in any currently recognized subfamily.

4.3. Extrafamilial Relationships

Clearly, the Cylindrodontidae are too specialized to be ancestral to any later rodents. It is also difficult to determine any definite sister group of the family.

Based on cranial foramina, the Chadronian cylindrodonts share more derived characters with the Ischyromyinae (Wahlert, 1974) than with any other protrogomorphous rodent. Clearly the uniserial enamel of the cylindrodonts was attained separate from that of the ischyromyids because the early and middle Eocene species of each family are primitively pauciserial. Dawson (1977) pointed out similarities of the upper cheek teeth of cylindrodonts with early Tertiary Asian ctenodactyloids. These similarities may be the

5. Problematical Taxa

5.1. *Sespemys*

The species *Sespemys thurstoni* from the Arikareean of California was originally referred to the Prosciurinae (Wilson, 1934, 1949c), but several authors have considered this species a cylindrodont (Burke, 1936; Wood, 1937, 1980; Stehlin and Schaub, 1951). It is believed here that *S. thurstoni* is a prosciurine aplodontid that is convergent with cylindrodonts. The occlusal pattern of the cheek teeth of *S. thurstoni* is closer to that of *Prosciurus*, but the massive nature of the tooth crown and large size of the lower incisor are similar to those of cylindrodonts. Stratigraphically, the reference of *Sespemys* to the Prosciurinae is much more consistent with the known record of that subfamily than with that of the Cylindrodontidae.

A mandible from the Whitneyan of Nebraska was identified by Wood (1937) as "cf. *Sespemys* sp." This specimen has been shown to represent an as yet unnamed genus and species of prosciurine (Korth, 1989a).

5.2. Tsaganomyinae

Wood (1974) thoroughly reviewed the systematics of the middle to late Oligocene Asian tsaganomyids and included them as a subfamily of the Cylindrodontidae. However, these rodents are bathyergid-like in that they have protrogomorphous skulls, hystricognathous mandibles, and procumbent incisors, though they predate the earliest bathyergid from Africa which is at least partially hystricomorphous (Lavocat, 1973). Recent studies have revealed that the incisor microstructure of tsaganomyines is a derived multiserial pattern as in hystricognath rodents and not uniserial as in cylindrodonts (Wahlert, personal communication). The Tsaganomyidae still remain as a problematical family (or subfamily), but they definitely cannot be referred to the Cylindrodontidae.

5.3. ?*Anomeomys lewisi*

Black (1974) named *Pareumys lewisi* from the Uintan of Wyoming based on a single maxillary specimen with worn cheek teeth. The unique feature of the species was an anteroposteriorly oriented loph in the center of the tooth. Later, Wang (1986) named the genus *Anomeomys* from the Eocene of China with this same accessory loph on the molars and noted the similarity with Black's species. *P. lewisi* is here questionably referred to this Asian genus pending recovery of additional unworn material. The appearance of an Asian species in North America is not inconsistent with the remainder of the Uintan fauna. Other immigrant genera, such as the lagomorph *Mytonolagus*, similarly occur at this level.

6. Classification

Cylindrodontidae Miller and Gidley, 1918
  Cylindrodontinae Miller and Gidley, 1918
    Mysops Leidy, 1871 (Br)
      M. minimus Leidy, 1871* (Br)
      M. parvus parvus (Marsh, 1872) (Br)
      M. fraternus Leidy, 1873 (Br)
      M. parvus plicatus (Troxell, 1923) (Br)
    Cylindrodon Douglass, 1902 (Ch)
      C. fontis Douglass, 1902* (ch)
      C. nebraskensis Hough and Alf, 1956 (Ch)
      C. collinus Russell, 1972 (Ch)
    Pseudocylindrodon Burke, 1935 (U–Ch)
      P. neglectus Burke, 1935* (Ch)
      P. medius Burke, 1938 (Ch)
      P. tobeyi Black, 1970 (D)
      P. texanus Wood, 1974 (Ch)
      P. citofluminis Storer, 1984 (U)
      P. lateriviae Storer, 1988 (D)
    Ardynomys Matthew and Granger, 1925 (Ch)
      A. saskatchewaensis (Lambe, 1908) (Ch)
      A. occidentalis Burke, 1936 (Ch)
  Jaywilsonomyinae Wood, 1974
    Pareumys Peterson, 1919 (U–D)
      P. milleri Peterson, 1919* (U)
      P. grangeri Burke, 1935 (U)
      ?*P. troxelli* Burke, 1935 (U)
      P. guensburgi Black, 1970 (D)
      P. boskeyi (Wood, 1973) (U)
    Jaywilsonomys Ferrusquia and Wood, 1969 (Ch)
      J. ojinagaensis Ferrusquia and Wood, 1969* (Ch)
      J. pintoensis Ferrusquia and Wood, 1969 (Ch)
    Anomeomys Wang, 1986
      ?*A. lewisi* (Black, 1974) (U)
  Uncertain subfamily
    Dawsonomys Gazin, 1961 (W)
      D. woodi Gazin, 1961* (W)

Chapter 8
Protoptychidae

1. Characteristic Morphology	77
1.1. Skull	77
1.2. Dentition	78
1.3. Skeleton	80
2. Evolutionary Changes in the Family	80
3. Fossil Record	81
4. Phylogeny	81
4.1. Origin	81
4.2. Intrafamilial Relationships	82
4.3. Extrafamilial Relationships	82
5. Problematical Taxon: *Presbymys*	83
6. Classification	83

1. Characteristic Morphology

1.1. Skull

Presbymys is only known from dental elements (Wilson, 1949d), thus the description of the skull and skeleton of protoptychids is limited to the morphology of *Protoptychus* (Fig. 8.1). The skull of *Protoptychus* has a narrow, elongated rostrum. The infraorbital foramen is enlarged and opens anteriorly indicating the invasion of the foramen by the masseter muscle. The ventral surface of the anterior zygoma is unmodified. This condition is hystricomorphous (Scott, 1895; Wahlert, 1973). Turnbull (1991) suggested that the enlargement and invasion of the infraorbital foramen in *Protoptychus* had some similarity to that of the earliest myomorphous rodents and therefore might also be considered myomorphous. The auditory bulla of *Protoptychus* is greatly inflated and the surrounding bones are modified to accommodate the inflation. The cranial foramina are not greatly modified from the primitive condition as seen in primitive ischyromyids (Wahlert, 1973; Turnbull, 1991), except for the enlargement of the infraorbital and incisive foramina in *Protoptychus*.

The mandible of *Protoptychus* is sciurognathous or " . . . weakly or incipiently hystricognathous" (Turnbull, 1991, p. 15). The masseteric scar is V-shaped. It ends anteriorly below the posterior margin of P_4. The ascending ramus is relatively thin anteroposteriorly and high. The coronoid process extends far above the condyle and is long and slender.

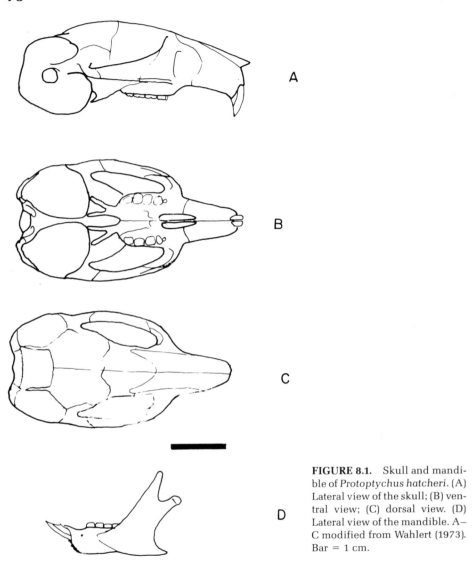

FIGURE 8.1. Skull and mandible of Protoptychus hatcheri. (A) Lateral view of the skull; (B) ventral view; (C) dorsal view. (D) Lateral view of the mandible. A–C modified from Wahlert (1973). Bar = 1 cm.

1.2. Dentition

The microstructure of the enamel of the incisors is pauciserial in *Protoptychus* (Wahlert, 1973). *Protoptychus hatcheri* retains the primitive dental formula of rodents but *P. smithi* and *Presbymys* lack P^3. The molars are lophate, but the individual major cusps of the cheek teeth are still recognizable (Fig. 8.2). P^3 is peglike in *P. hatcheri*. P^4 is submolariform and smaller

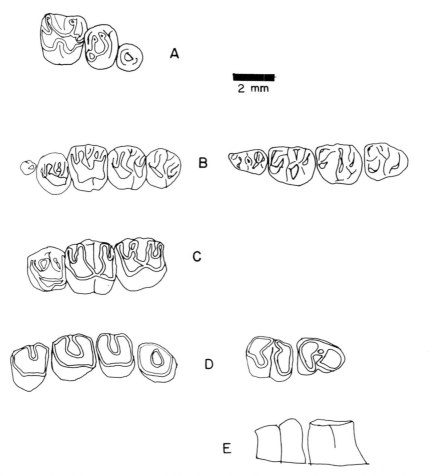

FIGURE 8.2. Dentitions of Protoptychidae. (A) Bridgerian unnamed sciuravid AMNH 12118, RP^3–M^1. (B) Uintan *Protoptychus hatcheri*, LP^3–M^3 and LP$_4$–M$_3$. (C) Uintan *P. smithi*, LP^4–M^2. (D) Duchesnean *Presbymys lophatus*, LP^4–M^3 and RP$_4$–M$_1$. (E) Lateral view of RP$_4$–M$_1$ of *P. lophatus*. A from Dawson (1962), B from Turnbull (1991), C from Wilson (1937b), D and E from Wilson (1949d).

than the molars. The hypocone is relatively smaller and the metaconule is larger than in the molars. The hypocone on the upper molars is subequal to that of the protocone, and the conules are not distinguishable from the proto- and metalophs. The anterior and posterior cingula are nearly as well developed as the major lophs. M^3 is smaller than the anterior molars and is reduced posteriorly. The lower cheek teeth are dominated by two major lophs, metalophulid I and the hypolophid. The hypoconulid is large. The mesoconid is lacking, and the connection between the hypoconid and protoconid is weak or absent. P_4 is smaller than the molars and is submolariform, the trigonid being much narrower (buccolingually) than the talonid. M_3 is reduced posteriorly similar to the condition of M^3. The cheek teeth of *Protoptychus* are mesodont in crown height and those of *Presbymys* are subhypsodont and rooted. The upper cheek teeth of *Presbymys* are unilaterally hypsodont, being higher crowned lingually.

1.3. Skeleton

In the postcranial skeleton of *Protoptychus* the forelimbs are somewhat reduced, the hindlimbs and pes are greatly enlarged and elongated with tibia and fibula fused, and the tail is lengthened. Turnbull (1991) showed that the proportions of the skeleton compared favorably with those of modern dipodoids and heteromyids that are adapted to ricochetal locomotion (Fig. 8.3).

2. Evolutionary Changes in the Family

The only major change from the Uintan *Protoptychus* to the Duchesnean *Presbymys* is in the crown height of the cheek teeth. Teeth of *Protoptychus* are mesodont and those of *Presbymys* are subhypsodont.

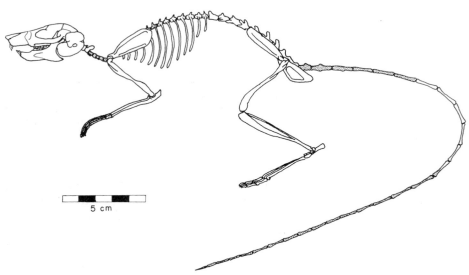

FIGURE 8.3. Skeleton of *Protoptychus hatcheri*. Reconstruction based on descriptions by Turnbull (1991). Shaded areas represent bones not known in the fossil record.

3. Fossil Record

All known specimens of *Protoptychus* have been reported from the Uintan of the Washakie Basin of Wyoming and the Uinta Basin of Utah (Turnbull, 1991). *Presbymys* is only known from the Duchesnean Pearson Ranch fauna of the Sespe Formation in southern California (Wilson, 1949d; Fig. 8.4).

4. Phylogeny (Fig. 8.5)

4.1. Origin

Protoptychus shares a number of dental similarities with ctenodactyloids: large hypocones on upper molars and lower molars with large hypoconulids strongly developed metalophulid I. The conules on the upper molars are small and the mesoconid is absent on the lower molars in contrast to those of early ctenodactyloids. Features of the cheek teeth of *Protoptychus* that are shared with ctenodactyloids may only be primitive. This may indicate that protoptychids developed independently from a ctenodactyloid stock separate from ischyromyids and sciuravids in North America.

The only pre-Uintan North American rodent that has a dentition similar to *Protoptychus* is an unnamed Bridgerian sciuravid AMNH 12118 (Dawson, 1962; see discussion in Chapter 6, Section 5.3) which has a large hypocone on the upper molar which is lophate though less lophate than those of *Protoptychus*. The upper premolars of AMNH 12118 are also specialized differently than in *Protoptychus*. It is remotely possible that this Bridgerian species was ancestral to *Protoptychus*.

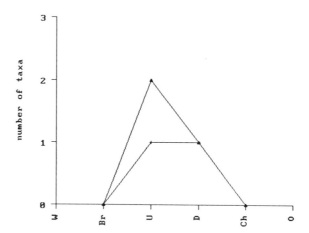

FIGURE 8.4. Occurrence of Protoptychidae in the Tertiary of North America (▲, species; + genera). Abbreviations as in Fig. 5.4.

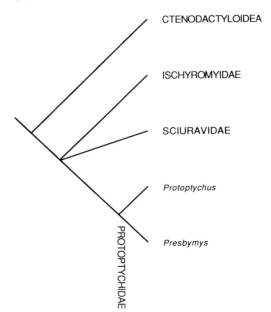

FIGURE 8.5. Cladogram of protoptychid relationships.

4.2. Intrafamilial Relationships

The only evident relationship within the Protoptychidae is the probable ancestry of *Presbymys* within *Protoptychus*. Wilson (1949d) noted the similarity in the cheek teeth of *Presbymys* and *Protoptychus*. The occlusal morphology of the former is nearly identical to that of the Uintan *Protoptychus*, especially *P. smithi*, differing only in having much higher crowned cheek teeth.

4.3. Extrafamilial Relationships

The Protoptychidae has been assigned to a number of different rodent groups throughout its history. Because of the inflation of the auditory bullae and the adaptation of the postcranial skeleton for ricochetal locomotion, many authors have included *Protoptychus* in the Dipodidae (Scott, 1895; Matthew, 1910), or as an offshoot of the Ischyromyidae, possibly heading toward the dipodoids (Wood, 1935a; Wilson, 1937b, 1949a; Simpson, 1945; Turnbull, 1991). Wahlert (1973), however, suggested that *Protoptychus* was involved with the ancestry of the South American hystricomorphous–hystricognathous Caviomorpha but showed some similarity in dental morphology to primitive cylindrodonts. Dawson (1977) noted the great similarity in the dentition of *Protoptychus* to that of the Oligocene ctenodactyloid

Yuomys from Asia. There is also a great deal of similarity in the occlusal morphology of the cheek teeth of *Protoptychus* to that of the problematical *Guanajuatomys* from central Mexico (see Chapter 23, Section 3.3).

The specializations of the skull and postcranial skeleton of *Protoptychus* appear to be convergent with those of modern dipodoids because *Protoptychus* retains pauciserial enamel of the incisors (not uniserial as in dipodoids) and a complete dental formula (P_4 and P^3 lost and P^4 reduced to peg or absent in dipodoids). The specializations of the occlusal morphology of the cheek teeth of the dipodoids are not present in *Protoptychus* (see Chapter 19).

The hystricomorphy of the skull of *Protoptychus* also differs from that of dipodoids in lacking the smaller accessory foramen contained within the infraorbital foramen of dipodoids.

5. Problematical Taxon: *Presbymys*

Wilson (1949d) originally allocated *Presbymys lophatus* questionably to the Cylindrodontinae of the Ischyromyidae based on the crown height (hypsodonty) and lophodonty of the cheek teeth. However, the distinctive occlusal pattern of the cheek teeth (especially the lowers) of *Presbymys* and the compressed lower incisor (not anteriorly broadened as in cylindrodonts) differentiate *Presbymys* from all known cylindrodonts. The similarity of the upper cheek teeth of *Presbymys lophatus* to those of *Protoptychus smithi* was noted by Wilson (1949d). With the description of the lower cheek teeth of *P. hatcheri* (Turnbull, 1991) comparison is now possible with those of *Presbymys lophatus*. Other than hypsodonty, there is little to separate the P_4 and M_1 of *Presbymys* with those of *Protoptychus*. The trigonid of P_4 is much narrower (buccolingually) than the talonid, and the cusps of the talonid on P_4 and M_1 are not directly connected with those of the trigonid (no ectolophid or mesoconid).

It appears that the Duchesnean *Presbymys lophatus* is the latest surviving protoptychid that has attained greater hypsodonty.

6. Classification

Protoptychidae Tullberg, 1899
    Protoptychus Scott, 1895 (U)
      P. hatcheri Scott, 1895* (U)
      P. smithi Wilson, 1937 (U)
    Presbymys Wilson, 1949 (D)
      P. lophatus Wilson, 1949* (D)

Chapter 9
Aplodontidae

1. Characteristic Morphology	85
1.1. Skull	85
1.2. Dentition	87
1.3. Skeleton	87
2. Evolutionary Changes in the Family	89
3. Fossil Record	89
4. Phylogeny	91
4.1. Origin	91
4.2. Intrafamilial Relationships	93
4.3. Extrafamilial Relationships	93
5. Problematical Taxa	94
5.1. *Eohaplomys*	94
5.2. *Sespemys*	94
5.3. *Cedromus*	95
5.4. *Mylagaulodon*	95
5.5. *Epeiromys*	95
5.6. *Horatiomys*	96
6. Classification	96

1. Characteristic Morphology

1.1. Skull

The zygomasseteric structure of aplodontids is the primitive protrogomorphous condition (Fig. 9.1). The rostrum is dorsoventrally deep and the incisive foramina are small. The diastemata are short especially in aplodontines. In later species, the neurocranium is low and broad. There is never a single sagittal crest, but always broadly separated lyrate crests that unite at the midline of the cranium posteriorly or parallel parasagittal crests. Where known, all fossil aplodontids possess a short, laterally directed postorbital process. In the Recent genus *Aplodontia*, this is reduced to a small shelf above the orbit. The palate is broad and flat or slightly convex ventrally. There is a wide variety in the range of morphology of the cranial foramina from the most primitive to Recent genera. Primitively, the pterygoid flanges are long, the masticatory and buccinator foramina are united, and a stapedial foramen, an accessory foramen ovale, and a sphenofrontal foramen are present (Korth and Emry, 1991). In *Aplodontia* and more advanced fossil genera, the pterygoid

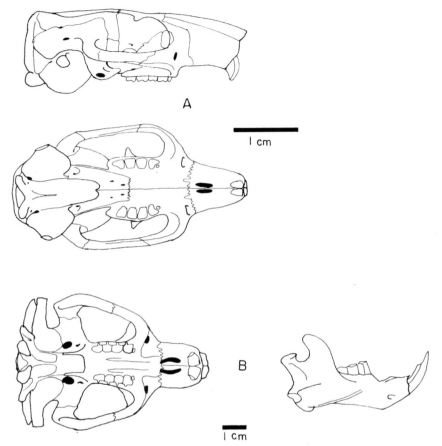

FIGURE 9.1. Skull and mandible of Aplodontidae. (A) Lateral and ventral views of the skull of Orellan prosciurine *Prosciurus relictus*. (B) Ventral view of skull and lateral view of mandible of the Recent aplodontine *Aplodontia rufa* (see also Fig. 2.3A and 2.6B). A (composite) from Wahlert (1974) and Korth and Emry (1991), B from Ellerman (1940). Skulls not to same scale.

area is much shortened (thus eliminating the accessory foramen ovale), the masticatory and buccinator foramina are separate, and the stapedial and sphenofrontal foramina are lost as a result of loss of the internal carotid artery. Much of this change in the skull is reflected in the increasing hypsodonty in the later aplodontines. Aplodontines possess an interpremaxillary foramen on the palatal surface of the premaxillary anterior to the incisive foramen, which is lacking in all other aplodontids. Early genera have inflated bullae but aplodontines have reduced bullae with an elongate external meatus. Where the skull is known, all aplodontids except the early prosciurine *Prosciurus* have septate bullae.

The mandible is deep and broad in meniscomyines and aplodontines, and much more gracile in prosciurines and allomyines. The masseteric fossa is bounded by ridges both dorsally and ventrally and extends anteriorly to below the lower premolar or anterior edge of the first molar. A unique feature of the mandible of aplodontines is the angular process which extends medially, forming a horizontal process. There is no indication of the medial extension of the angular process in other aplodontids in which the angle of the mandible is as in other primitive rodents.

1.2. Dentition

All aplodontids retain the primitive rodent dental formula. In early genera, the incisors are relatively narrow with convex anterior surfaces. In meniscomyines and aplodontines they are broader and less strongly convex. The microstructure of incisor enamel for all known aplodontids is uniserial in pattern (Wahlert, 1968).

The cheek teeth of aplodontids range from brachydont in prosciurines and allomyines to hypsodont in meniscomyines and aplodontines (Fig. 9.2). In several lineages of aplodontids, the last premolar is larger than the first molar. Primitively, the upper molars had large conules, a reduced hypocone, an accessory protocone crest (Rensberger, 1975), and a well-developed mesostyle that became enlarged and associated with a complete ectoloph and crescentic buccal cusps in several lineages. P^4 is molariform with an expanded anterobuccal area, developing an anterocone in most species. In hypsodont species, the occlusal pattern is lost in early wear, but the outline of the upper molars maintains a well-developed mesostyle.

The lower cheek teeth primitively possess a partial or complete hypolophid and an entoconid widely separated from the posterolophid. The lower premolar is generally molariform but the trigonid is always narrower relative to the talonid than in the molars. As in the upper cheek teeth, in more hypsodont species the occlusal pattern of the lower molars disappears in relatively early wear, leaving only a few isolated enamel lakes. There are dentine tracts (enamel failure) on the sides of the molars of many meniscomyines, a common occurrence in many rodents that achieve hypsodonty.

1.3. Skeleton

No postcranial skeletal material has been definitely referred to a Tertiary aplodontid. Thus, the skeletal specializations described here are based on the skeleton of the Recent *Aplodontia* which is the most specialized genus in cranial and dental characteristics. The specializations may be related to its fossorial habits, which are not known for fossil species.

The skeleton is generally robust with short limbs and well-developed flanges and tubercles for muscular attachments. The girdles are strongly

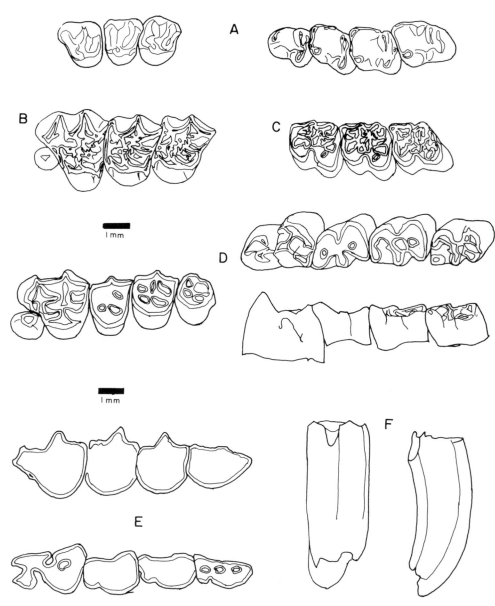

FIGURE 9.2. Dentitions of Tertiary Aplodontidae. (A) Orellan prosciurine *Prosciurus relictus*, LP^4–M^2 and LP_4–M_3. (B) Arikareean allomyine *Allomys simplicidens*, LP^3–M^2. (C) Arikareean allomyine *Allomys tessellatus*, LM_1–M_3. (D) Arikareean meniscomyine *Meniscomys hippodus*, LP^3–M^3, RP_4–M_3 and medial view of RP_4–M_3. (E) Barstovian aplodontine *Liodontia alexandrae*, LP^3–M_3 and LP_4–M_3. (F) Buccal (left) and posterior (right) view of upper molar of *L. alexandrae*. A and E from McGrew (1941a), B–D from Rensberger (1983), F from Shotwell (1958).

developed. The hind limb is less robust than the forelimb. The fibula does not fuse with the tibia.

2. Evolutionary Changes in the Family

The earliest and possibly ancestral aplodontids (Prosciurinae) have brachydont and cuspate cheek teeth. In later Tertiary aplodontids, the teeth achieve differing levels of hypsodonty, culminating in the totally hypsodont (ever-growing) cheek teeth of the aplodontines. Within the Allomyinae (Arikareean to ?Barstovian), there is very little indication of increase in crown height; rather there is an increase in the complexity of enamel (number and height of lophules; Rensberger, 1983). Within the Prosciurinae there is a tendency to enlarge the last premolars and develop an ectoloph on the upper molars. In all other (later) subfamilies, the last premolars are larger than the first molars and an ectoloph is developed with a pronounced mesostyle projecting buccally.

In prosciurines the skull and mandible are more generalized, not differing greatly from those in reithroparamyine ischyromyids. The dorsal margin of the mandibular diastema lacks the distinct ridge of ischyromyids. The auditory bulla is inflated and ossified and a stapedial foramen is present (= stapedial branch of the internal carotid artery), again quite similar to that of *Reithroparamys* (Meng, 1990). There is no indication of the broadening of the neurocranium and shortening of the pterygoid region and rostrum present in aplodontines. One specialization of *Prosciurus* not known for any other aplodontids (including other prosciurines) is the fusion of the buccinator and masticatory foramina in the alisphenoid bone. In allomyines and later prosciurines, these foramina are separated (as in *Aplodontia*) and the pterygoid region is shortened (Korth and Emry, 1991). The stapedial foramen (and artery) is also lost in all other known aplodontids. The auditory bulla in primitive aplondontids is inflated and that of aplodontines is much smaller with an elongate external meatus.

Thus, it appears that the general trend of evolutionary change in aplodontids is from a primitive, generalized skull and dentition to a more specialized morphology which includes higher crowned teeth and probably fossorial adaptations that resulted in shortening and broadening of the skull and a robust appendicular skeleton.

3. Fossil Record

The oldest known aplodontid is *Spurimus* from the Uintan of Wyoming (Black, 1971). The greatest diversity of the family in North America is in the Arikareean (Fig. 9.3). The earliest and most primitive subfamily is the Prosciurinae which appears in the Uintan and last occurs in the Arikareean. The greatest diversity of prosciurines is in the Orellan where there are four genera

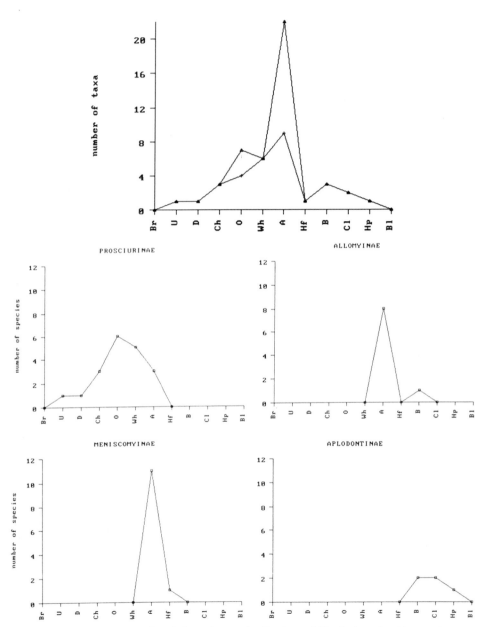

FIGURE 9.3. Occurrence of Aplodontidae in the Tertiary of North America (▲, species; +, genera). Abbreviations as in Fig. 5.4.

and seven species. The great diversity of aplodontids (over 20 species) in the Arikareean is related almost exclusively to two subfamilies, the Allomyinae and Meniscomyinae. Both of these subfamilies are short-lived and only one species from both families is known from the Hemingfordian (Rensberger, 1983). The problematical species "*Allomys*" *stirtoni* from the Barstovian of Nebraska (Klingener, 1968; Voorhies, 1990a) is the only allomyine that exists past the Arikareean.

Aplodontines first occur in North America in the Barstovian and are never very diverse (never more than two species) until the Blancan from which no aplodontids are known. Only one species of aplodontid currently exists, *Aplodontia rufa*.

Outside North America the record of aplodontids is spotty, and never continuous throughout any lengthy span of the Tertiary. In the Oligocene and into the earliest Miocene, the record of allomyine aplodontids is diverse in Europe, paralleling the development of the North American allomyines (Schmidt-Kittler and Vianey-Liaud, 1979). The only other Oligocene aplodontid reported from Europe is the prosciurine *Ephemeromys* from Germany (Wang and Heissig, 1984).

In Asia, Oligocene aplodontids are more rare but are equally as diverse as those from Europe. Three prosciurines and a primitive meniscomyine have been reported from the middle Oligocene of China (Rensberger and Li, 1986; Wang, 1987), and one prosciurine has been reported from Mongolia and the Soviet Union (Shevyreva, 1966; Kowalski, 1974; Wang, 1987). The supposed prosciurine *Prosciurus lohiculus* from the middle Oligocene of China (Matthew and Granger, 1923) was identified as a cylindrodontid and assigned a new genus by Wang (1986). The occurrence of diverse aplodontids in both Europe and Asia in the Oligocene demonstrates a worldwide radiation at that time.

In the later Tertiary, aplodontids outside North America become even more rare. A single species, *Ameniscomys selenoides*, an aplodontid of uncertain subfamilial affinities from the ?middle Miocene, is the only record of the Aplodontidae in Europe (Dehm, 1950; Rensberger, 1983). Likewise, only a single species of aplodontine, *Pseudaplodon asiatica*, is represented from the latest Miocene or earliest Pliocene of Asia (Schlosser, 1924; Shotwell, 1958).

4. Phylogeny (Fig. 9.4)

4.1. Origin

The origin of the aplodontids is most certainly within the Ischyromyidae. Several authors have included the Prosciurinae (the earliest subfamily of aplodontids) within the Ischyromyidae (Wilson, 1949a; Black, 1971) or Paramyidae (Wood, 1962). Emry and Korth (1989) suggested that the early Bridgerian ischyromyid *Reithroparamys huerfanensis* had dental and mandibular characters that would indicate an ancestral relationship with the

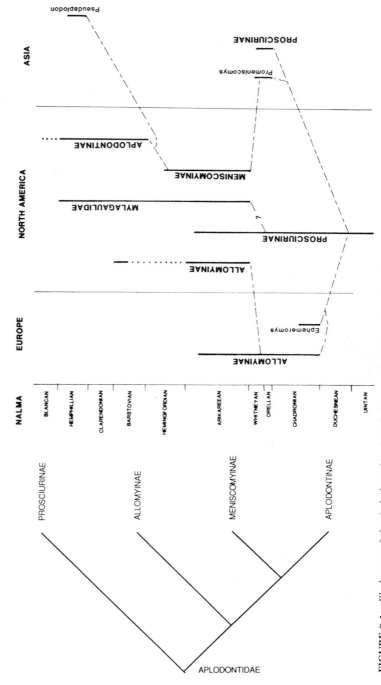

FIGURE 9.4. Phylogeny of the Aplodontidae. (A) Cladogram of aplodontid relationships. (B) Dendrogram of the Aplodontidae with age of occurrence (see Fig. 5.5 for explanation). Age equivalence of European and Asian taxa approximate.

earliest prosciurines. In a study of the auditory region of the skull of the Bridgerian *R. delicatissimus*, Meng (1990) suggested a close systematic relationship between the Reithroparamyinae (= his Reithroparamyidae) and the aplodontids. Because the earliest record of prosciurines (Uintan and Duchesnean) is based on isolated teeth only (Black, 1971; Storer, 1988), and because the similarity of these teeth with those of primitive ischyromyids is obvious, it is not a long morphological step from a reithroparamyine morphology to that of the earliest prosciurines.

4.2. Intrafamilial Relationships

The Prosciurinae have long been considered ancestral to the remainder of the Aplodontidae (Matthew, 1910; Wilson, 1949a; Rensberger, 1975). However, none of the known North American prosciurines can be shown to be directly ancestral to any of the later subfamilies. The most likely ancestor for the Allomyinae is *Ephemeromys* from the Oligocene of Europe (Wang and Heissig, 1984), a prosciurine which possessed crenulate enamel on the cheek teeth and a doubled metaconule on the upper molars, diagnostic characters of the Allomyinae (Rensberger, 1983). This implies that the allomyines originated in Europe and migrated into North America near the beginning of the Arikareean rather than evolving from one of the North American species known from the early or middle Oligocene.

Likewise, the first meniscomyine is *Promeniscomys* from the middle Oligocene of China (Wang, 1987) which is more brachydont than any other meniscomyine, but which has an enlarged P^4 and upper molars with a complete ectoloph and crescentic buccal cusps as in all later meniscomyines. This indicates an Asian origin for the Meniscomyinae. Whereas the dental pattern of *Promeniscomys* can be derived from that of a generalized prosciurine, no species is currently known that is intermediate in morphology between the Prosciurinae and *Promeniscomys*. The great radiation of meniscomyines in North America in the Arikareean also appears to be the result of immigration rather than evolution from a North American ancestor.

The Aplodontinae are clearly derived from the Meniscomyinae. The increase in crown height seen in the meniscomyines is merely continued in the aplodontines. Though no complete skull of a meniscomyine is known, the auditory bulla of meniscomyines is much more highly septate than that of prosciurines and allomyines (Rensberger, 1983), a feature of the bulla in aplodontines.

4.3. Extrafamilial Relationships

The only family of rodents that is likely derived from the Aplodontidae is the Mylagaulidae. The great similarity in the cranial morphology of these two families has led some authors to believe that the mylagaulids should be just a subfamily of the Aplodontidae (Wahlert, 1974). The earliest mylagaulid, *Promylagaulus* from the Arikareean, is already distinct from contemporaneous aplodontids (McGrew, 1941a; Rensberger, 1979). The early mylagaulids parallel the meniscomyines in enlargement of the last premolar and increase in crown height of the cheek teeth but maintain a distinct occlusal pattern of the cheek teeth that is unlike that of any aplodontid. It is quite possible that the Mylagaulidae are derived from a primitive aplodontid such as *Promeniscomys* or some early prosciurine (see Korth, 1989a), but the lineages leading to the Recent *Aplodontia* and to that of the mylagaulids have been separate at least since the Arikareean. Much of the similarity in the skull and skeleton of mylagaulids and *Aplodontia* appears to be convergences relating to their fossorial habits rather than to shared ancestors (Fagan, 1960).

Besides the Mylagaulidae, the family most closely related to the Aplodontidae are the squirrels, the Sciuridae. It is clear that the aplodontids did not give rise to the squirrels (or the inverse), but there are a number of cranial and dental morphologies that unite these families and indicate a common ancestry. Both the Sciuridae and Aplodontidae primitively have inflated auditory bullae, uniserial microstructure of the incisor enamel, simplified cheek teeth, fused buccinator and masticatory foramina in the skull, loss of the internal carotid artery (Korth and Emry, 1991), and numerous features of the basicranium that clearly unite the two families (Lavocat and Parent, 1985; Meng, 1990). The first sciurid is from the Chadronian, thus indicating that the common ancestor for the sciurids and aplodontids must be found somewhere in the Bridgerian or later Eocene.

5. Problematical Taxa

5.1. *Eohaplomys*

This genus from the Uintan of California (Stock, 1935) was long considered the basal aplodontid (Wilson, 1949a). However, Rensberger (1975) demonstrated that the prosciurine-like features of the dentitions of *Eohaplomys* were convergent with those of early aplodontids and did not reflect any close relationship. Most recently, Korth (1988a) noted the close similarity of *Eohaplomys* to European Eocene species of *Ailuravus*, and referred *Eohaplomys* to the Ailuravinae of the Ischyromyidae.

5.2. *Sespemys*

This genus was described from the Arikareean Upper Sespe beds of southern California (Wilson, 1934) and assigned to the Ischyromyidae (which included the Prosciurinae) on the basis of a number of similarities between *Sespemys* and *Prosciurus*. Later, Burke (1936) referred *Sespemys* to the

Cylindrodontidae which was followed by Wood (1974) in his review of the Cylindrodontidae. However, the similarities of Sespemys to the cylindrodonts appear to be in the general robustness of the mandible and dentition, but in the detail of the occlusal pattern of the cheek teeth, Sespemys much more closely resembles prosciurines (Wilson, 1949c). The horizon of occurrence of Sespemys (Arikareean) is also much more consistent with the fossil record of aplodontids than that of cylindrodontids which became extinct in North America in the Chadronian. It is most likely that Sespemys is a primitive aplodontid that is convergent with cylindrodonts in the relative size of the teeth and mandible.

5.3. *Cedromus*

Wilson (1949b) first named Cedromus wardi from the Orellan of Colorado. Since that time, this species has been considered a sciurid (Galbreath, 1953) or a prosciurine (Wood, 1962, 1980; Black, 1971). Most recently, complete cranial material of Cedromus has been recovered, and it is evident that this genus represents a sciurid (Korth and Emry, 1991).

Wood (1962, 1980) included "Sciurus" jeffersoni from the Chadronian in the genus Cedromus. However, studies of dentition (Black, 1965), cranial morphology (Korth and Emry, 1991), and the postcranial skeleton (Emry and Thorington, 1982) have demonstrated that this species is clearly a sciurine that is not Cedromus.

5.4. *Mylagaulodon*

Sinclair (1903) described Mylagaulodon angulatus from the Hemingfordian of Oregon based on a palate with a right P^4 and both P^3s. This species (and genus) has long been considered the basal mylagauline (McGrew, 1941a; Shotwell, 1958). The premolars of the holotype are not expanded anteroposteriorly as in mylagaulids and more closely resemble the premolars of meniscomyines, particularly Niglarodon (see Rensberger, 1981). It is likely that Mylagaulodon is an advanced meniscomyine rather than a mylagaulid. Identifications of Mylagaulodon from the Hemingfordian of the Great Plains (McGrew, 1941a; Skwara, 1988) appear to be incorrect, and the teeth supposedly referable to this genus are more likely specimens of the mylagaulid Mesogaulus (see Korth, 1992a).

5.5. *Epeiromys*

Korth (1989a) described Epeiromys spanios from the Orellan of Nebraska. Though this species is brachydont, as in prosciurines, it lacks the hypolophid and isolated entoconid of the lower molars that is characteristic of prosciurines. It is sciurid-like in the simplicity of the lower cheek teeth but differs from early sciurids in the position of the masseteric fossa on the mandible and relative proportions of the lower molars (longer than wide, rather than the reverse in sciurids). It is also possible that this species is a derived ischyromyid. Because *E. spanios* is known only from a single specimen, it is impossible to determine its affinities with any confidence. It is retained in the Aplodontidae based on the original arguments of its familial relationships (Korth, 1989a).

5.6. *Horatiomys*

Wood (1935b) first described *Horatiomys montanus*, based on a single mandible with one cheek tooth from the Arikareean of Montana, as belonging to the Cricetidae. Later, Black (1969) recognized that the mandible belonged to a very young individual and that the cheek tooth was deciduous. He referred the specimen to the Geomyidae. More recently, Rensberger (1981) was able to determine that the type specimen of *H. montanus* was a very young individual of a meniscomyine aplodontid. It was recovered from the same locality as species of *Niglarodon* and thus might be the senior synonym of the latter genus and one of its species. However, Rensberger failed to propose such a synonymy because of the fragmentary nature of the type specimen and because of the poor record of deciduous premolars of meniscomyines from the type locality. *H. montanus* is maintained here as a distinct species pending further study and recovery of fossil material from the type area.

6. Classification

Aplodontidae Trouessart, 1897
  Prosciurinae Wilson, 1949
    Prosciurus Matthew, 1903 (Ch–Wh)
      P. relictus (Cope, 1873) (O)
      P. vetustus Matthew, 1903* (Ch)
      P. parvus Korth, 1989 (O)
      P. magnus Korth, 1989 (O–Wh)
    Pelycomys Galbreath, 1953 (Ch–Wh)
      P. rugosus Galbreath, 1953* (Ch)
      P. placidus Galbreath, 1953 (O)
      P. brulanus Korth, 1986 (O)
      P. new species Korth, 1986 (Wh)
    Spurimus Black, 1971 (U–?Ch)
      S. scotti Black, 1971 (D–?Ch)
      S. selbyi Black, 1971 (U)

    Downsimus Macdonald, 1970 (A)
        D. chadwicki Macdonald, 1970* (A)
    Campestrallomys Korth, 1989 (O–A)
        C. dawsonae (Macdonald, 1963)* (A)
        C. annectens Korth, 1989 (O)
        C. siouxensis Korth, 1989 (Wh–?A)
    Oropyctis Korth, 1989 (Wh)
        O. pediasius Korth, 1989* (Wh)
    Haplomys Miller and Gidley, 1918 (Wh)
        H. liolophus (Cope, 1881)* (Wh)
    Pseudallomys Korth, 1992 (O)
        P. nexodens Korth, 1992* (O)
Allomyinae Marsh, 1877
    Parallomys Rensberger, 1983 (A)
        P. americanus Korth, 1992 (A)
    Allomys Marsh, 1877 (A, ?B)
        A. nitens Marsh, 1877* (A)
        A. cavatus (Cope, 1881) (A)
        ?*A. stirtoni* Klingener, 1968 (B)
        A. simplicidens Rensberger, 1983 (A)
        A. reticulatus Rensberger, 1983 (A)
        A. tessellatus Rensberger, 1983 (A)
    Alwoodia Rensberger, 1983 (A)
        A. harkseni (Macdonald, 1963) (A)
        A. magna Rensberger, 1983* (A)
Meniscomyinae Rensberger, 1981
    Meniscomys Cope, 1879 (A)
        M. hippodus Cope, 1879* (A)
        M. uhtoffi Rensberger, 1983 (A)
        M. editus Rensberger, 1983 (A)
    Niglarodon Black, 1961 (A)
        N. koerneri Black, 1961* (A)
        N. yeariani (Nichols, 1976) (A)
        N. petersonensis (Nichols, 1976) (A)
        N. progressus Rensberger, 1981 (A)
        N. blacki Rensberger, 1981 (A)
        N. loneyi Rensberger, 1981 (A)
    Horatiomys Wood, 1935 (A)
        H. montanus Wood, 1935* (A)
    Rudiomys Rensberger, 1983 (A)
        R. mcgrewi Rensberger, 1983* (A)
    Sewelleladon Shotwell, 1958 (Hf)
        S. predontia Shotwell, 1958* (Hf)
?Meniscomyinae
    Mylagaulodon Sinclair, 1903 (Hf)

        M. angulatus Sinclair, 1903* (Hf)
Aplodontinae Trouessart, 1897
    Liodontia Miller and Gidley, 1918 (B–HP)
        L. alexandrae (Furlong, 1910)* (B)
        L. furlongi Gazin, 1932 (Cl–Hp)
    Tardontia Shotwell, 1958 (B–Cl)
        T. occidentale (Macdonald, 1956)* (Cl)
        T. nevadans Shotwell, 1958 (B)
Subfamily uncertain
    Epeiromys Korth, 1989 (O)
        E. spanios Korth, 1989* (O)
    Sespemys Wilson, 1934 (Wh)
        S. thurstoni Wilson, 1934* (Wh)

Chapter 10
Mylagaulidae

1. Characteristic Morphology	99
1.1. Skull	99
1.2. Dentition	101
1.3. Skeleton	101
2. Evolutionary Changes in the Family	104
3. Fossil Record	105
4. Phylogeny	105
4.1. Origin	105
4.2. Intrafamilial Relationships	107
4.3. Extrafamilial Relationships	108
5. Problematical Taxon: "*Mesogaulus*" *novellus*	108
6. Classification	108

1. Characteristic Morphology

1.1. Skull

The skull of mylagaulids, from their earliest occurrence in the Arikareean, possesses a number of morphologies characteristic of fossorial rodents. The skull is low and broad (Fig. 10.1). The rostrum is short, dorsoventrally deep, and wide. The ventral surface of the zygoma and infraorbital foramen are unmodified, retaining the primitive protrogomorphous condition. The occipital forms a nearly vertical wall or is sloped slightly anteriorly. The zygoma is massive. A postorbital process projects laterally from the frontal bone above the orbit and a corresponding process rises dorsally from the jugal bone. A single sagittal crest or two parasagittal crests arise from the postorbital processes and converge posteriorly. Well-developed lambdoidal crests are always present. The palatal surface is broad, and the tooth rows converge posteriorly. The pterygoid region of the skull is greatly shortened. The auditory bulla shows some inflation and has an elongated external meatus as in aplodontine aplodontids. Some of the known genera have a pair of "horn cores" on the nasal bones. These structures are crescentic in cross section and rise well above the rest of the skull.

    The cranial foramina of mylagaulids are quite similar in arrangement to those of *Aplodontia* (Wahlert, 1974). The optic foramen is always small, the ethmoid foramen is within the frontal bone, and the buccinator and mastica-

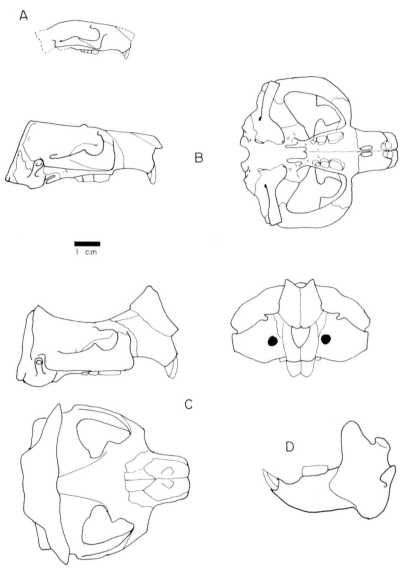

FIGURE 10.1. Skulls of mylagaulids. (A) Lateral view of Arikareean *Trilaccogaulus lemhiensis*. (B) Lateral and ventral (right) views of *Mylagaulus laevis*. (C) Lateral, anterior (right) and dorsal (below) views of *Ceratogaulus rhinocerus*. (D) Lateral view of mandible of *C. rhinocerus*. A From Nichols (1976), B from Wahlert (1974), C and D modified from Matthew (1902).

tory foramina are separate. There is a small interpremaxillary foramen present in all genera except the earliest (*Promylagaulus*).

The mandible, as in the skull, is massive (deep and broad). The coronoid process rises well above the tooth row and condyle and is broad anteroposteriorly. The angle is similarly broadened and there is a lateral process similar to that of aplodontine aplodontids. The fossa for the masseter terminates anteriorly below the premolar.

1.2. Dentition

The incisors are broad anteriorly, but maintain a convex anterior surface. On some specimens of *Mylagaulus* there are a few shallow furrows on the upper incisor. All known mylagaulids have a uniserial microstructure of the incisor enamel (Wahlert, 1968). In promylagaulines and primitive mylagaulines, the complete rodent dental formula is present. In *Mylagaulus* and later genera the dental formula is reduced in two ways. First, the simple peglike P^3 is lost. Second, the last premolar enlarges so much that its eruption pushes the first (and sometimes the second) molars out of the jaw, leaving only one or two molars remaining.

The last premolar is always larger than the molars in mylagaulids, and the molars decrease in size from M^1/M_1 to M^3/M_3. The occlusal surface of the cheek teeth disappears in early stages of wear, leaving behind an enamel outline of the tooth and differing patterns and numbers of small enamel "lakes" or fossettes (fossettids). Because of this occlusal pattern, the normal dental terminology does not apply to mylagaulids except in unworn specimens of the more primitive species (see Galbreath, 1984). The terminology used in descriptions of the teeth are limited to the number and position of the enamel lakes (Fig. 10.2; Storer, 1975; Rensberger, 1979; Munthe, 1988). As the size of the animal and proportionate size of the premolar enlarges, the number of enamel lakes and the complexity of their arrangement increase.

The most primitive mylagaulines have cheek teeth that are mesodont in height, showing some loss of enamel on the anterior and posterior walls of the molars leaving dentine tracts. The teeth are highly lophate, cusps being reduced to junctions of the lophs (Fig. 10.3). The crown height of the cheek teeth in the mylagaulines increases, but never attains complete hypsodonty. Often in mylagaulines, cementum is developed on the sides and within the lakes of the cheek teeth.

1.3. Skeleton

All mylagaulids for which the postcranial skeleton is known show a great deal of fossorial adaptation. The limbs, especially the forelimbs, are massive with exaggerated development of all ridges and processes for muscular

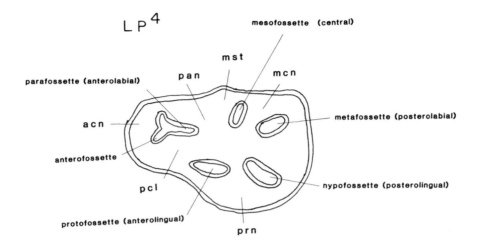

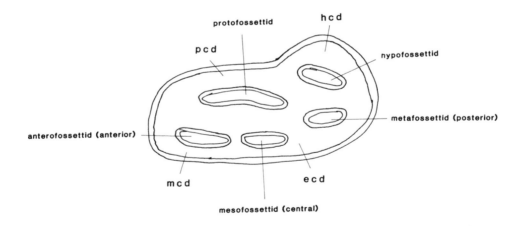

FIGURE 10.2. Nomenclature used for the fossettes (-ids) of mylagaluid premolars. After Munthe (1988), with alternate nomenclature in parentheses from Rensberger (1979). Abbreviations for cusps same as in Fig. 2.8.

Mylagaulidae 103

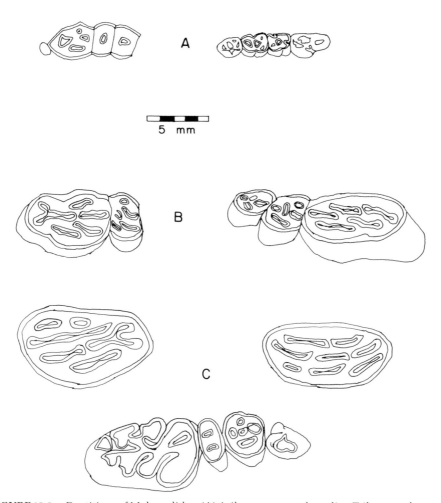

FIGURE 10.3. Dentitions of Mylagaulidae. (A) Arikareean promylagauline *Trilaccogaulus montanensis*, LP^3–M^2 and RP_4–M_3. (B) Hemingfordian mylagauline *Mylagaulus vetus*, LP^4–M^1. (C) Hemphillian mylagauline ?*Mylagaulus* sp., R^4, RP_4, and LP_4–M_3 (below). A from Nichols (1976), B from Munthe (1988), C from Wilson (1937a).

attachment. The pes is enlarged and the ungules massive. The girdles are also strengthened. The hind limbs do not have as much hypertrophy of the muscular attachments but are more massive than in *Aplodontia* and most other rodents. The fibula and tibia are never fused (Fig. 10.4).

2. Evolutionary Changes in the Family

The greatest change in the Mylagaulidae is the increase in hypsodonty of the cheek teeth and the relative enlargement of the last premolar. In promylagaulines the premolar is the largest cheek tooth, but it is only slightly larger than the first molar. In *Mylagaulus* the premolar has increased in size so much that when the permanent tooth replaces its deciduous precursor, it pushes the first molar out with the deciduous premolar. In some individuals of *Epigaulus*, none of the molars are retained after the eruption of the premolar.

Along with the enlargement of the premolar, the number and complexity of arrangement of the enamel lakes on the cheek teeth increase in mylagaulids. In Arikareean *Trilaccogaulus* and *Promylagaulus*, the number of enamel lakes in moderately worn premolars is three or four, and there is one lake in each molar. In Barstovian *Mylagaulus*, the lakes increase to six or seven in the premolars and four in the molars. In *Epigaulus* and the most derived species of *Mylagaulus*, there are as many as ten lakes in each premolar. Also associated with the increase in size of the premolar, there is an increase in overall size through time; the promylagaulines being the smallest

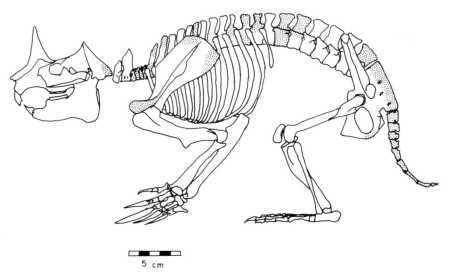

FIGURE 10.4. Skeleton of Hemphillian mylagaulid *Epigaulus hatcheri*, from Gidley (1907). Shaded areas indicate bones not represented in the fossil.

animals and the Hemphillian species the largest. However, Baskin (1980) recognized a lineage within the genus *Mylagaulus* that showed a reduction in size (dwarfism) but not in complexity of the dentition in the Clarendonian and Hemphillian.

3. Fossil Record

Mylagaulids are exclusive to North America and have never been very diverse throughout their fossil history. They first appear in the Arikareean, and their last occurrence is in the Hemphillian. Their greatest diversity was in the Hemingfordian where six species have been recognized (Fig. 10.5). In terms of generic diversity, three genera are known from the Arikareean and only two from any other age. As pointed out by Galbreath (1984), some of this lack of diversity is probably related to the fact that a systematic review of this family of rodents has not been undertaken since that of Matthew (1902), and several additional taxa may be recognizable ultimately.

The two subfamilies of the Mylagaulidae are mutually exclusive in their ages of occurrence. All of the promylagaulines are restricted to the Arikareean, and the mylagaulines first appear in the Hemingfordian and continue through the Hemphillian.

4. Phylogeny (Fig. 10.6)

4.1. Origin

The origin of the mylagaulids is uncertain. Several authors have noted the great similarity between *Promylagaulus*, the earliest mylagaulid, and contem-

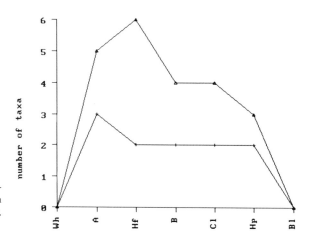

FIGURE 10.5. Occurrence of Mylagaulidae in the Tertiary of North America (▲, species; +, genera). Abbreviations as in Fig. 5.4.

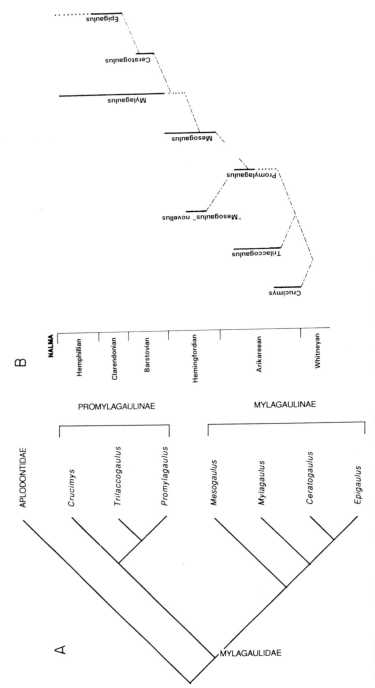

FIGURE 10.6. Phylogeny of the Mylagaulidae. (A) Cladogram of mylagaulid relationships. (B) Dendrogram of the Mylagaulidae with age of occurrence (see Fig. 5.5 for explanation).

poraneous meniscomyine aplodontids, thus suggesting a possible ancestry for the mylagaulids in *Meniscomys* or a related genus (McGrew, 1941a; Wilson, 1949a; Shotwell, 1958; Rensberger, 1979). Although the hypsodonty of the cheek teeth and enlargement of the premolar are shared by *Promylagaulus* and meniscomyines, the details of the occlusal morphology of the cheek teeth differ (Rensberger, 1979, 1980). With the recognition of a lower crowned promylagauline, *Crucimys*, it appears likely that the hypsodonty of the cheek teeth was developed in parallel in the meniscomyines and promylagaulines (Rensberger, 1980).

Because both meniscomyines and promylagaulines appear at the beginning of the Arikareean, it is likely that any ancestor would have existed earlier, in the Whitneyan. Korth (1989a) suggested that a Whitneyan ?prosciurine, *Oropyctis*, could represent an ancestral mylagaulid, but that the morphological gap between *Oropyctis* and the Arikareean promylagaulines was great. Mylagaulids probably originated from an advanced prosciurine (aplodontid) or prosciurine-like rodent sometime in the late middle Oligocene (Whitneyan) and rapidly developed the distinctive dental and skeletal morphology characteristic of the family.

4.2. Intrafamilial Relationships

The later, more advanced mylagaulines most certainly were derived from a more primitive promylagauline which was a suitable ancestor both morphologically and temporally. Rensberger (1979, 1980) could not find any genus of promylagauline generalized enough morphologically from which to derive the later mylagaulines. However, Korth (1992a) included most of the species recognized by Rensberger (1979) as *Promylagaulus* (on which he based his conclusions) in a new genus *Trilaccogaulus*.

Among the mylagaulines, *Mesogaulus* is clearly ancestral to the later and more advanced *Mylagaulus*. *Mesogaulus* has no derived morphologies that would bar it from an ancestral position. Fossorial adaptations of the skull and skeleton of *Mylagaulus* are present in *Mesogaulus* but are not as well developed (Fagan, 1960; Galbreath, 1984; Munthe, 1988).

The genera *Ceratogaulus* and *Epigaulus* are the only known mylagaulids with "horn cores" on the nasal bones. Clearly, this feature unites these two genera, and separates them from all other mylagaulids. However, it is impossible to determine if these "horns" are truly diagnostic of separate genera, or are simply the result of sexual dimorphism. There is a great deal of dental similarity between the Barstovian *Ceratogaulus* and contemporaneous species of *Mylagaulus*, and also between species of *Epigaulus* and later species of *Mylagaulus*. Matthew (1902) first suggested the possibility of sexual dimorphism in mylagaulids, but rejected it because he could find no record of such marked dimorphism in the entire order.

If *Ceratogaulus* and *Epigaulus* are accepted as distinct genera, then it is clear that the older *Ceratogaulus* is ancestral to *Epigaulus* based on smaller size, less complex occlusal pattern of the cheek teeth, and smaller nasal horns.

4.3. Extrafamilial Relationships

The Mylagaulidae are far too derived to be ancestral to any later family of rodents. The most closely related family is clearly the Aplodontidae. The similarity of these families is so great that a number of authors have suggested that the mylagaulids be included only as a subfamily of the Aplodontidae (Wilson, 1949a; Wahlert, 1974). However, these two families have been separate since at least the middle of the Oligocene and clearly developed many of their fossorial adaptations independently (Fagan, 1960). Thus, the Mylagaulidae is considered here a separate family but is included in the superfamily Aplodontoidea.

5. Problematical Taxon: "*Mesogaulus*" *novellus*

Originally referred to *Mylagaulus* from the Hemingfordian of Nebraska (Matthew, 1924), this species is one of the best represented species of mylagaulids (Black and Wood, 1956; Munthe, 1988). Black and Wood (1956) transferred this species to *Mesogaulus* based on the simpler pattern and lower crown of the cheek teeth with respect to *Mylagaulus*. However, Galbreath (1984) noted a number of differences in the occlusal morphology of the premolars between other species of *Mesogaulus* and "*M.*" *novellus* which indicated to him that the latter belonged in a different (and possibly new) genus. Most recently, Korth (1992a) proposed a close relationship between *Promylagaulus riggsi* and "*M.*" *novellus*, excluding the latter from *Mesogaulus*. This species is not referable to any currently recognized genus of mylagaulid.

As stated previously, a detailed systematic review of this family is much needed. A number of synonymies and perhaps the recognition of some new genera may be in order. No new genus for "*M.*" *novellus* is named and no formal synonymies are proposed here, pending such a review of the family.

6. Classification

Mylagaulidae Cope, 1881
   Promylagaulinae Rensberger, 1980
     Promylagaulus McGrew, 1941 (A)
       P. riggsi McGrew, 1941* (A)
     Trilaccogaulus Korth, 1992 (A)

        T. lemhiensis (Nichols, 1976)* (A)
        T. ovatus (Rensberger, 1979) (A)
        T. montanensis (Rensberger, 1979) (A)
    Crucimys Rensberger, 1980 (A)
        C. milleri (Macdonald, 1970)* (A)
Mylagaulinae Cope, 1881
    Mylagaulus Cope, 1878 (Hf–Hp)
        M. sesquipedalis Cope, 1878* (?Cl)
        M. monodon Cope, 1881 (?B–?Hp)
        M. laevis Matthew, 1902 (Hf–?B)
        M. vetus Matthew, 1924 (Hf)
        M. douglassi McKenna, 1955 (B)
        M. kinseyi Webb, 1966 (Hp)
        M. elassos Baskin, 1980 (Cl)
    Mesogaulus Riggs, 1899 (Hf–B)
        M. ballensis Riggs, 1899* (Hf)
        M. paniensis (Matthew, 1902) (Hf)
        M. pristinus (Douglass, 1903) (B)
        M. proximus (Douglass, 1903) (B)
        ?*M. novellus* (Matthew, 1924) (Hf)
        M. praecursor Cook and Gregory, 1941 (Hf)
    Ceratogaulus Matthew, 1902 (B)
        C. rhinocerus Matthew, 1902* (B)
    Epigaulus Gidley, 1907 (Cl–?Hp)
        E. hatcheri Gidley, 1907 (?Hp)
        E. minor Hibbard and Phillis, 1945 (Cl)

Chapter 11
Sciuridae

1. Characteristic Morphology	111
1.1. Skull	111
1.2. Dentition	113
1.3. Skeleton	113
2. Evolutionary Changes in the Family	116
3. Fossil Record	117
4. Phylogeny	119
4.1. Origin	119
4.2. Intrafamilial Relationships	121
4.3. Extrafamilial Relationships	121
5. Problematical Taxon: ?*Protosciurus jeffersoni*	122
6. Classification	122

1. Characteristic Morphology

1.1. Skull

The skull in sciurids is high and broad. All sciurines and petauristines except the Chadronian ?*Protosciurus jeffersoni* have a sciuromorphous zygomasseteric structure with the infraorbital foramen small, laterally compressed, and low on the rostrum with a bony flange extending lateral and ventral to it (Fig. 11.1). The zygoma of ?*Protosciurus jeffersoni* is protrogomorphous with a primitive infraorbital foramen. The Cedromurinae have a modified protrogomorphous zygoma, in which the ventral surface of the anterior root of the zygoma is broader than in other protrogomorphous rodents and is clearly tilted anterodorsally which has caused some lateral compression of the infraorbital foramen. Korth and Emry (1991) have suggested that this might be a modified myomorphy with the possibility of invasion of the infraorbital foramen by the masseter.

    The nasal bones extend farther posteriorly on the skull roof than the surrounding premaxillaries. There is a distinct postorbital process on the frontals that extends laterally and posteriorly. Low lyrate crests on the parietals merge at the occipital crest. The auditory bulla is inflated and septate. Within the bulla the stapedial artery is always contained within a bony tube across the promontorium. The pterygoid flanges of the skull are elongate and

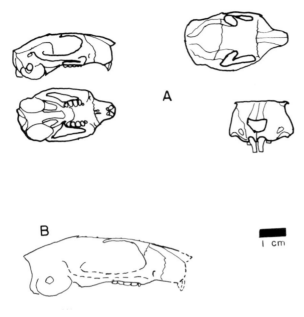

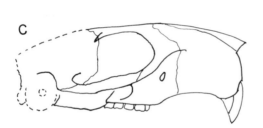

FIGURE 11.1. Skulls of Sciuridae. (A) Lateral, ventral, dorsal, and anterior views of Orellan cedromurine *Cedromus wilsoni*. (B) Lateral view of Arikareean sciurine *Protosciurus condoni*. (C) Lateral view of marmotine *Palaearctomys montanus*. A from Korth and Emry (1991), B and C from Black (1963).

extend posteriorly to the bulla. The palate is broad and flat or gently concave ventrally. The basisphenoid and basioccipital are slightly tiled anterodorsally, making a slight angle with the plane of the palate.

There are several characteristic morphologies of the cranial foramina in sciurids: incisive foramina are short; posterior palatine foramina are contained within the palatine bone; alisphenoid canal opens ventrally within the foramen ovale in the roof of the pterygoid fossa; carotid canal is lacking (internal carotid artery lost) in many species; postglenoid foramen is always large, and the petrosal bone is exposed through it; large accessory foramen ovale on the alisphenoid bone; the masticatory and buccinator foramina are always fused; and the ethmoid foramen is surrounded entirely by the frontal bone.

1.2. Dentition

The incisors of sciurids are laterally compressed and gently convex anteriorly. Ornamentation of the anterior surface of the incisors is rare, but does occur in some genera of marmotines, most notably the Blancan *Paenemarmota* (Hibbard and Schultz, 1948; Repenning, 1962). The microstructure of the enamel is uniserial.

Sciurids generally retain the complete primitive dental formula of rodents with a small, peglike P^3. In some marmotines, P^3 is enlarged, but it never attains the size or occlusal pattern of P^4 or the molars. Only one North American genus, *Tamias*, has reduced the dental formula through the loss of P^3. A few Recent African genera have similarly lost this tooth (see Ellerman, 1940). The upper cheek teeth are generally simplified in occlusal pattern and brachydont (Fig. 11.2). The conules and hypocones of the upper cheek teeth are reduced or absent. P^4 is generally molariform, often with an anterior expansion at the anterobuccal corner of the tooth (anterocone). In the most primitive species, ?*Protosciurus jeffersoni* and *Protosciurus mengi*, the metaconules of the upper molars are doubled (Korth, 1987a). There is a distinct protocone crest that runs anterobuccally from the protocone on the teeth of ?*P. jeffersoni*, a primitive character lost in all later sciurids. Cedromurines and some primitive marmotines have an ectoloph on the upper molars (Black, 1963; Korth and Emry, 1991).

The lower cheek teeth are similarly simplified. The cusps are bulbous and marginally placed with a shallow talonid basin. The trigonid is small and generally posteriorly closed by the metalophulid II. The entoconid is progressively incorporated into the posterolophid and becomes virtually indistinct in many genera. The lower molars are characteristically wider buccolingually than long. In the most primitive sciurine and the cedromurines, there is a partial hypolophid on the lower cheek teeth, a primitive condition. The earliest sciurids have lower molars that are two-rooted, later sciurids have four-rooted lower cheek teeth.

In the Sciurini, the lophs of the cheek teeth are generally low and rounded. In the tamiines and marmotines, the cheek teeth become more lophate and the cusps are essentially unrecognizable. In marmotines the cheek teeth attain mesodonty. Also in marmotines the last molar is often considerably larger than the anterior molars.

The petauristines closely resemble the Sciurini in general pattern, but in all petauristines the enamel on the cheek teeth is highly crenulated. The accessory lophules that produce this crenulation vary in degrees of height and size among the genera of the Petauristinae.

1.3. Skeleton

The skeleton of sciurids are essentially primitive for rodents, following the general pattern of Eocene ischyromyids. The tibia and fibula are never

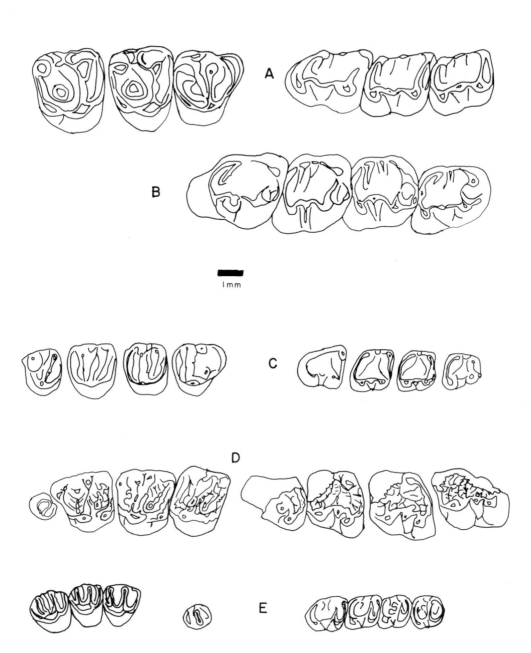

FIGURE 11.2. Dentitions of Sciuridae. (A) Orellan cedromurine *Cedromus wardi*, RP^4–M^2 and RM$_1$–M$_3$. (B) Chadronian sciurine "*Protosciurus*" *jeffersoni*, RP$_4$–M$_3$. (C) Hemingfordian marmotine *Protospermophilus kelloggi*, LP^4–M^3 (composite, and RP$_4$–M$_3$ (composite). (D) Clarendonian petauristine *Petauristodon matthewsi*, LP^3–M^2 and LP$_4$–M$_3$. (E) Blancan marmotine *Paenemarmota barbouri* RP^3, M^1–M^3, and RP$_4$–M$_3$. A–D to same scale (top); E to bottom scale. A (lowers) from Wood (1962), A (uppers) and B from Wood (1937), C from Black (1963), D from James (1963), E from Repenning (1962).

Sciuridae

fused, though the scaphoid and lunate bones of the carpus are fused in all species but the earliest sciurid, the Chadronian ?*Protosciurus jeffersoni*. The skeleton of ?*P. jeffersoni* has already attained the arboreal adaptations of the Sciurini (Fig. 11.3). The skeleton is generally less robust than that in primitive ischyromyids, and the hind limb is shorter relative to the forelimb. The tree squirrels (Sciurini) have the most gracile skeleton of the Sciurinae. The limb bones and digits are more slender than those of ground squirrels. The only difference between the skeleton of the Chadronian genus and later genera is a matter of degree. ?*P. jeffersoni* is only slightly more robust than later sciurines and clearly less robust than the ground squirrels (Emry and Thorington, 1982).

The marmotines develop more robust limbs and digits for their more terrestrial and often burrowing habits. The tamiines are intermediate in developing some of the terrestrial adaptations of the maromotines, while retaining some of the arboreal adaptations of the sciurines.

One unique feature of the skeleton of sciurids not known in any other rodents (except the north African anomalurid *Anomalurus*) is the presence of a subscapular spine on the scapula (Emry and Thorington, 1982).

No postcranial skeletal material is currently known for the fossil North American petauristines. The skeletons of modern flying squirrels are more gracile than those of sciurines with the major limb bones being much more slender and delicate. The cartilaginous struts that some flying squirrels develop on the forelimb to support their gliding membrane most likely would not be preserved in the fossil record.

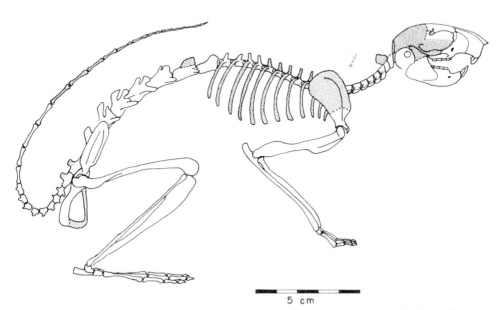

FIGURE 11.3. Reconstruction of the skeleton of the Chadronian "*Protosciurus*" *jeffersoni* (based on description from Emry and Thorington, 1982).

2. Evolutionary Changes in the Family

Once sciuromorphy was attained within the sciurids, little change in the skull occurred in any of the tribes and subfamilies. The skulls of marmotines are characteristically lower and broader than those of sciurines, probably a fossorial adaptation.

The skeleton of the earliest Sciurine from the Chadronian of North America is nearly identical to that of Recent *Sciurus* (Emry and Thorington, 1982). Vianey-Liaud (1974) described the earliest sciurid skeleton from Europe (middle Oligocene) and found some shared characters with the ground squirrels, but noted that the proportions of the limbs more nearly approached those of tree squirrels. The earliest known postcranial elements of the skeleton of a marmotine, *Ammospermophilus fossilis* from the Clarendonian (James, 1963) are somewhat longer and more slender than those of Recent marmotines, more closely approaching the morphology of the sciurines. Similarly, the earliest record of *Spermophilus* limb and hind foot bones of the Hemphillian *Spermophilus* (*Otospermophilus*) *shotwelli*, only differ from these same elements of Recent *Spermophilus* in being slightly longer (Black, 1963). This implies that the skeleton of the marmotines was modified from the primitive sciurine condition by shortening the limb and foot bones and increasing their robustness. There also has been a shortening of the tail in marmotines from the long-tailed sciurine condition since the Chadronian. However, no complete sequences of caudal vertebrae are known for Tertiary marmotines, so the time and rate at which such a shortening occurred cannot be determined.

Because there is no fossil postcranial material for any Tertiary petauristines, cedromurines, or tamiines, nothing can be definitely stated about the modification of the skeleton of these squirrels except by comparison with Recent species. The tamiines developed some of the terrestrial features of the marmotines (shorter, broader limb bones, shorter tail) but maintained more slender foot bones as in sciurines. The petauristines developed longer, more slender limb bones as an adaptation to gliding.

Dentally, the change in Sciurini from their earliest occurrence is one of simplification. The upper molars are always brachydont and bunodont with the metaloph and protoloph low and rounded. The hypocone is much reduced or lost, and the conules are essentially indistinguishable. Primitively, in Chadronian and Orellan Sciurini, the metaconule is doubled on the upper molars and the hypocone is relatively larger (Emry and Thorington, 1982; Korth, 1987a). A protocone crest is present in the Chadronian species. These features are lost in later Sciurini.

The lower molars are also simplified in Sciurini. The talonid basin is simple and the cusps marginal. The entoconid is nearly indistinguishable. Again, in the earliest sciurines, the entoconid is a distinguishable cusp and a partial hypolophid is sometimes present.

The marmotines and tamiines developed lophodonty on their cheek teeth. The difference in the tamiines is markedly less than that in the marmotines. Tamiines simplified the cusps and basins of the teeth as did the later sciurines, but the lophs of the teeth are slightly higher and better defined in chipmunks. In marmotines the cheek teeth became higher crowned and the lophs greatly elevated. There is also a distinct anteroposterior compression of the teeth, especially in the most derived marmotines, Cynomys and Marmota.

The Petauristinae are remarkably conservative in their dentition. The cheek teeth have a basic pattern of a later sciurine (simplified) with crenulate enamel. This pattern varies little throughout the history of this group, the greatest difference being in size and height of the crenulations.

3. Fossil Record

Sciurids first appear in the Chadronian and are known throughout the Tertiary of North America. The greatest number of species (17) occurred in the Blancan, but the greatest number of genera (5) is known from the Hemingfordian to the Clarendonian (Fig. 11.4). The earliest tribe, the Sciurini, though still present in North America today, only has a fossil record from the Chadronian to the Arikareean. There is no record of sciurines in North America between the Hemingfordian and the Pleistocene. The shortest-lived subfamily is the Cedromurinae which is only known from the Orellan and Whitneyan (Korth and Emry, 1991).

Overall, it appears that the early history of the sciurids in North America is dominated by the Sciurini and Cedromurinae (Chadronian to Arikareean), only to be replaced by the rapidly diversifying marmotines. The great diversity of sciurids in the Hemphillian and Blancan is related to the marmotines, mainly the species of Spermophilus which rapidly diversified during this time. The absence of Sciurini in the latter Tertiary suggests that the modern genera of this tribe were reintroduced during the Pleistocene or Holocene. The petauristines and tamiines were never a large part of the sciurid diversity in the Tertiary, but their earliest occurrences roughly coincide with the appearance of marmotines and the waning of the sciurines.

In Europe the first record of sciurids is nearly simultaneous with that in North America, in the earliest Oligocene (after the "grande coupure"). However, the Petauristinae and Sciurinae appeared at the same time (de Bruijn and Unay, 1989). The Sciurini persist in Europe into modern times. However, the marmotines that make up the bulk of the diversity of the squirrels in the latest Tertiary in North America did not appear in Europe until later in the Pliocene (Black and Kowalski, 1974). The tamiines, only a small part of the sciurid fauna in Europe, as in North America, appeared in the earlier Miocene (Orleanian; Savage and Russell, 1983). The dominant subfamily of squirrels in Europe, especially in the Miocene and Pliocene, are the petauristines which are more diverse than the sciurines during this time.

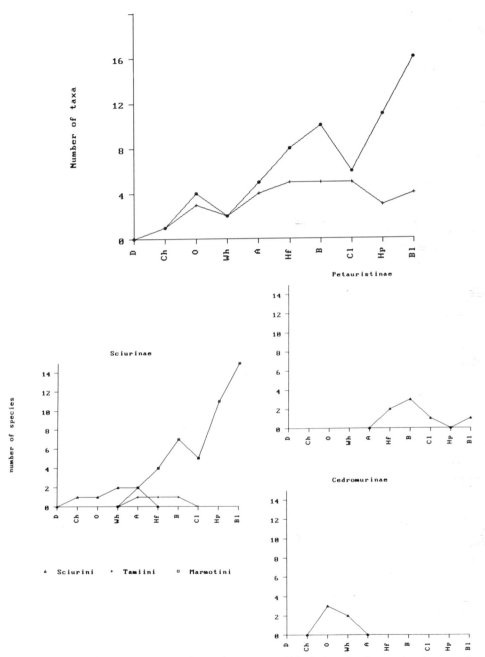

FIGURE 11.4. Occurrence of Sciuridae in the Tertiary of North America (●, species; +, genera). Abbreviations as in Fig. 5.4.

Although sciurids currently occupy all of the continents except Australia and Antarctica, the fossil record from all other continents outside of Europe and North America is nearly nonexistent. Only a single tamiine, *Eutamias*, from the later Miocene of Asia (Munthe, 1980), and the sciurines, *Atlantoxerus* (Lavocat, 1961, 1978; Jaeger, 1976), *Vulcanisciurus* (Lavocat, 1973, 1978), and *Kubwaxerus* (Cifelli et al., 1986), from the Miocene of Africa, have been reported from the Tertiary outside of Europe and North America.

Clearly, the marmotines developed and diversified in North America, while the petauristines reached greater diversity in Europe. The current widespread distribution of sciurids around the world was accomplished during the Pleistocene and Holocene.

4. Phylogeny (Fig. 11.5)

4.1. Origin

Early workers viewed the North American ischyromyids (including paramyids) as the ancestors of the sciurids. The genera *Prosciurus* (Matthew, 1910) and *Cedromus* (Wilson, 1949a) were suggested as possible ancestors. However, these two genera have been shown to be an aplodontid (Rensberger, 1975) and a sciurid (Korth and Emry, 1991), respectively. The Uintan paramyine *Uriscus* (known only from a single specimen) was proposed as an ancestral morphotype for sciurids based on the occlusal outline and relative proportions of the lower molars (Wood, 1962). Later authors agreed with this suggestion (Black, 1963, 1971; Black and Sutton, 1984; Korth, 1984) although the subfamily of the Ischyromyidae to which *Uriscus* belonged was not certain.

Recent studies of the basicranium and dentition of the Bridgerian reithroparamyine ischyromyid *Reithroparamys* have cited a number of shared derived characters with it and sciurids (Emry and Korth, 1989; Meng, 1990). The dental features shared by *Reithroparamys* and the Chadronian ?*Prosciurus jeffersoni* (earliest sciurid) are: (1) a protocone crest on the upper molars, (2) doubled metaconule on upper molars, and (3) at least partial hypolophid on the lower molars. The shared cranial features are: (1) the presence of an inflated ossified bulla and (2) the stapedial artery enclosed within a bony tube across the promontorium (Meng, 1990). The mandible of one species of *Reithroparamys*, *R. huerfanensis*, has a masseteric fossa that extends father anteriorly than in ischyromyids, more similar to the condition in of sciurids (Wood, 1962). The known postcranial skeleton of *Reithroparamys* is similar to that of paramyines but is less robust (see Wood, 1962, Figs. 35, 44, 46) similar to the differences noted between ?*P. jeffersoni* and the paramyines (Emry and Thorington, 1982). All of these features make it very likely that a *Reithroparamys*-like ischyromyid was the ancestor of the Sciuridae.

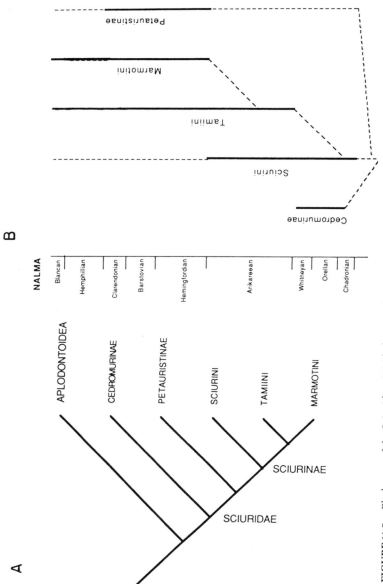

FIGURE 11.5. Phylogeny of the Sciuridae. (A) Cladogram of sciurid relationships. (B) Dendrogram of the Sciuridae with age of occurrence (see Fig. 5.5 for explanation).

4.2. Intrafamilial Relationships

The Orellan and Whitneyan cedromurines retain primitive characters of the skull (modified protrogomorphy) and dentition (hypolophids on lower molars) that are only present in the most primitive Sciurinae. It is evident that the cedromurines are an aberrant side branch of the Sciuridae and have nothing to do with the evolution of the sciuromorphous sciurids (Korth and Emry, 1991).

If sciuromorphy were attained only once in the Sciuridae, all Sciurinae and Petauristinae must have been derived from a protrogomorphous ?*P. jeffersoni*-like ancestor. Because the evidence of the postcranial skeleton is so strong for the inclusion of the earliest squirrels in the Sciurini (Vianey-Liaud, 1974; Emry and Thorington, 1982), this tribe represents the primitive condition for sciurines from which all later tribes were derived. The earliest ground squirrels (Marmotini) for which the postcranial skeleton is known differ from Recent marmotines only in having longer limb and foot bones, similar to the condition in sciurines. This means that the first squirrels were adapted to an arboreal habit, then later descended from the trees to develop the more robust skeletons of the ground squirrels. This scheme is also consistent temporally with the occurrence of Sciurini predating that of marmotines and tamiines (see above discussion of the fossil record).

Dentally, the early sciurines are also suitable as ancestral morphotypes for the later squirrels. The greater lophodonty of the marmotines and tamiines is easily attained from the basic simplified pattern of the sciurines. Similarly, the crenulate enamel of the petauristines is not known in any sciurines, but the occlusal pattern of the cheek teeth of the petauristines is not greatly different from that of sciurines with shallow basins and marginal cusps.

The precise relationships of the Petauristinae are unclear. The almost simultaneous appearance of petauristines along with the earliest sciurines in Europe has led some authors (Heissig, 1979; de Bruijn and Unay, 1989) to suggest that a separate family for the flying squirrels should be established, arguing that they did not share a recent common ancestor with the Sciurinae. Unfortunately, none of the early petauristines from Europe or North America are known from postcranial material, so it is not known if they had developed the specializations of the skeleton for gliding that are characteristic of Recent petauristines.

No definite conclusions can be drawn on the currently known material of fossil petauristines. If better cranial and postcranial material became available, the relationships of the early flying squirrels could be determined.

4.3. Extrafamilial Relationships

Recent studies of the cranial and dental morphology of early sciurids and aplodontids have demonstrated a close relationship between the Sciuridae and Aplodontidae (Lavocat and Parent, 1985; Meng, 1990; Korth and Emry, 1991; see also Chapter 9, Section 4.3).

The only other suggestion of a family of rodents that might be closely related to the Sciuridae was made by Bugge (1974). He found that the arterial circulation in the skull showed a great deal of similarity between sciurids and the myomorphous Old World dormice (Gliridae). In his classification of rodents, Bugge (1974) listed only two Recent families in the suborder Sciuromorpha, the Sciuridae and Gliridae. Nearly all previous authors had included the dormice in the suborder Myomorpha (Simpson, 1945) or as a sister group to the muroid rodents (Wahlert, 1978). Hartenberger (1971) demonstrated that the earliest glirid (*Eogliravus*) was derived from Eocene species of the European reithroparamyine ischyromyid *Microparamys* based on dental morphology. Both of these lines of evidence imply that, at least at some level, the glirids and sciurids shared a common ancestor (apparently within the Reithroparamyinae). However, a much closer study of the recent and fossil glirids is necessary before any definite conclusion can be drawn as to the relationship of these two families.

5. Problematical Taxon: ?*Protosciurus jeffersoni*

Douglass (1902) first described *Sciurus jeffersoni* from the Chadronian of Montana, noting its sciurid affinities. Since that time this species has been included in a number of genera of either sciurid or aplodontid. Osborn and Matthew (1909) and Wood (1937) included this species in *Prosciurus*. Wood (1962, 1980) included it in the genus *Cedromus* which he considered to be a prosciurine. Black (1965) described the upper dentition of this species and demonstrated that it was not a prosciurine and was distinct from *Cedromus*. He included *jeffersoni* questionably in the sciurid genus *Protosciurus*. Emry and Thorington (1982) described the skeleton of this species and demonstrated that it was clearly a sciurine.

"*Sciurus*" *jeffersoni* is distinct from all other sciurids in the possession of primitive features of the dentition that are common with a sciurid–aplodontid ancestor (doubled metaconule and protocone spur on the upper molars, partial hypolophid on the lower molars; Korth and Emry, 1991) and a protrogomorphous zygomasseteric structure (other *Protosciurus* show the beginnings of sciuromorphy; Black, 1963; Korth, 1987a). This species most likely represents a new genus of primitive sciurid. The naming of this genus will not be done here. The species *jeffersoni* is included below questionably in *Protosciurus* following the arguments of Black (1965).

6. Classification

Sciuridae Gray, 1821
    Sciurinae Gray, 1821

Sciurini Gray, 1821
   Protosciurus Black, 1963 (O, A–Hf)
     ?*P. jeffersoni* (Douglass, 1902)* (Ch)
     P. tecuyensis (Bryant, 1945) (Hf)
     P. condoni Black, 1963* (A)
     P. mengi Black, 1963 (O)
     P. rachelae Black, 1963 (A)
   Miosciurus Black, 1963 (A)
     M. ballovianus (Cope, 1881)* (A)
Tamiini Black, 1963
   Nototamias Pratt and Morgan, 1989 (A–Hf)
     N. hulberti Pratt and Morgan, 1989* (Hf)
     N. quadratus Korth, 1992 (A)
   Eutamias Trouessart, 1880
     E. ateles Hall, 1930 (B–Cl)
Marmotini Pocock, 1923
   Ammospermophilus Merriam, 1892 (Cl, Pl–R)
     A. fossilis James, 1963 (Cl)
     A. junturensis (Shotwell and Russell, 1963) (Cl)
   Arctomyoides Bryant, 1945 (B)
     A. arctomyoides (Douglass, 1903)* (B)
   Spermophilus Cuvier, 1825 (B–R)
   S. (Spermophilus) Cuvier, 1825 (Hp–R)
     S. (S.) tuitus (Hay, 1921) (Bl)
     S. (S.) bensoni (Gidley, 1922) (Bl–Pl)
     S. (S.) howelli (Hibbard, 1941) (Bl)
     S. (S.) meadensis (Hibbard, 1941) (Bl–?Pl)
     S. (S.) cragini (Hibbard, 1941) (Bl)
     S. (S.) rexroadensis (Hibbard, 1941) (Bl)
     S. (S.) mckayensis (Shotwell, 1956) (Hp)
     S. (S.) magheei (Strain, 1966) (Bl)
     S. (S.) meltoni Hibbard, 1972 (Bl)
     S. (S.) johnsoni Hibbard, 1972 (Bl)
   S. (Otospermophilus) Brandt, 1844 (B–R)
     S. (O.) gidleyi (Merriam, Stock, and Moody, 1925) (Hp)
     S. (O.) tephrus (Gazin, 1932) (B)
     S. (O.) argonautus (Stirton and Goeriy, 1942) (Hp)
     S. (O.) fricki (Hibbard, 1942) (Hp)
     S. (O.) primitivus (Bryant, 1945) (B)
     S. (O.) pattersoni (Wilson, 1949) (Hp)
     S. (O.) wilsoni (Shotwell, 1956) (Hp)
     S. (O.) matthewi (Black, 1963) (Cl)
     S. (O.) shotwelli (Black, 1963) (Hp)
     S. (O.) finlayensis (Strain, 1966) (Bl)
     S. (O.) boothi (Hibbard, 1972) (Bl)
   Spermophilus (?) *matachicensis* (Wilson, 1949) (Hp)

    Paenemarmota Hibbard and Schultz, 1948 (Hp–Bl)
        ?*P. nevadensis* (Kellogg, 1910) (Hp)
        P. barbouri Hibbard and Schultz, 1948* (Bl)
        P. sawrockensis (Hibbard, 1964) (Hp)
    Protospermophilus Gazin, 1930 (A–B)
        P. vortmani (Cope, 1879) (A)
        P. quatalensis Gazin, 1930* (B)
        P. malheurensis (Gazin, 1932) (B)
        P. angusticeps (Matthew and Mook, 1933) (Hf)
        P. oregonensis (Downs, 1956) (B)
        P. kelloggi Black, 1963 (Hf)
    Miospermophilus Black, 1963 (Hf)
        M. bryanti (Wilson, 1960)* (Hf)
        M. wyomingensis Black, 1963 (Hf)
    Marmota Frisch, 1755 (Cl–R)
        M. vetus (Marsh, 1871) (Cl)
        M. minor (Kellogg, 1910) (Hp)
        M. arizonae Hay, 1921 (Bl)
    Cynomys Rafinesque, 1817 (Bl–R)
        C. vetus Hibbard, 1942 (Bl)
        C. hibbardi Eshelman, 1975 (Bl)
Petaruistinae Miller, 1912
    Cryptopterus Mein, 1970 (Bl)
        C. webbi Robertson, 1976 (Bl)
    Petauristodon Engesser, 1978 (Hf–Cl)
        P. matthewsi (James, 1963)* (Cl)
        P. uphami (James, 1963) (B)
        P. jamesi (Lindsay, 1972) (B)
        P. minimus (Lindsay, 1972) (B)
        P. pattersoni Pratt and Morgan, 1989 (Hf)
    Sciurion Skwara, 1986 (Hf)
        S. campestre Skwara, 1986* (Hf)
Cedromurinae Korth and Emry, 1991
    Cedromus Wilson, 1949 (O–?Wh)
        C. wardi Wilson, 1949* (O)
        C. wilsoni Korth and Emry, 1991 (O)
        C. sp. Korth and Emry, 1991 (Wh)
    Oligospermophilus Korth, 1987 (O–?Wh)
        O. douglassi (Korth, 1981)* (O)

Chapter 12
Eutypomyidae

1. Characteristic Morphology 125
   1.1. Skull 125
   1.2. Dentition 127
   1.3. Skeleton 127
2. Evolutionary Changes in the Family 127
3. Fossil Record 129
4. Phylogeny 129
   4.1. Origin 129
   4.2. Intrafamilial Relationships 131
   4.3. Extrafamilial Relationships 131
5. Problematical Taxa 132
   5.1. *Anchitheriomys* and *Amblycastor* 132
   5.2. "*Eutypomys*" *magnus* 132
6. Classification 132

1. Characteristic Morphology

1.1. Skull

In the Oligocene and Miocene, eutypomyids are intermediate to large size rodents. The primitive Wasatchian to Duchesnean species are small. No cranial material is known for species earlier than the Chadronian. The skull is low with a markedly elongate rostrum which is manifest in elongated nasal and premaxillary bones and upper diastema (Fig. 12.1). The neurocranium is also elongated. The skull is fully sciuromorphous (not known in pre-Chadronian species). The infraorbital foramen is relatively low on the rostrum as in sciurids. There is a pronounced process ventral to the infraorbital foramen for the attachment of the masseter superficialis. The incisive foramina are short.

Wahlert (1977, p. 6) pointed out two specializations in the cranial foramina of eutypomyids that were shared with castorids: (1) dorsal palatine foramen separated from sphenopalatine foramen and situated in orbitosphenoid–maxillary suture or maxilla and (2) interorbital foramen posterior to optic foramen. The sphenopalatine foramen is contained entirely in the maxillary, a feature elsewhere known only in castorids and eomyids. The posterior maxillary foramen is enclosed laterally.

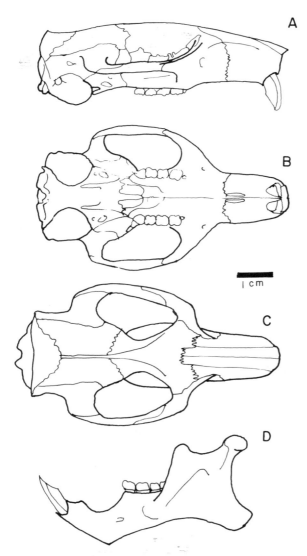

FIGURE 12.1. Skull and mandible of *Eutypomys thompsoni*. (A) Lateral view of the skull; (B) ventral view; (C) dorsal view. (D) Lateral view of left mandible. A and B from Wahlert (1977), C and D from Wood (1937).

The maxillary tooth rows are parallel, not converging as in castorids. The bullae are moderately inflated.

The mandible is shallow for the size of the animal and is clearly sciurognathous. The masseteric scar extends anteriorly to the posterior margin of P_4 on all but the Wasatchian *Mattimys* where it extends only to below the posterior half of M_2. The mental foramen is high on the mandible below the posterior half of the diastema. The coronoid process is broad.

1.2. Dentition

The incisor enamel of *Eutypomys* is uniserial (Wahlert, 1968). The enamel microstructure for pre-Chadronian species is not known. Incisors are procumbent where known; the upper incisor extending beyond the anterior edge of the nasals and the lower incisor having a large radius of curvature. In *Amblycastor* and *Anchitheriomys*, the largest of the eutypomyids, the incisors have numerous parallel grooves and ridges that run the length of the tooth, similar to some large castorids. The incisors are broad anteriorly. The anterior surface is convex on I_1 and nearly flattened on I^1.

The primitive rodent dental formula is present in *Eutypomys* and earlier genera. The Hemingfordian and later species of *Amblycastor* lose P^3. Cheek teeth are brachydont to mesodont in crown height. The last premolar (except in the earliest species) is molariform and equal to or larger in size than the molars. The occlusal pattern of the cheek teeth is dominated by two transverse lophs (metalophulid II and hypolophid on the lower cheek teeth; protoloph and metaloph on the upper cheek teeth), well-developed anterior and posterior cingula, and an irregular pattern of lophules within the basins (Fig. 12.2). In higher crowned species, the occlusal morphology becomes a complex pattern of enamel lakes in later stages of wear.

1.3. Skeleton

Matthew (1905) reported the first postcranial material of *Eutypomys*. Wood (1937, Pl. XXX) fully described and figured all of the known elements (several vertebrae, humerus, proximal and distal tibia, and nearly complete tarsus and pes). He noted that in general, the vertebrae and limb bones were not far from that of primitive ischyromyids (fibula not fused to tibia, no apparent adaptations for specialized movement) and more nearly approached that of sciurids than castorids. However, he noted that the limb bones were slender as in those of sciurids, but the muscular attachments were larger. Wood (1937) concluded that the pes, with elongate calcaneum, metatarsals, and phalanges, was unique among rodents.

2. Evolutionary Changes in the Family

The major changes within the Eutypomyidae were the result of marked size increase over time. The cheek teeth became more complex in occlusal

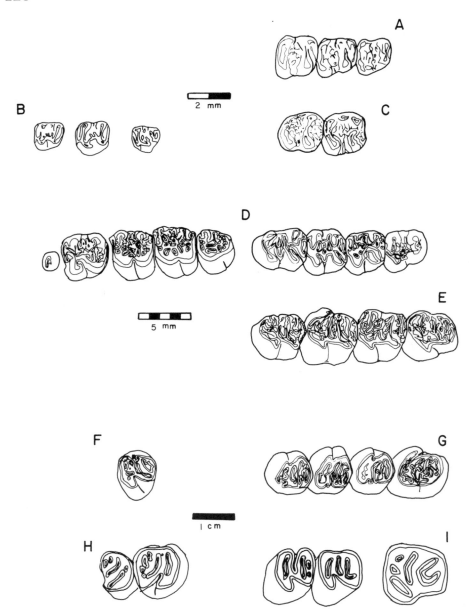

FIGURE 12.2. Dentitions of Eutypomyidae. (A) Wasatchian *Mattimys kalicola*, RP_4–M_2. (B) Duchesnean *Janimus dawsonae*, LP^4, M^1 or M^2, RM^3 (reversed). (C) Uintan *Janimus rhinophilus*, RM_2–M_3. (D) Orellan *Eutypomys thompsoni*, LP^3–M^3 and RP_4–M_3. (E) Orellan "*Eutypomys*" *magnus*, RP_4–M_3. (F, G) Asian Miocene *Anchitheriomys tungurensis*, LM^1 or M^2 and LP_4–M_3. (H, I) Barstovian *Amblycastor fluminis*, RP^4–M^1 and RP_4 and RM_1–M_4. A from Matthew (1918b), B from Storer (1987), C from Dawson (1966), D and E from Wood (1937), G from Stirton (1934), F H, and I from Stirton (1935).

pattern and increased in crown height. Similarly, the elongation of the skull and procumbency of the incisors are much greater in Miocene *Amblycastor* and *Anchitheriomys* than in earlier genera.

3. Fossil Record

Mattimys kalicola from the Wasatchian of Wyoming was the earliest eutypomyid (Matthew, 1918a; Korth, 1984). The last occurrence in North America was *Amblycastor fluminis* from the Barstovian of Nebraska (Matthew, 1918b; Voorhies, 1990b). Eutypomyids were usually rare in faunas from the Tertiary and were never very diverse. During the Duchesnean they reached their peak of diversity with the recognition of two genera and four species (Fig. 12.3).

The only eutypomyids known from outside of North America are two Miocene species of *Anchitheriomys*, *A. wiedemanni* and *A. tungurensis*, from Asia.

4. Phylogeny (Fig. 12.4)

4.1. Origin

The distinctive morphology of the anterior cingulum on the lower molars of the Wasatchian and Uintan species of eutypomyids is shared with the reithroparamyine genus *Microparamys*. Dawson (1966) believed that the ancestor of the Uintan *Janimus* could be found among the earlier microparamyines. Black (1971) included *Janimus* in the Reithroparamyinae. Korth (1984) derived *Janimus* and later *Eutypomys* from the late Wasatchian *Mattimys* which, in turn, was derived from a *Microparamys*-like ancestor within the Reithroparamyinae.

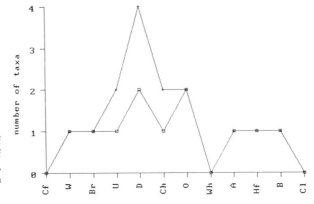

FIGURE 12.3. Occurrence of Eutypomyidae in the Tertiary of North America (□, genus; +, species). Abbreviations as in Fig. 5.4

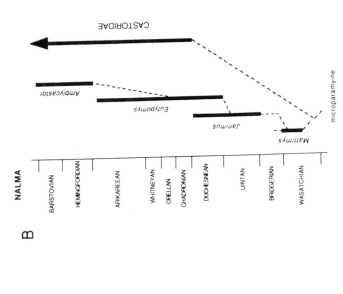
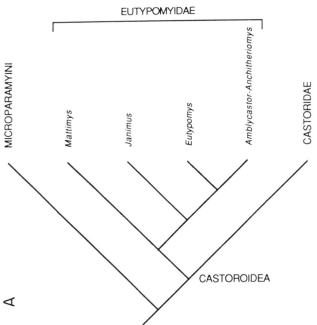

FIGURE 12.4. Phylogeny of the Eutypomyidae. (A) Cladogram of eutypomyid relationships. (B) Dendrogram of the Eutypomyidae with age of occurrence (see Fig. 5.5 for explanation).

It is most likely that the Eutypomyidae were derived from a microparamyine ancestor, probably in the early Wasatchian (see Chapter 5, Section 4.3).

4.2. Intrafamilial Relationships

Whereas the increase in crown height and complexity of the cheek teeth and the cranial elongation and procumbency of the incisors were progressive through time in eutypomyids, it is likely that the earliest genera were ancestral to the later genera, thus forming a progression of ancestry from the Wasatchian and Bridgerian *Mattimys* to the Uintan and Duchesnean *Janimus* to the Duchesnean to Arikareean *Eutypomys* to the possibly Orellan to later Miocene *Anchitheriomys* and *Amblycastor*. The only deviation from this pattern would be *Amblycastor* from the Hemingfordian and Barstovian. *Amblycastor* lacks the complexity of the occlusal pattern of the cheek teeth present in all other genera. This suggests that *Amblycastor* was derived from a *Eutypomys*-like ancestor in the Arikareean or earlier that began to simplify the occlusal pattern of the teeth.

The Asian Miocene species of *Anchitheriomys* could be derived morphologically from the Orellan "*Eutypomys*" *magnus*. The greater crown height of the cheek teeth of the latter makes it distinct from other species of *Eutypomys* and makes it more likely referable to the otherwise Asian *Anchitheriomys*.

4.3. Extrafamilial Relationships

In the first description of *Eutypomys*, Matthew (1905) suggested a close relationship with the castorids. The Eutypomyidae have been included in the superfamily Castoroidea in virtually all subsequent classifications of rodents. However, a number of sciurid-like characters have been recognized in the eutypomyids (Matthew, 1910; Wood, 1937; Wilson, 1949a). In his study of the cranial foramina of eutypomyids, Wahlert (1977) supported the close relationship between eutypomyids and castorids, separating these families only on dental morphology of the cheek teeth.

Several authors have noted that characters of the skull and dentition of eutypomyids are also shared with eomyids but none of them held there to be any particularly close relationship between the eomyids and eutypomyids (Wood, 1937; Wilson, 1949a; Wahlert, 1977). Stirton (1935), in his review of the Tertiary castorids, included *Eutypomys thompsoni* in his phylogeny of the Castoridae as being near a lineage that led to the Mylagaulidae.

Based on cranial and dental characters, it appears that eutypomyids do form a sister group to the castorids. Although the earliest possible eutypomyid (Wasatchian) predates the earliest castorid (Chadronian), there is no evidence that the Eutypomyidae are part of the ancestry of the castorids.

5. Problematical Taxa

5.1. *Anchitheriomys* and *Amblycastor*

Matthew (1918b) described *Amblycastor fluminis* from the Barstovian of Nebraska and referred it to the Castoridae. A second species, *A. tungurensis* from the Miocene of Mongolia, was named by Stirton (1934). The major difference between the Asian species and *A. fluminis* was the amount of plications in the enamel fossettids of the cheek teeth. Wood (1937) noted the similarity in the complex occlusal pattern of the cheek teeth of *Eutypomys* to that of *A. tungurensis*, but believed the latter to be a castorid and the similarity to reflect parallelism.

Stehlin and Schaub (1951) first suggested the synonymy of *Anchitheriomys* (recognized at that time only from the late Miocene of Asia; Roger, 1885) and *Amblycastor*. Wilson (1960) described a new species from the Hemingfordian of Colorado that he did not name but referred to ?*Anchitheriomys*. However, Wilson (1960) did note that even in the unworn cheek teeth of his new species the minor lophules or plications of the occlusal pattern of the Miocene Asian species were lacking. He included Stirton's species as "*Amblycastor*" *tungurensis*, questioning its inclusion in *Amblycastor*. Most recently, Voorhies (1990b) included *Amblycastor fluminis* in *Anchitheriomys* following Stehlin and Schaub's (1951) synonymy.

The complications of the occlusal pattern of the cheek teeth present in *A. wiedemanni* (the type species of *Anchitheriomys*) and *A. tungurensis* are not present in the North American Hemingfordian and Barstovian species of eutypomyids. Therefore, the North American species are maintained here in *Amblycastor* while the Asian species are referred to *Anchitheriomys*.

5.2. "*Eutypomys*" *magnus*

Wood (1937) named *Eutypomys magnus* from the Orellan of South Dakota and noted that it differed from all other species of the genus by its much larger size and greater crown height. In both of these features, *E. magnus* more closely resembles the Asian *Anchitheriomys*. This species is here questionably referred to *Anchitheriomys* because of the similarity of the cheek teeth. It may prove that ?*A. magnus* represents a new genus intermediate between *Eutypomys* and the much younger *Anchitheriomys* but ?*A. magnus* is currently known only from the type specimen, a lower jaw.

6. Classification

Eutypomyidae Miller and Gidley, 1918
    Mattimys Korth, 1984 (W–?Br)

    M. kalicola (Matthew, 1918)* (W)
Janimus Dawson, 1966 (U–D)
    J. rhinophilus Dawson, 1966* (U)
    J. mirus Storer, 1984 (U)
    J. dawsonae Storer, 1987 (D)
Eutypomys Matthew, 1905 (D–O, A)
    E. thompsoni Matthew, 1905* (O)
    E. parvus Lambe, 1908 (Ch)
    E. montanensis Wood and Konizeski, 1965 (A)
    E. inexpectatus Wood, 1974 (Ch)
    E. acares Storer, 1988 (D)
    E. obliquidens Storer, 1988 (D)
Microeutypomys Walton, 1993 (U–D)
    M. tilliei (Storer, 1988) (D)
    M. karenae Walton, 1993* (U)
Anchitheriomys Roger, 1885 (O)
    ?*A. magnus* (Wood, 1937) (O)
Amblycastor Matthew, 1918 (Hf–B)
    A. fluminis (Matthew, 1918) (B)
    A. new species Wilson, 1960 (Hf)

Chapter 13
Castoridae

1. Characteristic Morphology	135
1.1. Skull	135
1.2. Dentition	137
1.3. Skeleton	140
2. Evolutionary Changes in the Family	141
3. Fossil Record	141
4. Phylogeny	142
4.1. Origin	142
4.2. Intrafamilial Relationships	144
4.3. Extrafamilial Relationships	145
5. Problematical Taxa	146
5.1. *Hystricops*	146
5.2. *Monosaulax*	146
5.3. *Anchitheriomys* and *Amblycastor*	147
6. Classification	147

1. Characteristic Morphology

1.1. Skull

The size of castorids (beavers) ranges from intermediate to the largest rodents ever, culminating in the Pleistocene *Castoroides* which had a total body length of approximately 7 feet. The skulls of castorids, partly because of their larger size, have a dorsoventrally deep rostrum and broad zygomatic arch (Fig. 13.1). The palaeocastorine beavers had shorter rostra and lower, broader neurocrania as an adaptation for their fossorial mode of life. Martin (1987) proposed that the highly fossorially adapted palaeocastorines may have had a nasal horn as in advanced mylagaulids based on the presence of thickened nasal bones and numerous minute nutritive foramina in the same area. All known castorids are fully sciuromorphous. The infraorbital foramen is small, laterally compressed, and situated about middepth of the rostrum. Ventral to the infraorbital foramen is a small bony knob for the attachment of the masseter superficialis as in sciurids.

There is a strongly developed sagittal crest along the center of the neurocranium or two parallel, narrowly spaced parasagittal crests that are continuous with well-developed occipital crests. Often, the dorsal surface of the parietal bones is rugose. The auditory bullae are not inflated but there is an

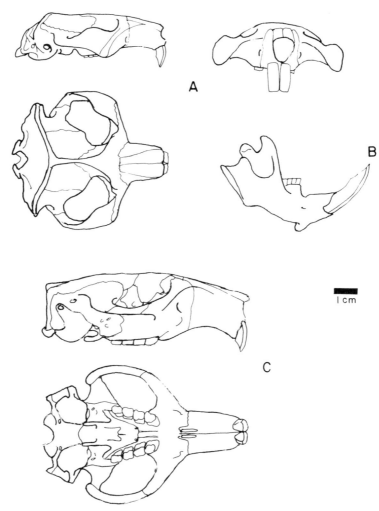

FIGURE 13.1. Skulls of castorids. (A) Lateral, dorsal, and anterior views of the skull of the Arikareean palaeocastorine *Euhapsis platyceps*. (B) Lateral view of the mandible of the Arikareean palaeocastorine *Palaeocastor fossor*. (C) Lateral and ventral views of the skull of the Hemphillian castoroidine *Dipoides stirtoni*. A (lateral and dorsal views) and B from Stirton (1935), A (anterior view) from Martin (1987), C from Wahlert (1972).

elongate external meatus that extends dorsally and slightly posteriorly. The upper tooth rows diverge posteriorly in all but the most primitive genera in which they are parallel.

There are a number of distinctive features of the cranial foramina of castorids (Olson, 1940; Wahlert, 1972, 1977): the incisive foramina are small; the sphenopalatine foramen is anteriorly situated (dorsal to P^4) and entirely surrounded by the maxillary bone (elsewhere only known in eomyids and eutypomyids); interorbital foramen posterior to the optic foramen; posterior palatine foramen in maxillary–palatine suture on the palate; and the stapedial and sphenofrontal foramina are lacking (loss of stapedial artery). In *Agnotocastor*, the earliest and most primitive castorid, the posterior palatine foramina are contained within the palatine bone and the stapedial and sphenofrontal foramina are present (primitive condition for rodents). An accessory foramen ovale is present in some genera, but is lost in later genera because of the enlargement of the pterygoid flange (Wahlert, 1972). Martin (1987) noted a reduction in the relative size of the optic foramen in some of the highly fossorial palaeocastorines and suggested that this indicated total blindness in these genera.

The mandible of castorids is dorsoventrally deep and mediolaterally broad. The diastema is deep, extending ventrally nearly half the depth of the mandible. There is a distinctive chin process (symphyseal flange). The coronoid process is large and curved posteriorly. The mandible is sciurognathous with a broad angle that frequently has a laterally curved ventral margin. The base of the lower incisor is marked on the lateral side of the ascending ramus by a bulbous lateral expansion of bone or a well-developed shelf of bone. The mental foramen is commonly multiple with one main foramen and several smaller foramina.

1.2. Dentition

The incisors of castorids are large and broadened anteriorly. In the palaeocastorines the anterior surface is flat, in other genera the anterior surface is gently convex. On the largest genera, notably *Castoroides*, the anterior surface of the enamel on the incisor is covered with multiple parallel grooves and ridges. The microstructure of the enamel of the incisors is uniserial (Wahlert, 1968). In the palaeocastorines the incisors were procumbent (Martin, 1987).

Castorids lose P^3 in all but *Agnotocastor*, and otherwise maintain the maximum dental formula for rodents. The most primitive castorids had cheek teeth that were mesodont in crown height and in later Tertiary genera the cheek teeth attained full hypsodonty. P^4 and P_4 are the largest of the cheek teeth, and the molars decrease in size from M_1^1 to M_3^3. The cusps on the cheek teeth of castorids are difficult to identify even in the earliest genera (Fig. 13.2). In the most primitive genera the occlusal pattern of the cheek teeth consisted of an enamel outline and several smaller isolated enamel lakes (Fig. 13.3). In later Tertiary castorids the occlusal pattern became a series of deep reentrant valleys of enamel. The castorids with higher crowned cheek teeth often have cement between these folds of enamel. The most derived castoroidines had cheek teeth that become a series of compressed enamel ovals with dentine in the center and cement between each oval. Interstitial wear is heavy on the

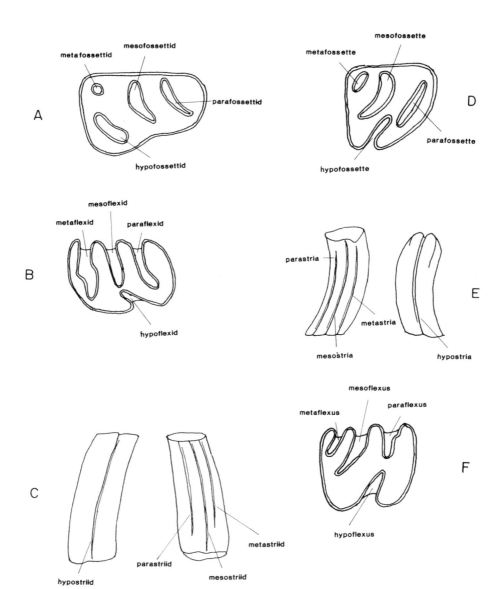

FIGURE 13.2. Specialized dental nomenclature of castorids. (A) Worn RP_4 of *Eucastor*, occlusal view. (B) RP_4 of *Castor*, occlusal view. (C) Same as B, buccal view (left) and lingual view (right). (D) Worn RP^4 of *Eucastor*, occlusal view. (E) RP^4 of *Castor*, buccal view (left) and lingual view (right). (F) Same as E, occlusal view. All panels not drawn to scale.

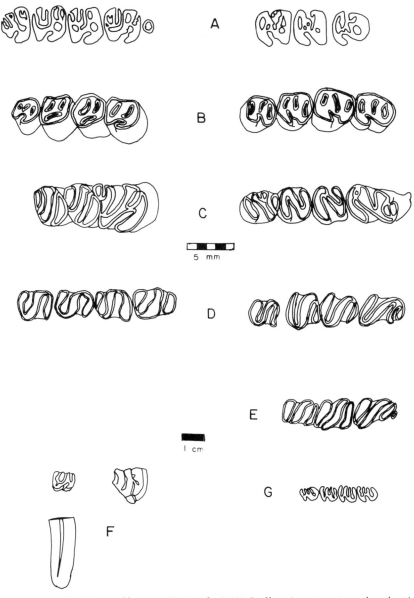

FIGURE 13.3. Dentitions of beavers (Castoridae). (A) Orellan *Agnotocastor coloradensis*, RP³–M³ (composite) and RP$_4$–M$_2$. (B) Arikareean palaeocastorine *Palaeocastor fossor*, RP⁴–M³ and RP$_4$–M$_3$. (C) Clarendonian castoroidine *Eucastor dividerus*, RP⁴–M² and RP$_4$–M$_3$. (D) Hempillian castoroidine *Dipoides vallicula*, RP⁴–M³ and LP$_4$–M$_3$. (E) Blancan castoroidine *Procastoroides idahoensis*, LP$_4$–M$_2$. (F) Blancan castorid *Castor californicus*, RP⁴ and M² with lingual view of M². (G) Blancan castorid *Castor accessor*, RP$_4$–M$_3$. A–D to same scale (above) and E–G to same scale (below). A from Galbreath (1953), B, D, F from Stirton (1935), C, E, G from Shotwell (1970).

cheek teeth evidenced by anteroposterior shortening of specimens that show advanced age.

1.3. Skeleton (Fig. 13.4)

The skeletons of castorids are generally robust for rodents, partially because of their larger size. The skeleton does not greatly differ from that of ischyromyids of similar size except in the proportions of the limbs. The forelimb of castorids is slightly longer than the hindlimb. The tibia and fibula remain as distinct bones in all castorids.

The relative proportions of the limbs in the aquatic castorines differ from those of the other beavers. The femur is shorter and the elements of the lower part of the limbs are elongated (radius and ulna of the forelimb; tibia and fibula of the hindlimb). The pes in *Castor* is also much enlarged relative to the manus and the pes of other castorids. Clearly, this is an adaptation for propulsion in the water by the hindlimbs.

The palaeocastorine beavers showed a number of fossorial adaptations in their skeleton (Peterson, 1905). The forelimb was massive, the scapula was long and narrow, and the manus was relatively enlarged with large ungual phalanges. The neck was shorter than in other beavers and the skull was proportionally larger compared with the body than in other beavers.

In *Castor* the caudal vertebrae are dorsoventrally compressed and the transverse processes are relatively large on all but the terminal vertebra, clearly for the support of the broad, fleshy tail used for swimming. There is no evidence of this in any of the palaeocastorine or castoroidine beavers (Shotwell, 1970). Martin (1987) used the relative shortening of the tail as a diagnostic character of the Palaeocastorinae.

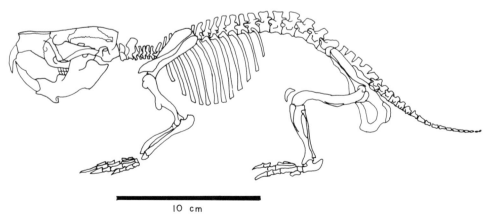

FIGURE 13.4. Reconstruction of the skeleton of the Arikareean palaeocastorine *Palaeocastor fossor*. Taken from Peterson (1905).

Castoridae 141

2. Evolutionary Changes in the Family

The two most obvious changes in the castorids through time are the increase in size (especially in the later Tertiary castoroidines) and hypsodonty of the cheek teeth. The earliest castorid, *Agnotocastor*, had cheek teeth that were already mesodont in crown height (Wilson, 1949b, Fig. 2; Emry, 1972a, Fig. 1; Korth, 1988c, Fig. 1.3). In all lineages of castorids including the Palaeocastorinae, there was an increase in crown height of the cheek teeth in the later species. Castorines and especially castoroidine genera with higher-crowned (subhypsodont–hypsodont) cheek teeth develop cement between the bands of enamel and dentine on the cheek teeth.

The incisors develop numerous longitudinal grooves and ridges in the largest genera of castoroidines (*Procastoroides*, *Castoroides*). The more fossorially adapted palaeocastorines developed broader, shorter, and lower skulls and wider incisors (Martin, 1987).

3. Fossil Record

The first occurrence of castorids in North America was *Agnotocastor galushai* from the Chadronian of Wyoming (Emry, 1972a). The greatest diversity of castorids in North America was in the Arikareean (Fig. 13.5) involving the radiation of the palaeocastorine beavers which were restricted to North America except perhaps for the middle Oligocene *Propalaeocastor* of Asia (Borisoglebskaya, 1967). A second time of greater diversity was the Blancan

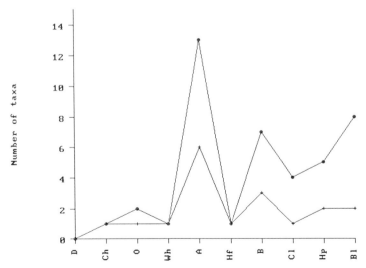

FIGURE 13.5. Occurrence of Castoridae in the Tertiary of North America (●, species; +, genera). Abbreviations as in Fig. 5.4.

and into the Pleistocene which primarily involved the radiation of the castoroidine beavers. Only one species of castorid survived the Pleistocene in North America, *Castor canadensis*.

Castorids are also well represented in the fossil record from both Europe and Asia. The earliest occurrence of castorids in Europe was from the early Oligocene, contemporaneous with the first occurrence in North America (Misonne, 1957). However, the known species of castorids from Europe were restricted to the genus *Stenofiber* Geoffroy Saint-Hilaire (1833) through the Oligocene until the late Miocene (Astracian) when additional genera appeared. The greatest diversity of castorids in Europe was coincident with the second radiation peak in North America in the Pliocene and involves castorine as well as castoroidine species.

The first record of castorids in Asia was also in the early Oligocene (Lychev, 1978). The earliest Asian species was more similar to those in North America than in Europe, being represented by the otherwise North American genus *Agnotocastor*. Later beavers from Asia were either of the same genera as those from Europe or distinctive indigenous forms. Castorids were never very diverse in Asia, only being represented by three or fewer genera at any one time. The only genera that have been recognized from all three continents are *Dipoides* and *Castor*. Both of these genera first appeared in North America in the Hemphillian and at the same time in both Europe and Asia (Savage and Russell, 1983). As in North America, only one species of castorid survived the Pleistocene in Eurasia, *Castor fiber*.

It is difficult to trace the migrations of castorids from one continent to another. The simultaneous first occurrence of castorids in Europe, Asia, and North America makes it difficult to determine the continent of origin since there are no obvious ancestors of the castorids on either continent. It is likely that the first castorids in Asia migrated from North America in the middle Oligocene because of the generic similarity of the genera present on both continents. The occurrence of *Dipoides* in both Europe and Asia also might be the result of migration from North America. While the first occurrence of *Dipoides* on all of these continents was simultaneous, the most likely ancestor to *Dipoides*, *Eucastor*, is restricted to North America.

The first occurrence of *Castor* was also nearly simultaneous on all three continents, hence its continent of origin is also uncertain. Stirton (1935) viewed the origin of the modern beavers from a European ancestry through the genus *Stenofiber*.

4. Phylogeny (Fig. 13.6)

4.1. Origin

Matthew (1910) derived the castorids from Bridgerian *Sciuravus* in his phylogeny, but no Eocene ancestor of the castorids has ever been described,

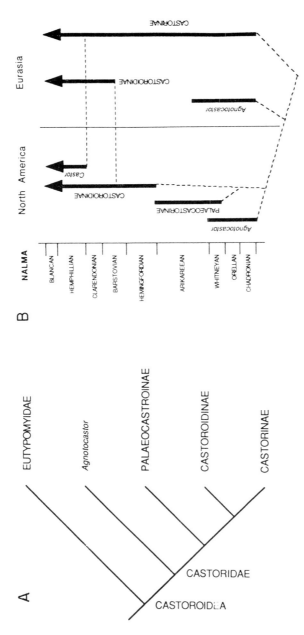

FIGURE 13.6. Phylogeny of the Castoridae. (A) Cladogram of castorid relationships. (B) Dendrogram of the Castoridae with age of occurrence (see Fig. 5.5 for explanation).

making this suggestion questionable (Stirton, 1935; Wilson, 1949a). Although the eutypomyids appear to be closely related to the castorids, and predate them in the known fossil record (eutypomyids as early as the Wasatchian; castorids earliest occurrence in the Chadronian), the eutypomyids cannot be considered as ancestral to the castorids because characters of the skull and dentition of eutypomyids are derived in another direction (e.g., elongation of the rostrum, complex enamel pattern on the cheek teeth). There is no indication of the increase in crown height in the earliest castorids in any of the pre-Chadronian eutypomyids.

If the cranial similarities of eutypomyids and castorids (Wahlert, 1977) are an indication of true relationship, and the eutypomyids are derived from a microparamyine ischyromyid (Chapter 12, Section 4.1), then it is likely that the castorids, too, can be ultimately traced back to a microparamyine ancestor. Wahlert (1972) followed this same reasoning and concluded that the ancestry of castorids was to be found within the early paramyids. Wilson (1949b) commented that there was superficial resemblance between the early beavers and the European "anomaluroids" (= theridomyids). However, there are no pre-Chadronian or Eocene castorids or "pre-castorids" known from any continent, so any suggestion as to an ancestor is highly speculative at this time.

4.2. Intrafamilial Relationships

Matthew (1910) recognized two lineages of castorids that were distinct since at least the Miocene which he referred to the Castorinae and Castoroidinae, the former leading to the Recent *Castor* and the latter culminating in the Pleistocene genus *Castoroides*. In the last review of the Castoridae of the world, Stirton (1935, p. 395) divided the known genera into two categories: (1) those with a flattened anterior surface on the incisors and (2) those with convex incisors. In the first group he included the palaeocastorine beavers, one species of *Stenofiber*, the European *Palaeomys*, and the Recent *Castor*. He also included *Agnotocastor* with question. The second group consisted of the remainder of the species of *Stenofiber*, *Monosaulax*, *Eucastor*, *Dipoides*, *Amblycastor*, *Trogontherium*, and *Castoroides*. Though not formally defined, these two groups have been used as two subfamilies of the castorids, the Castorinae which included *Castor* and the European lineage of *Stenofiber* and *Paleomys*; and the Castoroidinae which included the remaining North American genera (except *Agnotocastor* and the palaeocastorines) and the Eurasian *Trogontherium* (Simpson, 1945; Wilson, 1949a).

Agnotocastor first appeared almost simultaneously in both Asia and North America. It is distinctive among other castorids on the basis of primitive characters of the cranial foramina (Wahlert, 1972) and cheek teeth (mesodont crown height, more complex pattern of enamel fossettes, presence of P^3). Because of these less advanced characters and absence of more derived castorid features, *Agnotocastor* cannot be allocated to any recognized subfamily. Its primitive characteristics of the skull and dentition make it a likely ancestral morphotype for the later castorids. Its simultaneous appearance in the fossil record with the European castorine *Stenofiber* prohibits *Agnotocastor* from being recognized as the actual ancestor of all other beavers based on chronology.

Martin (1987) erected the subfamily Palaeocastorinae for the North American genera *Palaeocastor* Leidy (1869), *Capitanka* Macdonald (1963), *Capacikala* Macdonald (1963) and *Euhapsis* Peterson (1905), all of which occur only in the Arikareean of the Great Plains. He also recognized two new genera and three new tribes of palaeocastorines. He diagnosed this highly fossorial subfamily of beavers as having flattened incisors and shortened, rounded tails. Martin believed that this radiation of burrowing beavers was an aberrant side branch of the castorids that were not responsible for any later radiations of the family (see Peterson, 1905; Martin and Bennett, 1977, for discussions of *Daemonelix* burrows).

A series of North American genera considered as members of the Castoroidinae have been accepted as a direct lineage by several authors (Matthew, 1910; Stirton, 1935; Wilson, 1949a, 1960; Shotwell, 1970). Beginning with the Hemingfordian *Monosaulax*, this lineage ultimately became extinct at the end of the Pleistocene where it was represented by the genus *Castoroides*. This lineage consisted of five subsequent genera, *Monosaulax–Eucastor–Dipoides–Procastoroides–Castoroides*. The later Tertiary and Pleistocene Eurasian genus *Trogontherium* Fischer (1809) also has been included in this grouping (Stirton, 1935; Simpson, 1945).

The European genus *Stenofiber* remains the earliest member of the castorines. All of the Eurasian genera of castorids excluding *Trogontherium* and those also known from North America as part of the Castoroidinae (see Savage and Russell, 1983, for occurrences of *Dipoides* and *Monosaulax* in Europe and Asia) are considered to belong to the Castorinae. Stirton (1935) viewed the lineage that led to the Recent *Castor* to be from European stock as a *Stenofiber–Palaeomys–Castor* lineage.

4.3. Extrafamilial Relationships

The Castoridae "sister-group" relationship with the Eutypomyidae has been long understood and these two families have been classified under the superfamily Castoroidea by virtually all authors (Stirton, 1935; Wilson, 1949a,b; Wahlert, 1977). Beyond this relationship there is little indication of any other related rodent groups. In early classifications of rodents the castorids were included in the suborder Sciuromorpha along with the sciurids mainly because of their sciuromorphous zygomasseteric structure and nearctic occurrence (e.g., Brandt, 1855; Miller and Gidley, 1918; Simpson, 1945; Wilson, 1949a). Wood (1955a) first proposed the suborder Castorimorpha because he believed that other than sciuromorphy there were no characteristics of the castorids that allied them to the other Sciuromorpha. This conclusion has been corroborated by several authors using other characters of the morphology and physiology of fossil and Recent rodents (Bugge, 1974; Lavocat and Parent, 1985).

5. Problematical Taxa

5.1. *Hystricops*

Leidy (1858) first named *Hystrix (Hystricops) venustus* from the Barstovian of Nebraska based on two isolated cheek teeth. He believed that these teeth belonged to the Hystricidae, the family of Old World porcupines. In his catalog of fossil vertebrates, Hay (1902) transferred this species to *Erethizon*, a genus of Recent New World porcupine. Matthew (1902) was the first to suggest that *Hystricops* was more likely a castorid.

Shotwell and Russell (1963) identified some isolated cheek teeth from the Clarendonian of Oregon as ?*Hystricops* sp., and in the same publication, Shotwell (1963) named a second species of *Hystricops*, *H. browni*, from the Hemphillian of Oregon based on two associated cheek teeth. In an abstract, Stout and Stone (1971) referred *Monosaulax senrudi* from the Barstovian of Montana (Wood, 1945) to *Hystricops* which was followed by Cassiliano (1980). This was significant because *M. senrudi* was known from a nearly complete mandible with lower dentition. However, the relative size of the premolar (in relation to the molars), depth of the hypoflexid on both the premolar and molars, and height of the mesostriid on the premolar of *M. senrudi* are different from those of *H. venustus* and more closely approach the morphologies in species of *Monosaulax*. Therefore, this species is retained in *Monosaulax*. Voorhies (1990a) referred a number of specimens (isolated teeth and one mandibular fragment) to *Hystricops* that had been previously referred to *Amblycastor* from the Barstovian of Colorado (Matthew, 1902; Stirton, 1935) and Saskatchewan (Storer, 1975).

The systematic position of *Hystricops* is uncertain because the record is so poor. This genus and its two recognized species are known from only several isolated cheek teeth, hence the assignment of this genus to any recognized subfamily or lineage of castorid is impossible at the present time.

5.2. *Monosaulax*

Stirton (1935) erected the genus *Monosaulax* and designated *Stenofiber pansus* Cope (1874) as its type species. Stout (in Skinner and Taylor, 1967) suggested that this species was synonymous with *Eucastor tortus* Leidy (1858), the type species of *Eucastor*, and therefore the generic name *Monosaulax* was invalid. Stout used "*Eucastor*" in the place of *Monosaulax* as the generic name. Later, Voorhies (1990a) used "*Monosaulax*" as the generic reference pending a more complete study of the available material in order to determine if *M. pansus* and *E. tortus* were truly synonymous.

5.3. *Anchitheriomys* and *Amblycastor*

All previous authors have considered these genera (or single genus) as castorids. However, the general skull morphology, elongation of the rostrum, and lower crown height and complexity of the cheek teeth of these genera are similar to the characteristics of eutypomyids (Voorhies, 1990a). These genera differ only in their larger size and later occurrence from other eutypomyids. *Anchitheriomys* and *Amblycastor* are here considered eutypomyids. The generic validity of *Anchitheriomys* and *Amblycastor* has been discussed earlier (Chapter 12, Section 5.1).

6. Classification

Castoridae Gray, 1821
  Castorinae Gray, 1821
      Castor Linnaeus, 178 (Hp–R)
        C. californicus Kellogg, 1911 (?Hp–Pl)
        C. accessor Hay, 1927 (Bl–Pl)
  Palaeocastorinae Martin, 1987
    Tribe Palaeocastorini Martin, 1987
      Palaeocastor Leidy, 1869 (?Wh–A)
        P. nebrascensis* (Leidy, 1856) (?Wh, ?A)
        P. peninsulatus (Cope, 1881) (A)
        P. fossor (Peterson, 1905) (A)
        P. simplicidens (Matthew, 1907) (A)
      Capitanka Macdonald, 1963 (A)
        C. cankpoepi Macdonald, 1963* (A)
        C. magnus (Romer and McCormick, 1928) (A)
    Tribe Capacikalini Martin, 1987
      Capacikala Macdonald, 1963 (A)
        C. gradatus (Cope, 1879)* (A)
      Pseudopalaeocastor Martin, 1987 (A)
        P. barbouri (Peterson, 1905)* (A)
    Tribe Euhapsini Martin, 1987
      Euhapsis Peterson, 1905 (A)
        E. platyceps Peterson, 1905* (A)
        E. ellicottae Martin, 1987 (A)
        E. breugerorum Martin, 1987 (A)
      Fossorcastor Martin, 1987 (A)

    F. brachyceps (Matthew, 1907) (A)
    F. greeni Martin, 1987* (A)
Castoroidinae Trouessart, 1880
  Monosaulax Stirton, 1935 (Hf–B)
    M. pansus (Cope, 1874)* (B)
    M. complexus (Douglass, 1901) (?B)
    M. hesperus (Douglass, 1901) (?B)
    M. curtus (Matthew and Cook, 1909) (B)
    ?*M. senrudi* (Wood, 1945) (?B)
    M. n. sp. Wilson, 1960 (Hf)
    M. typicus Shotwell, 1968 (B)
    M. progressus Shotwell, 1968 (B)
  Eucastor Leidy, 1858 (?B–Cl)
    E. tortus Leidy, 1858* (B)
    E. lecontei (Merriam, 1869) (?Cl)
    E. dividerus Stirton, 1935 (Cl)
    E. planus Stirton, 1935 (?Cl)
    E. malheurensis Shotwell and Russell, 1963 (Cl)
  Dipoides Jager, 1835 (Hp–Bl)
    D. stirtoni Wilson, 1934 (Hp–Bl)
    D. williamsi Stirton, 1936 (Hp)
    D. rexroadensis Hibbard and Riggs, 1949 (Bl)
    D. wilsoni Hibbard, 1949 (Hp)
    D. smithi Shotwell, 1955 (Hp)
    D. intermedius Zakrzewski, 1969 (Bl)
    D. vallicula Shotwell, 1970 (Hp)
  Procastoroides Barbour and Schultz, 1937 (Bl–Pl)
    P. sweeti Barbour and Schultz, 1937* (Bl–Pl)
    P. idahoensis Shotwell, 1970 (Bl–Pl)
  Castoroides Foster, 1838 (Bl–Pl)
    C. ohioensis Foster, 1838* (Bl–Pl)
Subfamily uncertain
  Agnotocastor Stirton, 1935 (Ch–Wh)
    A. praetereadens Stirton, 1935* (Wh)
    A. coloradensis Wilson, 1949 (O)
    A. galushai Emry, 1972 (Ch)
    A. readingi Korth, 1988 (O)
  Hystricops Leidy, 1858 (B–Hp)
    H. venustus Leidy, 1858* (B)
    H. browni Shotwell, 1963 (Hp)

Chapter 14
Eomyidae

---

1. Characteristic Morphology	149
1.1. Skull	149
1.2. Dentition	151
1.3. Skeleton	151
2. Evolutionary Changes in the Family	151
3. Fossil Record	154
4. Phylogeny	155
4.1. Origin	155
4.2. Intrafamilial Relationships	157
4.3. Extrafamilial Relationships	158
5. Problematical Taxa	158
5.1. *Adjidaumo–Eomys–Omegodus*	158
5.2. *Kansasimys–Ronquillomys–Comancheomys*	159
5.3. *Meliakrouniomys*	160
6. Classification	161

---

1. Characteristic Morphology

1.1. Skull

The Eomyidae were small rodents (except the Hemphillian species). The skull was fully sciuromorphous in all post-Chadronian species (Fig. 14.1) and not known for the Uintan and Duchesnean species. The infraorbital foramen was low on the rostrum with a small bony knob ventral to it for the attachment of the masseter superficialis as in sciurids. There was never a sagittal crest on the skull; there were either no markings at all, weakly formed lyrate parasagittal crests, or strongly developed, subparallel parasagittal crests (the latter present only in the Hemphillian *Kansasimys*; Wahlert, 1978). In one genus, *Aulolithomys*, there was a small postorbital process on the frontals (Wood, 1974). The bullae were ossified and moderately inflated.

Wahlert (1978, Table 2) cited a number of characters of the cranial foramina of eomyids that were also shared by the geomyoids: entrance to transverse canal separated from alisphenoid canal; carotid canal short; temporal foramina rare or absent; ratio of incisive foramen length to diastemal length low; infraorbital canal long; sphenopalatine foramen far anterior; and sphenopterygoid foramen present. The buccinator, masticatory, and accessory foramen ovale were all fused (also present in florentiamyids and helisco-

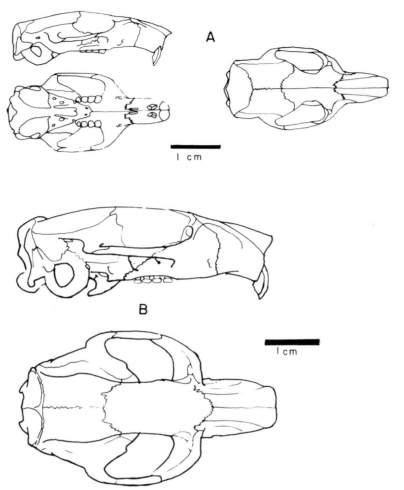

FIGURE 14.1. Skulls of eomyids. (A) Lateral, ventral, and dorsal views of the skull of Orellan *Paradjidaumo trilophus*. (B) Lateral and dorsal views of the skull of Hemphillian *Kansasimys dubius*. All from Wahlert (1978).

myids among the geomyoids). The sphenopalatine foramen was entirely enclosed by the maxilla (only elsewhere known in eutypomyids).

The mandible was slender and sciurognathous. The coronoid process was small and not laterally deflected as in heteromyids and geomyids. A small bulbous projection or shelf was on the ascending ramus even with the condyle which housed the base of the incisor. The masseteric fossa extended anteriorly on the mandible to below P_4. The ridges marking the fossa formed a V- or U-shape at their anterior end. In *Meliakrouniomys* (Harris and Wood, 1969; Wood, 1974; Emry, 1972b), *Metadjidaumo* (Korth and Tabrum, in press),

Eomyidae and *Paranamatomys* (Ostrander, 1983; Korth, 1992b), the masseteric scar ended anteriorly in a small shelf anterior to the tooth row similar to that of heteromyids. The mental foramen was high on the mandible near the border of the diastema, just posterior to its center.

1.2. Dentition

The incisor enamel of eomyids is uniserial but is unique (Fig. 14.2) in having a two-part internal layer (portio interna; Wahlert and von Koenigswald, 1985). The incisor of *Paradjidaumo* had a single longitudinal ridge along the anterolateral corner of the enamel. The only other reported ornamentation of the incisors was that of *Aulolithomys* which had a single ridge near the center of the anterior enamel surface (Black, 1965).

The Yoderimyinae retained the primitive rodent dental formula. All other known eomyids lacked P^3. The premolars were molariform, but almost always smaller than the first molars. The cheek teeth were cuspate to lophate and brachydont to mesodont.

The occlusal pattern of the cheek teeth (Fig. 14.3) primitively consisted of five transverse lophs on unworn teeth: anterior and posterior cingula, mesoloph or mesolophid, and two major transverse lophs (protoloph and metaloph on uppers, metalophid and hypolophid on lowers) that were joined at their lingual ends to an anteroposteriorly running loph (endoloph) on the upper cheek teeth and buccal ends on lower cheek teeth (ectolophid). With wear, the pattern became that of three connected transverse lophs in a lowercase "omega" pattern (Stehlin and Schaub, 1951).

1.3. Skeleton

Cope (1884) described parts of the pelvis, humerus, tibia, femur, and calcaneum of Orellan eomyids. Wood (1937) redescribed these specimens, his conclusions differing only slightly from those of Cope (1884). In general shape the limb bones of eomyids have been described as slender and long. Both Wood (1937) and Cope (1884) noted that the features of the postcranials more closely resembled those of sciurids than those of muroids in detail. Cope (1884, p. 822) described the fibula as being separate from the tibia (the primitive condition). However, Wilson (1949b, p. 47) noted that in European species the fibula and tibia were fused.

2. Evolutionary Changes in the Family

The changes in the North American lineages of eomyids are few. In a number of lineages the cheek teeth became more lophodont in later genera

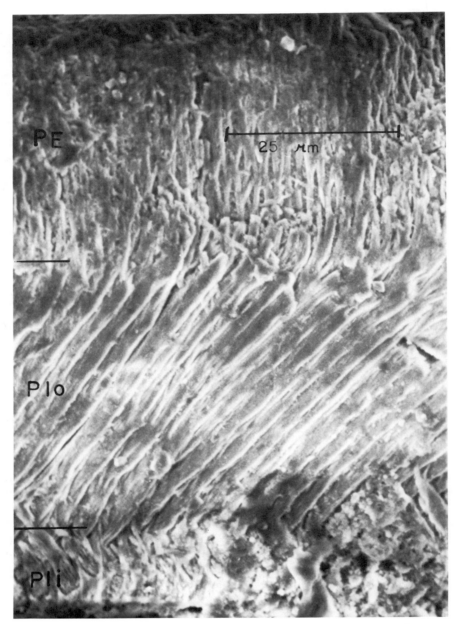

FIGURE 14.2. Enamel microstructure of eomyids. Scanning electron micrograph of transverse section of LI[1] of *Litoyoderimys lustrorum*. Abbreviations: PE, portio externa; PIo, outer layer of portio interna; PIi, inner layer of portio interna. Anterior to top of page.

Eomyidae

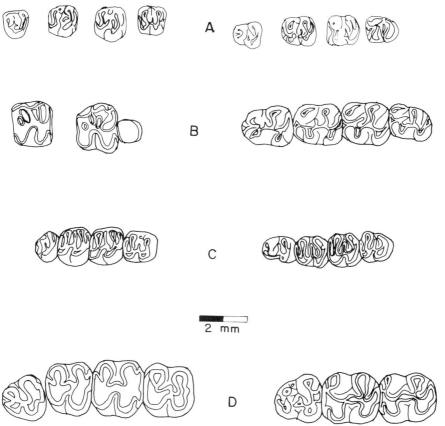

FIGURE 14.3. Dentitions of eomyids. (A) Uintan namatomyine *Metanoiamys fantasmus*, RP^4–M^3 (composite) and LP$_4$–M$_3$ (composite). (B) Chadronian yoderimyine *Yoderimys bumpi*, LP^3–P^4 and M^2 (reversed) and LP$_4$–M$_3$. (C) Chadronian and Orellan *Paradjidaumo trilophus*, RP^4–M^3 and LP$_4$–M$_3$. (D) Hemphillian *Kansasimys dubius*, RP^4–M^3 and LP$_4$–M$_2$. A from Lindsay (1968), B from Wood (1955b), C from Black (1965), D (uppers) from Wahlert (1978) and (lowers) Wood (1936a).

(*Zemiodontomys* in the yoderimyines; *Montanamus* in the namatomyines; and *Pseudotheridomys* in the eomyines). In the later Chadronian and Arikareean genera of yoderimyines, P^3 was lost but dP^3 is retained (Korth, 1992a; Emry and Korth, 1994). Among eomyines the crown height of the cheek teeth was increased in *Paradjidaumo* and a number of European genera (see Fahlbusch, 1973, 1979).

There was no general trend for size increase in any lineage of eomyid. The latest species of *Leptodontomys* from the Hemphillian were very near the size of the Uintan and Duchesnean species of eomyids. The Hemphillian species of *Kansasimys* were the largest eomyids, but no gradual increase in size is discernible from any earlier eomyids that might lead to *Kansasimys*.

Among the Yoderimyinae, there was an increase in complexity of the occlusal pattern and degree of lophodonty of the cheek teeth through the Chadronian (Emry and Korth, 1994).

3. Fossil Record

The earliest record of any eomyids was from the Uintan of southern California (Lindsay, 1968) and Saskatchewan (Storer, 1984). Both Namatomyini and Eomyini were represented from Saskatchewan. The earliest record of Yoderimyinae was from the Chadronian (Emry and Korth, 1994).

The time of greatest diversity for eomyids was the Chadronian where 13 genera and 20 species have been recognized (Fig. 14.4). The number of species and genera greatly decreased by Orellan times and remained low until the Hemphillian when eomyids became extinct in North America.

The large Hemphillian *Kansasimys* had a relatively primitive eomyid dental morphology but a uniquely derived skull (Wood, 1936a; Wahlert, 1978). There was no known Clarendonian eomyids from North America that could have been ancestral to *Kansasimys* suggesting migration of this genus from another continent. However, no possible ancestor is known among the European eomyids of appropriate age, and the origin of this lineage is thus unknown.

In Europe, the first record of eomyids was in the early Oligocene, and

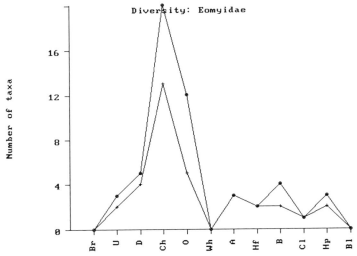

FIGURE 14.4. Occurrence of Eomyidae in the Tertiary of North America (●, species; +, genera). Abbreviations as in Fig. 5.4.

they persisted until the Pleistocene (Fahlbusch, 1979). The greatest diversity of eomyids in Europe was in the late Oligocene and early to middle Miocene (Fahlbusch, 1973).

The genus *Pseudotheridomys* appeared in North America as early as the Arikareean (Macdonald, 1972; Korth, 1992a) and persisted into the Barstovian (Shotwell, 1967a; Barnosky, 1986a). In Europe this genus appeared slightly earlier (Fahlbusch, 1979). There are no earlier North American species that appear to be ancestral to *Pseudotheridomys* suggesting the immigration of this genus from Europe into North America in the middle to late Oligocene (Wilson, 1960; Black, 1965; Fahlbusch, 1973, 1979; Engesser, 1979).

Fahlbusch (1973) suggested there were two migration events between Europe and North America in eomyid history. The first was from North America to Europe in the middle Oligocene (the genus *Eomys*). The second episode was from Europe to North America in the latest Oligocene (the genus *Pseudotheridomys*).

Wang and Emry (1991) described the first eomyids from Asia. They recognized four genera and species from the middle to late Oligocene of Mongolia that were more similar to European species than any North American species. The late Eocene *Zelomys* from China, originally described as a sciuravid (Wang and Li, 1990), had a cheek tooth morphology similar to that of the Chadronian *Namatomys*. The zygomasseteric structure is not known for *Namatomys* or any other namatomyine, but all other known eomyids are fully sciuromorphous. *Zelomys* is protrogomorphous. Its horizon of occurrence is approximately equal to the first appearance of "*Namatomys*" in southern California. It is most likely that *Zelomys* is part of the early radiation of eomyids (Namatomyini) some of which were protrogomorphous.

4. Phylogeny (Fig. 14.5)

4.1. Origin

Several authors have demonstrated that the likely ancestral stock of the eomyids could be found in North American Eocene sciuravids (Matthew, 1910; Wilson, 1949a; Fahlbusch, 1979; Dawson and Krishtalka, 1984; Storer, 1987). The occlusal pattern of the cheek teeth of early eomyids was very similar to that of sciuravids, the latter lacking the longitudinal loph connecting the two major transverse lophs (ectolophid on the lowers; endoloph on the uppers). Sciuravids were slightly less lophate than the earliest eomyids. All known sciuravids possessed a P^3, unlike most eomyids, but this tooth was retained in most yoderimyines (Emry and Korth, 1994) and was possibly lost in one species of sciuravid (Dawson, 1968a).

The zygomasseteric structure of all eomyids was sciurognathous, though it is not known for the Uintan and Duchesnean species, and was protrogomorphous for all known sciuravids. The masseteric scar on the mandible of

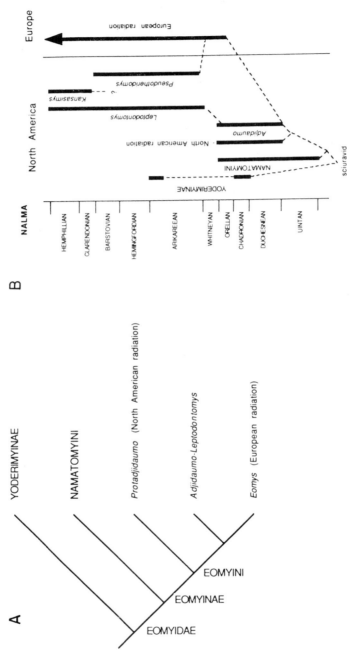

FIGURE 14.5. Phylogeny of the Eomyidae. (A) Cladogram of eomyid relationships. (B) Dendrogram of the Eomyidae with age of occurrence (see Fig. 5.5 for explanation).

all eomyids, including the Uintan "*Namatomys*," extends anteriorly to below P_4. In sciuravids it ended more posteriorly, below the first or second molars. The enamel microstructure of eomyids was uniserial and that of sciuravids was pauciserial (Wahlert, 1968).

While there were clear differences between the eomyids and sciuravids, the similarity of morphology of the cheek teeth of sciuravids to that of eomyids was striking, and the former was the most likely candidate for the ancestor of the latter.

4.2. Intrafamilial Relationships

The first author to discuss relationships within the eomyids was Burke (1934) who recognized two separate lineages, one containing the Duchesnean *Protadjidaumo* and the Chadronian and later *Adjidaumo*, the second containing only the species of *Paradjidaumo*. These separate lineages were based on the crown height of the cheek teeth (higher in *Paradjidaumo*) and connection of the anterior cingulum on the lower molars (free buccal end in *Protadjidaumo* and *Adjidaumo*). Burke (1934) viewed *Protadjidaumo* as likely ancestral to *Adjidaumo*, but too specialized to be ancestral to *Paradjidaumo*. Black (1965) viewed *Protadjidaumo* as too derived to be ancestral to *Adjidaumo*.

Later, Wood (1955b) named a new subfamily of eomyids, the Yoderimyinae, that represented a separate lineage from all other eomyids (Eomyinae). This was based on the primitive dental formula and the possession of an anteroconid on P_4 of yoderimyines.

In a discussion of both European and North American eomyids, Fahlbusch (1973) distinguished three groups of eomyids: the *Adjidaumo*–*Eomys* group, most primitive forms that were likely ancestral to later eomyids; the *Leptodontomys* group, later Tertiary eomyids that retained the primitive dental morphology of the earlier eomyids; and the *Pseudotheridomys* group, later Tertiary eomyids that have a more complex, lophate cheek tooth morphology.

Wahlert (1978) separated the Hemphillian *Kansasimys* from all other eomyines based on cranial features, but did not assign it to another subfamily.

The most detailed analysis of relationships within the eomyids was presented by Storer (1987). Storer not only recognized the two subfamilies of eomyids, but defined two lineages within the Eomyinae. The first and most primitive of these lineages was the "*Namatomys* group." This group was defined by the presence of an anteroconid on P_4, increase in crown height resulting from increase in cusps alone, anterior cingulum attached to the anterior arm of the protoconid on the lower molars, and the primitive retention of small protocone (-id) and hypocone (-id) and anteroposteriorly compressed buccal cusps on the upper molars and lingual cusps on the lower molars. This lineage contained the Uintan and Duchesnean species referred to "*Namatomys*" and the Chadronian genera *Namatomys* and *Montanamus*.

Later, Korth (1992b) referred the members of this group to the tribe Namatomyini and included another new genus.

Within the *Adjidaumo–Paradjidaumo* lineage, Storer (1987, Fig. 2) recognized that morphology of the cheek teeth of *Adjidaumo* was unique among all other Uintan to Chadronian eomyids and formed a separate lineage, though he did not name it. Fahlbusch (1973, 1979) derived all later eomyids from Europe (including *Pseudotheridomys*) from the Oligocene species of *Eomys* which, in turn, is part of the *Adjidaumo* lineage.

The Uintan and Duchesnean "*Namatomys*" remains the most primitive eomyid, sharing many features of dentition with sciuravids. These namatomyines may well represent the basal stock from which all later eomyids were derived, although the reduction in the dental formula in known species bars them from being ancestral to the yoderimyines.

By Duchesnean times, the Eomyini had differentiated into two separate lineages. The lineage that contained most Chadronian North American eomyids began with *Protadjidaumo*. Ancestors for many of the Chadronian genera can be found in the species of *Protadjidaumo* (Storer, 1987). This lineage was the basis for the major radiation of eomyids in the Chadronian in North America, but became extinct in the Orellan. The *Adjidaumo* lineage remained fairly conservative in North America, retaining the primitive dental morphology into Hemphillian times (Korth and Bailey, 1992). In Europe, the radiation of eomyids in the late Oligocene and Miocene was all part of the *Adjidaumo–Eomys* lineage.

4.3. Extrafamilial Relationships

Since Wilson's (1949a,b) recognition of the geomyoid affinities of eomyids, virtually all others have recognized this relationship (Wahlert, 1978; Dawson and Krishtalka, 1984). While the eomyids predate other geomyoids in the fossil record, it is not evident that they are ancestral to any later family of geomyoid (Wilson, 1949b, 1960; Wahlert, 1978). Similarly, the inclusion of eomyids along with the remainder of the geomyoids and the myodont rodents (muroids and dipodoids) in the Myomorpha, as first suggested by Wilson (1949b), has been nearly universally accepted.

5. Problematical Taxa

5.1. *Adjidaumo–Eomys–Omegodus*

The earliest, most primitive, and most diverse eomyid in Europe was *Eomys* Schlosser (1884) known from the middle Oligocene to the Miocene. Similarly, the genus *Adjidaumo* Hay (1899) was one of the most common North American eomyids from Chadronian to Orellan times (one species is currently known from the Duchesnean; Storer, 1987). Wood (1937) noted the similarity of these genera and included *Adjidaumo* along with the other members of the "Adjidaumidae" (Burke, 1934) in the European Eomyidae. Stehlin and Schaub (1951) found very little difference between the lower cheek teeth of *Eomys* and those of *Adjidaumo*. Fahlbusch (1973) noted the similarity between species of these two genera in the morphology of the cheek teeth and mandible. He did, however, cite one difference in the sutures of the skull. He questioned whether these genera should be recognized as distinct and concluded that they were best kept as separate genera because of their distinctive occurrence (*Adjidaumo* in North America; *Eomys* in Europe), phylogenetic position, and to avoid the difficulties of synonymy relating to the fact that both genera had been frequently cited in the literature for several decades. Wood (1980) stated that the two were probably congeneric but failed to synonymize them pending a detailed review of the family.

Wahlert (1978) suggested that the two genera were likely congeneric, but listed *Eomys* as a junior synonym of *Omegodus* Pomel (1853) without discussion. Later, Stucky (1992) listed *Adjidaumo* as a junior synonym of *Omegodus* in a table without discussion.

Engesser (1979) pointed out several general differences between the cheek teeth of North American eomyids and those of European eomyids. Wang and Emry (1991) specifically noted that these features (lingual extent of the valleys on the upper molars; extend of valley between the hypolophid and posterolophid on the lower molars) were consistent in described species of *Adjidaumo* and *Eomys* and thus these genera could be morphologically distinguished from one another and synonymy was not justified.

As to the validity of the genus *Omegodus*, Fahlbusch (1970, p. 14) clearly demonstrated that the generic name *Omegodus* was a *nomen obscurum* and correctly included it in the synonymy of *Eomys*. Hence, the reference of either European or North American species to *Omegodus* is erroneous.

5.2. *Kansasimys–Ronquillomys–Comancheomys*

Wood (1936a) named and described *Kansasimys dubius* from the Hemphillian of Kansas based on a single mandible with all of the lower cheek teeth. He was unsure about the familial assignment of this genus, noting the greatest similarity of the cheek teeth with the Eocene *Sciuravus*. Stehlin and Schaub (1951) first recognized the eomyid affinities of this genus. Wahlert (1978) later described the skull and upper dentition of *K. dubius*.

Jacobs (1977a) named a second large eomyid from the Hemphillian of Arizona, *Ronquillomys wilsoni*. Jacobs (1977a, p. 508) cited only one difference between *Ronquillomys* and *Kansasimys* other than size: "The endolophulid of the lower molars of *Ronquillomys* is directed posteromedially from the entoconid rather than anteromedially, as in *Kansasimys*." Korth (1980) listed *Ronquillomys* as a junior synonym of *Kansasimys* without explanation.

The next described large eomyid from the Hemphillian was *Comancheomys rogersi* from Texas (Dalquest, 1983) based on three isolated cheek teeth. In describing the holotype of *C. rogersi*, Dalquest (1983, Fig. 6C) mistakenly identified it as a lower molar. It is clearly an upper molar, and is quite similar to that of both *K. dubius* (Wahlert, 1978, Fig. 7A) and *R. wilsoni* (Jacobs, 1977a, Fig. 3). Because of this mistake in the identity of the holotype, a number of characters of *C. rogersi* have been misinterpreted. Later, Dalquest and Patrick (1989) described two additional lower molars of *C. rogersi* along with a single molar that they referred to *K. dubius*. They noted that the only difference between these two species was the greater lophodonty of *C. rogersi*. However, the lophodonty recognized in the molars of the latter appears to result from a greater stage of wear on the specimens (Dalquest and Patrick, 1989, Fig. 3). In size, the specimens from Texas are nearly identical to those of *K. dubius* from Kansas (Wood, 1936a; Wahlert, 1978). The morphology of the lower molar used by Jacobs (1977a) to separate his species from *K. dubius* is shared by the Texas molars.

It appears that all of the previously described large eomyids from the Hemphillian are referable to the genus *Kansasimys*, though there may be as many as four species (see also Voorhies, 1990b).

5.3. *Meliakrouniomys*

Harris and Wood (1969) described *Meliakrouniomys wilsoni*, an unusual eomyid from the Chadronian of Texas, based on a single mandible and lower dentition. It differed from all other eomyids in lacking the mesoconid, mesolophid, and ectolophid on the lower molars, making the occlusal pattern of the cheek teeth that of two subparallel lophs The masseteric scar on the mandible extended anterior to P_4 similar to heteromyids and more anterior than in eomyids. They described the phylogenetic position of this genus as "... an eomyid on the way to becoming a heteromyid" (Harris and Wood, 1969, p. 5).

A second species of this genus, *M. skinneri*, was later described from the Chadronian of Wyoming (Emry, 1972b). Emry (1972b) referred the genus to the Heteromyidae based on the morphology of the rostrum. Wood (1974, 1980) maintained this genus in the Eomyidae because the morphologies cited by Emry (1972b) as being heteromyid were also present in some eomyids. The lack of the central loph of the cheek teeth also occurs in some European eomyid genera (see Stehlin and Schaub, 1951; Engesser, 1990, for appropriate figures), and the heteromyid-like masseteric scar on the mandible is present on the eomyids *Paranamatomys* (Korth, 1992b) and *Metadjidaumo* (Korth and Tabrum, in press).

One test of the familial relationships of *Meliakrouniomys* might be to examine the microstructure of the incisor enamel to determine if it had the unique structure of eomyids (Wahlert and von Koenigswald, 1985). However, specimens of the known species of Meliakrouniomys are rare enough that their destruction through sectioning is currently unwarranted.

Wood (1974) pointed out the great similarity between Meliakrouniomys and the problematical geomyoid rodent Griphomys from the Uintan of California (Wilson, 1940). Later, Wood (1980) included Griphomys in the Eomyidae because of this similarity. Meliakrouniomys is here considered an eomyid based on the arguments of Wood (1974, 1980). However, the systematic position of Griphomys is still uncertain (see Chapter 23).

6. Classification

Eomyidae Depéret and Douxami, 1902
  Eomyinae Depéret and Douxami, 1902
    Tribe Eomyini Depéret and Douxami, 1902
      Adjidaumo Hay, 1899 (D–O)
        A. minutus (Cope, 1873)* (O)
        A. minimus (Matthew, 1903) (Ch)
        A. craigi Storer, 1987 (D)
        A. intermedius Korth, 1989 (O)
        A. maximus Korth, 1989 (O)
      Pseudotheridomys Schlosser, 1926 (A–B)
        P. hesperus Wilson, 1960 (Hf)
        P. pagei Shotwell, 1967 (B)
        P. cuyamensis Lindsay, 1974 (Hf)
      Paradjidaumo Burke, 1934 (D–O)
        P. trilophus (Cope, 1873)* (Ch–O)
        P. alberti Russell, 1954 (Ch)
        P. hansonorum (Russell, 1972) (Ch)
        P. hypsodus Setoguchi, 1978 (O)
        P. validus Korth, 1980 (O)
        P. reynoldsi Kelly, 1992 (D)
      Protadjidaumo Burke, 1934 (U–D)
        P. typus Burke, 1934* (D)
        P. altilophus Storer, 1984 (U)
        P. pauli Storer, 1987 (D)
      Kansasimys Wood, 1936 (Hp)
        K. dupius Wood, 1936* (Hp)
        K. wilsoni (Jacobs, 1977) (Hp)
      Centimanomys Galbreath, 1955 (Ch)
        C. major Galbreath, 1955* (Ch)
        C. galbreathi Martin and Ostrander, 1986 (Ch)
      Leptodontomys Shotwell, 1956 (A–Hp)
        L. douglassi (Burke, 1934) (A)
        L. oregonensis Shotwell, 1956* (Hp)

L. quartzi (Shotwell, 1967) (B)
  L. russelli (Storer, 1970) (B)
  L. stirtoni (Lindsay, 1972) (B)
 Aulolithomys Black, 1965 (Ch)
  A. bounities Black, 1965* (Ch)
 Meliakrouniomys Harris and Wood, 1969 (Ch)
  M. wilsoni Harris and Wood, 1969* (Ch)
  M. skinneri Emry, 1972 (Ch)
 Viejadjidaumo Wood, 1974 (Ch)
  V. magniscopuli Wood, 1974* (Ch)
 Metadjidaumo Setoguchi, 1978 (O)
  M. hendryi Setoguchi, 1978* (O)
 Cupressimus Storer, 1978 (Ch)
  C. barbarae Storer, 1978* (Ch)
 Orelladjidaumo Korth, 1989 (O)
  ?*O. cedrus* Korth, 1981 (O)
  O. xylodes Korth, 1989* (O)
 Aguafriamys Wilson and Runkel, 1991 (D)
  A. raineyi Wilson and Runkel, 1991* (D)
 Tribe Namatomyini Korth, 1992
  Namatomys Black, 1965 (Ch)
   N. lloydi Black, 1965* (Ch)
   "*N.*" *fantasmus* (Lindsay, 1968) (U)
   "*N.*" *fugitivus* (Storer, 1984) (U)
   "*N.*" *lacus* (Storer, 1987) (D)
  Montanamus Ostrander, 1983 (Ch)
   M. bjorki Ostrander, 1983* (Ch)
  Paranamatomys Korth, 1992 (Ch)
   P. storeri (Ostrander, 1983)* (Ch)
Yoderimyinae Wood, 1955
 Yoderimys Wood, 1955 (Ch)
  Y. bumpi Wood, 1955* (Ch)
  Y. stewarti (Russell, 1972) (Ch)
  Y. yarmeri Wilson and Runkel, 1991 (Ch)
 Litoyoderimys Emry and Korth, 1993 (Ch)
  L. lustrorum (Wood, 1974)* (Ch)
  L. auogoleus Emry and Korth, 1993 (Ch)
 Zemiodontomys Emry and Korth, 1993 (Ch)
  Z. burkei (Black, 1965)* (Ch)
 Arikareeomys Korth, 1992 (A)
  A. skinneri Korth, 1992* (A)

Chapter 15
Heliscomyidae

1. Characteristic Morphology	163
1.1. Skull	163
1.2. Dentition	164
1.3. Skeleton	167
2. Evolutionary Changes in the Family	167
3. Fossil Record	167
4. Phylogeny	167
4.1. Origin	167
4.2. Intrafamilial Relationships	168
4.3. Extrafamilial Relationships	170
5. Problematical Taxon: *Akmaiomys incohatus*	170
6. Classification	171

1. Characteristic Morphology

1.1. Skull

Heliscomyids were among the smallest rodents from North America, both fossil and Recent. The skull was sciuromorphous (Fig. 15.1) and the rostrum was narrow (laterally). There is no indication of sagittal or lyrate crests on the frontal bones. There is a distinct parapterygoid fossa on the posterior palatal surface as in all other geomyoids.

Heliscomyids had several distinct characteristics of the cranial foramina (Korth *et al.*, 1991): incisive foramen enlarged and intersected by the maxillary–premaxillary suture at its center; no rostral perforation anterior to infraorbital foramen as in heteromyids; the buccinator and masticatory foramina fused within the accessory foramen ovale as in eomyids and florentiamyids; optic foramen large; and the posterior maxillary notch remains open posterolaterally (laterally enclosed on all other geomyoids).

The mandible was dorsoventrally slender in *Heliscomys* and much deeper in *Apletotomeus* and *Akmaiomys*. The diastema was shallow and short. The anterior end of the masseteric fossa on the mandible was a small ridge that extended anterior to the tooth row as in heteromyids, and nearly reached the dorsal margin of the diastema. The mental foramen was dorsal and slightly anterior to the masseteric fossa. The angle of the jaw was not flared laterally or medially.

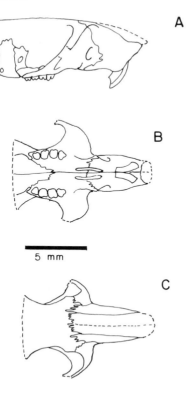

FIGURE 15.1. Partial skull and mandible of Chadronian *Heliscomys ostranderi*. (A) Lateral view; (B) ventral view; (C) dorsal view. (D) Lateral view of mandible. All from Korth et al. (1991).

1.2. Dentition

The dental formula for heliscomyids, and all later geomyoids (Heteromyidae, Florentiamyidae, Geomyidae) was $\frac{1013}{1013}$. The incisors of *Heliscomys* were small, delicate, and laterally compressed with a narrow, gently convex anterior enamel surface. The lower incisor of *Apletotomeus* had a similar cross-sectional shape but was a much more robust tooth. No studies of the microstructure of the enamel of the incisors of heliscomyids have been reported.

The cheek teeth were extremely brachydont. The molars consisted of four major cusps arranged in two transverse rows and much smaller stylar cusps along the buccal cingulum in lower molars and the lingual cingulum on

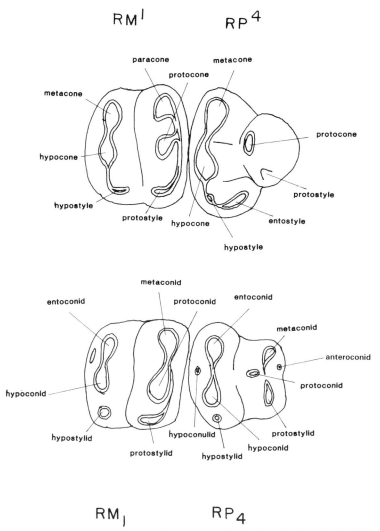

FIGURE 15.2. Schematic diagram of generalized geomyoid cheek teeth showing dental nomenclature. Anterior to the right.

upper molars (Figs. 15.2, 15.3). The central transverse valley of the molars was about the same depth as the longitudinal valleys that separate the major cusps. The anterior cingulum on the molars was generally continuous along the entire width of the tooth. The posterior cingulum was only present on the first or possibly the second molar as a small cusp in the valley between two of the major cusps. The third molars were always reduced, mainly in the posterior half of the tooth.

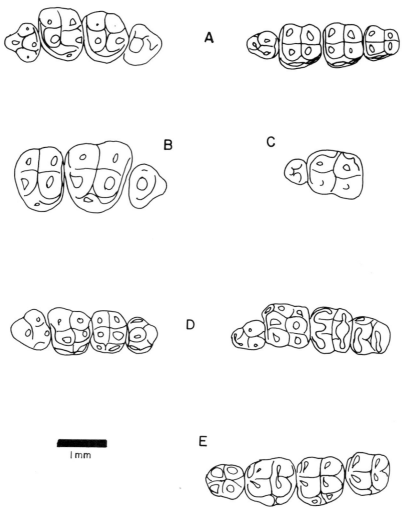

FIGURE 15.3. Dentitions of Heliscomyidae. (A) Orellan *Heliscomys vetus*, LP4–M3 and LP$_4$–M$_3$. (B) Orellan *H. mcgrewi*, RP4–M2. (C) Arikareean *H. woodi*, RP$_4$–M$_1$. (D) Orellan *H. hatcheri*, LP4–M3 and RP$_4$–M$_3$. (E) Orellan *Apletotomeus crassus*, LP$_4$–M$_3$. A (uppers) and B from Korth (1989b), A (lowers) from Wood (1935), C from McGrew (1941b), D from Wood (1939), E from Reeder (1960).

The posterior half of the upper premolar resembled that of the upper molars with a row of two larger cusps and a smaller stylar cusp. The protoloph consisted of a large protocone and variably occurring smaller cuspules on the buccal and lingual slopes of this cusp. The lower premolar had three or four major cusps (two in the hypolophid, one or two in the metalophid) and a variable number of smaller cuspules on either the hypolophid or metalophid.

1.3. Skeleton

No postcranial skeletal material has ever been reported for any heliscomyid.

2. Evolutionary Changes in the Family

There were two distinct lineages within the genus *Heliscomys* (Korth et al., 1991). One lineage involved the relative reduction of the premolars and simplification of the molars ranging from the Orellan to the Arikareean; the *H. vetus–H. mcgrewi–H. woodi* lineage. The other lineage involved the relative enlargement and increased complexity of the premolars. This clade also involved the greater development of stylar cusps on the upper cheek teeth. This lineage ranged from the Chadronian (*H. ostranderi*) to the Barstovian (*H. subtilis*), but lacked representative species from the Whitneyan and Arikareean.

The complexity of the lower premolar and enlarged size of the mandible of *Apletotomeus* and *Akmaiomys* (Reeder, 1960) does not fit into the *Heliscomys* lineages and represents an independent clade.

3. Fossil Record

The first record of *Heliscomys* was from the Duchesnean of Saskatchewan (Storer, 1987) and California (Kelly, 1992). In all instances, specimens from the Duchesnean were only single isolated cheek teeth, on the basis of which no specific identification could be made. The time of greatest diversity was the Orellan, from which three genera and five species are known (Fig. 15.4). At all other times there was no more than one species known, and all species have been referred to *Heliscomys*.

No recognized specimens of *Heliscomys* or any other heliscomyid have been described from the Whitneyan. This may reflect the sparsity of small mammals from the Whitneyan and may be a bias of collecting rather than a true representation of the fossil record. There is no record of heliscomyids from any continent other than North America.

4. Phylogeny (Fig. 15.5)

4.1. Origin

The Heliscomyidae, along with the remainder of the Geomyoidea, are believed to have evolved from a sciuravid stock (Wilson, 1949a) but there were no pre-Chadronian rodents known that could be interpreted as definitely ancestral (Wood, 1935a; Wilson, 1949a). Wood (1980) suggested that a morpho-

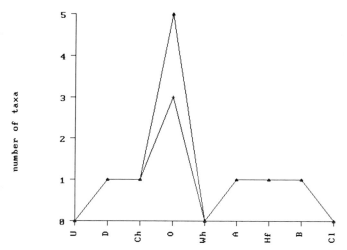

FIGURE 15.4. Occurrence of Heliscomyidae in the Tertiary of North America (▲, species; +, genera). Abbreviations as in Fig. 5.4.

logical sequence (based on cheek tooth morphology) that led to *Heliscomys* might be *Viejadjidaumo–Griphomys* or *Meliakrouniomys–Heliscomys*. This sequence was clearly not based on chronology because *Heliscomys* is known as early as the Duchesnean and *Meliakrouniomys* and *Viejadjidaumo* were Chadronian. However, this sequence does suggest the evolution from an eomyid-like occlusal pattern of the cheek teeth with its five transverse lophs to a more simplified, cuspate bilophed pattern. Alternatively, heliscomyids (and all other geomyoids) may have been derived directly from a sciuravid ancestor (see Chapter 6, Section 4.3).

The last common ancestor of heliscomyids and later geomyoids must have existed as early as the Uintan in order for the heliscomyids to have become separate by the Duchesnean. Again, there are no Bridgerian or Uintan rodents currently known that show a transitional molar morphology between that of eomyids or sciuravids and heliscomyids.

4.2. Intrafamilial Relationships

As noted in Section 2 of this chapter, there were two clades within the genus *Heliscomys*, the *Heliscomys vetus* lineage (reduction of premolars) and the *H. ostranderi* lineage (enlargement of premolars; additional stylar cusps) that were separate throughout their known occurrence. The common ancestor of these two clades may well have been represented in the poorly known Duchesnean representatives of *Heliscomys*.

There were no species of heliscomyid that were intermediate between

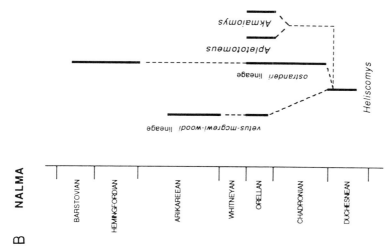
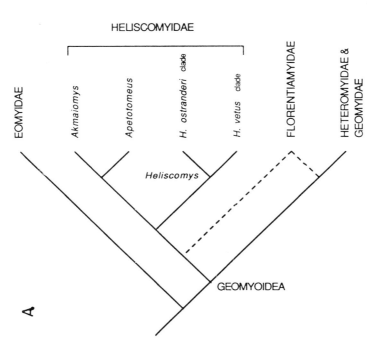

FIGURE 15.5. Phylogeny of the Heliscomyidae. (A) Cladogram of heliscomyid relationships. (B) Dendrogram of the Heliscomyidae with age of occurrence (see Fig. 5.5 for explanation).

Heliscomys and the two other Orellan genera which had a much heavier mandible and a more complex P_4. This may be, in part, related to the rare fossil record of the latter genera which are known only from a total of four specimens from the Orellan (Reeder, 1960; Korth, 1989b). If *Apletotomeus* and *Akmaiomys* were derived from within the genus *Heliscomys*, it would be from the *H. ostranderi* lineage because of the increased complexity of the lower premolar in that lineage.

4.3. Extrafamilial Relationships

In the first description of *Heliscomys*, Cope (1884) referred the genus to the Myomorpha. All subsequent reviews of this genus included it in the Heteromyidae (Wood, 1931, 1935a). Wood (1935a) was unable to determine to which subfamily of the heteromyidae *Heliscomys* belonged, thus he left it separate from the other heteromyids without naming a separate subfamily, but maintained that *Heliscomys* was the ancestral stock from which all of the heteromyids were derived. Rensberger (1973a) viewed *Heliscomys* as a morphological ancestor to later geomyoids (florentiamyids and geomyids) as well. However, because of the simplification of the premolar (reduced to three cusps) in the known species of *Heliscomys* at the time, Wilson (1949a) and Galbreath (1953) felt that this genus could not have been ancestral to the later heteromyids though they retained it within the family. With the discovery of referable cranial material, Korth et al. (1991) were able to establish a distinct family for *Heliscomys* that was clearly not ancestral to any other geomyoid rodents because of several features of the skull that were derived beyond other geomyoids (enlarged incisive foramina; mental foramen nearly within the diastema, dorsal to the masseter attachment).

Korth et al. (1991) cited several morphological characters of the skull and dentition that allied the Heliscomyidae with other geomyoids (sciuromorphy; parapterygoid fossa; infraorbital foramen sunken into rostrum; molars six-cusped and bilophate). They placed it as a sister group to all other geomyoids. Korth (1993a) noted that the relatively large optic foramen and fusion of the buccinator and masticatory foramina within the accessory foramen ovale of *Heliscomys* were also shared by florentiamyids. He placed the Heliscomyidae and Florentiamyidae as a clade which, in turn, was the sister group of a heteromyid–geomyid clade.

Clearly, the Heliscomyidae are a distinct, early radiation of the Geomyoidea that were not ancestral to any later geomyoids.

5. Problematical Taxon: *Akmaiomys incohatus*

Reeder (1960) named *Akmaiomys incohatus* based on a single partial mandible with P_4–M_1 from the Orellan of Colorado, originally identified as

Heliscomys by Galbreath (1953). He (Reeder) noted that it was unique because of its complex P_4 (five cusps) and deep mandible. He distinguished it from *Apletotomeus* by its relatively smaller incisor. Black (1965) suggested that *A. incohatus* was a junior synonym of *Heliscomys vetus*. Wood (1980) recognized it as a distinct species but transferred it to the genus *Proheteromys*. Korth (1989b) suggested that *A. incohatus* was the lower dentition of the florentiamyid *Ecclesimus tenuiceps* which was known only from upper dentitions.

The cheek tooth morphology of *A. incohatus* was clearly within the range of morphology of other heliscomyids, and the position of the mental foramen on the mandible (dorsal to the anterior extension of the insertion for the masseter muscle) was a derived feature of heliscomyids (Korth *et al.*, 1991). Because of its poor representation in the fossil record, little more can be determined about this species. At present, it is best to include *A. incohatus* as a distinct species and genus within the Heliscomyidae until more complete material allows for a better assessment of its systematic position.

6. Classification

Heliscomyidae Korth, Wahlert, and Emry, 1991
    Heliscomys Cope 1873 (D–B)
        H. vetus Cope, 1873* (O)
        H. gregoryi Wood, 1933 (O)
        H. hatcheri Wood, 1935 (O)
        H. senex Wood, 1935 (O)
        H. woodi McGrew, 1941 (A)
        H. subtilis (Lindsay, 1972) (B)
        H. ostranderi Korth, Wahlert, and Emry, 1990 (Ch)
    Apletotomeus Reeder, 1960 (O)
        A. crassus Reeder, 1960* (O)
    Akmaiomys Reeder, 1960 (O)
        A. incohatus Reeder, 1960* (O)

Chapter 16
Heteromyidae

1. Characteristic Morphology 173
   1.1. Skull 173
   1.2. Dentition 175
   1.3. Skeleton 177
2. Evolutionary Changes in the Family 177
3. Fossil Record 178
4. Phylogeny 179
   4.1. Origin 179
   4.2. Intrafamilial Relationships 179
   4.3. Extrafamilial Relationships 182
5. Problematical Taxa 182
   5.1. "*Diprionomys*" *agrarius* 182
   5.2. *Mookomys* 183
   5.3. *Proheteromys* 183
   5.4. *Dikkomys* and *Lignimus* 184
   5.5. *Schizodontomys* 185
   5.6. *Prodipodomys? mascallensis* 185
6. Classification 186

1. Characteristic Morphology

1.1. Skull

With only a few exceptions in the fossil record, heteromyids (pocket mice and kangaroo rats) are small animals. Their skulls are generally delicate in construction (Fig. 16.1) with thin, delicate zygoma and are fully sciuromorphous. The infraorbital foramen is dorsal to the center of the diastema and positioned just ventral to middepth of the rostrum opening laterally. Continuous with the infraorbital foramen is a large rostral perforation not present in any other rodents. As with all geomyoid rodents, there is a distinct parapterygoid fossa on the palate. The attachment of the temporalis muscle is restricted to the lateral sides of the parietal bones, producing no well-defined sagittal or lyrate crests. The rostrum is generally tapered anteriorly. Dorsally, the premaxillary bones extend farther posteriorly than the nasals on the skull roof.

Wahlert (1985, p. 14) cited several other unique features of the cranial foramina of heteromyids: (1) interpremaxillary foramen absent, (2) posterior end of optic foramen narrowly separated from the orbital fissure, (3) mastoid

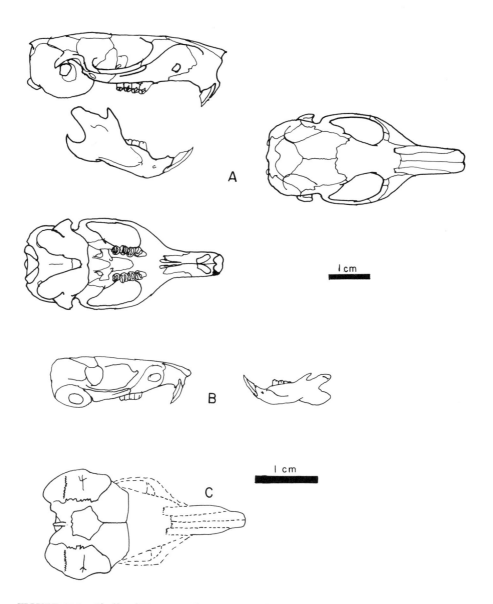

FIGURE 16.1. Skulls of Heteromyidae. (A) Lateral, ventral, and dorsal views of skull, lateral view of mandible of Hemingfordian harrymyine *Harrymys irvini*. (B) Lateral view of skull and mandible of Barstovian perognathine *Cupidinimus nebraskensis*. (C) Dorsal view (reconstruction) of Clarendonian dipodomyine *Eodipodomys celtiservator*. A from Wahlert (1991), B from Wood (1935a), C from Voorhies (1975). B and C to same scale (below), A to separate scale (above).

foramen minute or lacking, and (4) incisive foramina reduced (except in dipodomyines and harrymyines).

Among the subfamilies of Heteromyidae, there are also some unique features of the skull (Wahlert, 1985, 1991). In Heteromyinae and Harrymyinae, the buccinator and masticatory foramina are fused and the stapedial and sphenofrontal foramina are lacking (stapedial artery lost). In Perognathinae and Dipodomyinae the bullae are greatly inflated posteriorly and dorsally (especially in the latter) with anteroventral processes nearly touching, the optic foramen is large, and nonossification commonly occurs dorsal to the orbitosphenoid (sometimes including the ethmoid foramen). The Harrymyinae also show bullar inflation but this is restricted to the ventral part of the bulla and the bullae meet anteroventrally. The maxillary–premaxillary suture intersects the incisive foramina at different points in the different subfamilies. The intersection is near the center of the incisive foramina in harrymyines, at the anterior end in dipodomyines, and at the posterior end in perognathines and heteromyines.

The mandible of heteromyids is usually slender and sometimes very delicate. The anterior end of the masseteric fossa on the mandible extends anteriorly to a point anterior to the tooth row and terminates in a small ridge (anterior to the fusion of the dorsal and ventral ridges of the scar). The mental foramen is anterior and ventral to the masseteric scar. The coronoid process is small, delicate, and flared laterally as in geomyids. The angular process tapers posteriorly to a point that is generally directed slightly dorsally. The base of the incisor is marked by a lateral bulge on the ascending ramus posterior and dorsal to the tooth row.

1.2. Dentition

The dental formula for all heteromyids, as in all geomyoids, is $\frac{1013}{1013}$. The incisors are usually narrow (buccolingually) and gently convex anteriorly. The upper incisors are commonly sulcate, with a narrow central groove on the anterior surface (Wood, 1935a; Korth and Reynolds, 1991). The microstructure of the incisors is uniserial (Wahlert, 1968).

The cheek teeth of heteromyids vary in crown height from brachydont in perognathines and heteromyines to hypsodont in dipodomyines (Fig. 16.2). The occlusal pattern of all of the molars of heteromyids consists of two transverse rows of three cusps each. On the molars the stylar cusps are subequal in size to the major four cusps of the teeth. The first and second molars are nearly identical in size, the third molar is nearly always reduced in size and number of cusps. In harrymyines the hypolophid of the lower molars formed an anteriorly pointing V-shape that joins the metalophid posteriorly near its center.

The upper premolar has a complete three-cusped metaloph (metacone, hypocone, entostyle or hypostyle) and a protoloph dominated by a large

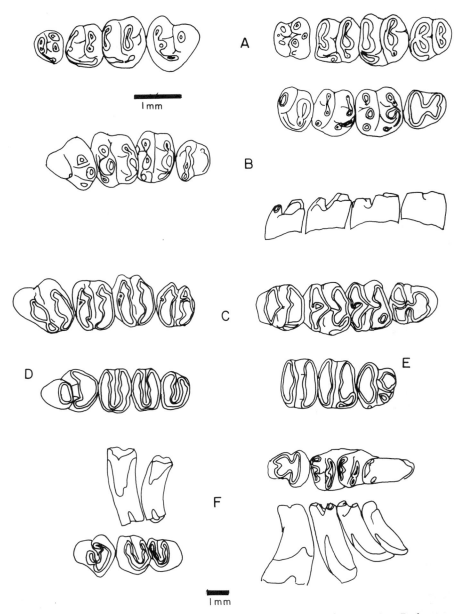

FIGURE 16.2. Dentitions of Heteromyidae. (A) Hemingfordian heteromyine *Proheteromys sulculus*, RP^4–M^3 (composite) and LP$_4$–M$_3$ (composite, M$_3$ reversed). (B) Barstovian perognathine *Perognathus furlongi*, LP^4–M^3 and RP$_4$–M$_3$ (composite) with lateral view. (C) Hemingfordian harrymyine *Harrymys irvini*, LP^4–M^3 and RP$_4$–M$_3$. (D) Barstovian heteromyine *Peridiomys oregonesis*, LP^4–M^3. (E) Barstovian heteromyine *Peridiomys rusticus*, RP$_4$–M$_2$. (F) Blancan dipodomyine *Dipodomys hibbardi*, LP^4–M^2 (composite) with lateral view of molars and RP$_4$–M$_3$ (composite, P$_4$ reversed) with lateral view of cheek teeth. A and B to same scale (above), and C–F to same sale (bottom). A from Wilson (1960), B from Lindsay (1972), C from Wahlert (1991), D and E from Wood (1935a), F from Tomida (1987).

central protocone. All other cusps of the protoloph are minute and on the slopes of the protocone. The lower premolar is usually smaller than the first molars. The hypolophid is three-cusped (entoconid, hypoconid, hypostylid) but the metalophid is more commonly two-cusped (metaconid, protostylid) with numerous accessory cuspules in varying occurrence. In heteromyines the union of the lophs on the lower premolar is either at the buccal or lingual ends, usually resulting in a central enamel pit. Other heteromyids have a central union of the metalophid and hypolophid on the lower premolar.

In mesodont heteromyids, there are enamel "chevrons" (Barnosky, 1986b) on the lateral sides of the cheek teeth that increase in height as the crown height increases. In hypsodont dipodomyines (*Eodipodomys*, *Dipodomys*), roots are retained but the occlusal surface is lost in early stages of wear and the dentine tracts on the lateral sides of the teeth run nearly the entire height of the tooth, resulting in an occlusal pattern of two crescentic ridges of enamel on the anterior and posterior sides of the tooth that are not continuous with one another. In the subhypsodont species of *Prodipodomys*, the cheek teeth were high crowned but there was no enamel failure. In this case, the occlusal pattern was eliminated at early stages of wear and only an enamel outline of the cheek teeth remained. The majority of the occlusal surface is exposed dentine in all dipodomyines.

1.3. Skeleton

The skeleton of heteromyids is very light and delicate (Fig. 16.3), and where known in the fossil record, had at least some modification for ricochetal locomotion (Wood, 1935a). All limb bones are slender and the tail is usually long. The hindlimbs are much longer than the forelimbs, the former being somewhat reduced and the latter being elongate. The tibia and fibula are always fused. Along with the lengthening of the hindlimb, the pes is often also elongated (particularly in the metatarsals). The greatest differential in limb length is among the highly ricochetal dipodomyines (kangaroo rats).

2. Evolutionary Changes in the Family

The greatest change in heteromyids through time was the increase in crown height of the cheek teeth. Early heteromyines and perognathines have brachydont cheek teeth which appeared to increase to mesodonty in several lineages. Within the Perognathinae there was nearly a continuous sequence of genera and species that showed increased crown height and enamel failure (Barnosky, 1986b). The earliest dipodomyine, *Eodipodomys* from the Clarendonian, already had the enamel crescents on the occlusal surface of the cheek teeth present in Recent dipodomyines.

Another major change was in the adaptation of the skeleton for saltatorial locomotion. There are no reported postcranial elements for the earliest

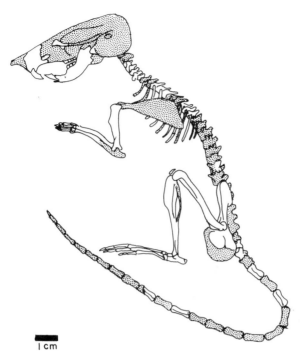

FIGURE 16.3. Reconstruction of the skeleton of problematical Clarendonian heteromyine "*Diprionomys*" *agrarius*. From Wood (1935a).

heteromyid (*Proheteromys*), but the earliest reported skeletons of both perognathines (*Cupidinimus*) and heteromyines ("*Diprionomys*" *agrarius*) from the Barstovian and Clarendonian, respectively, had reduced forelimbs and elongate hindlimbs (Wood, 1935a). The limb proportions of Recent dipodomyines can be attained by little modification of these fossil forms.

Finally, the inflation (enlargement) of the auditory bulla appeared to increase from more primitive heteromyines through perognathines through the most derived dipodomyines.

3. Fossil Record

Heteromyids were exclusively North American throughout their fossil history. The earliest heteromyid was the heteromyine *Proheteromys* from the Arikareean (Macdonald, 1963, 1970; Korth, 1992a). However, Wahlert (1993) and Green (in press) have suggested that *Proheteromys* may have been the basal geomyid. The Whitneyan *P. nebraskensis* (Wood, 1937) has been questionably identified as a florentiamyid (Korth, 1989b) and although Wood (1980) referred two Orellan species to this genus, these two species appear to be heliscomyids (Korth *et al.*, 1991; Chapter 15, Section 5).

The greatest Tertiary diversity of heteromyids was during the Hemingfordian and Barstovian (Fig. 16.4). Not surprisingly, the earliest radiation appears to be of heteromyines which peaked in the Arikareean and Hemingfordian, based almost entirely on the diversity of the genus *Proheteromys*. The Barstovian peak is mainly related to the great diversity of perognathines at that time, which diminished to lower levels by the end of the Tertiary. The dipodomyines first appeared in the Clarendonian and increased rapidly through the end of the Tertiary. Curiously, the record of heteromyines disappears in North America in the latest Tertiary (Hemphillian–Blancan) only to reappear in the Quaternary. This may reflect a migration of heteromyines to more tropical climates in Central America, from which they reinvaded North America in Quaternary times.

4. Phylogeny (Fig. 16.5)

4.1. Origin

The origin of heteromyids, as with the heliscomyids and other geomyoids, is difficult to determine (see Chapter 15, Section 4.1). If the earliest geomyoids, the Heliscomyidae, are not ancestral to the later geomyoids (Wilson, 1949a; Korth *et al.*, 1991), some common ancestor of the geomyoids must have existed in the later Eocene or early Oligocene. Wilson (1949a) suggested that the upper cheek teeth of the Bridgerian sciuravid *Taxymys* could be modified into the geomyoid pattern because it has basically bilophate. He saw the evolution of the geomyoid cheek tooth pattern through the Uintan and Duchesnean *Griphomys* which had fully bilophate cheek teeth (Chapter 23, Section 2). Wood (1980) viewed the origin of heteromyids through an eomyid ancestor. Fahlbusch (1985, Fig. 1) suggested that the ancestry of heteromyids could have come directly from a sciuravid ancestor or through an early eomyid. However, there remains no known fossil taxon that has a cheek tooth and cranial morphology that is intermediate between the primitive rodent morphology (or the morphology of any other group of rodents) and that of geomyoids.

4.2. Intrafamilial Relationships

Wood (1935a) was able to identify three subfamilies of the Heteromyidae in the Tertiary. His subdivisions of the family and generic assignments have been generally followed by most later workers. Korth (1979) removed *Cupidinimus* from the Dipodomyinae and included it in the Perognathinae. The only major change in Wood's (1935a) arrangement of the subfamilies was by Wahlert (1993) who included all of the Tertiary genera previously included in the Heteromyinae in the other subfamilies, retaining only the Recent genera

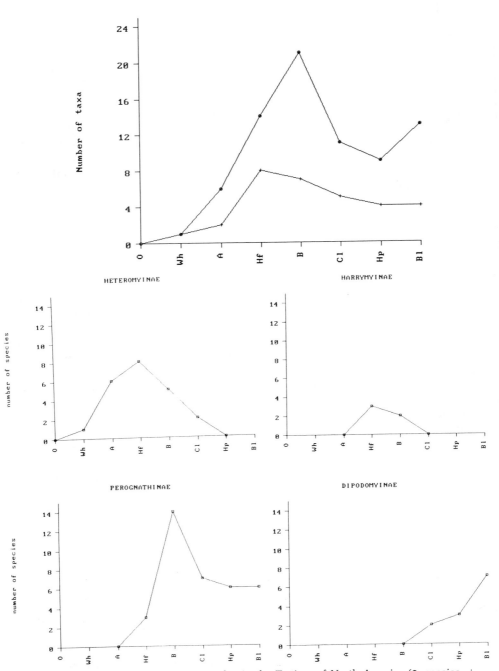

FIGURE 16.4. Occurrence of Heteromyidae in the Tertiary of North America (●, species, +, genera). Abbreviations as in Fig. 5.4.

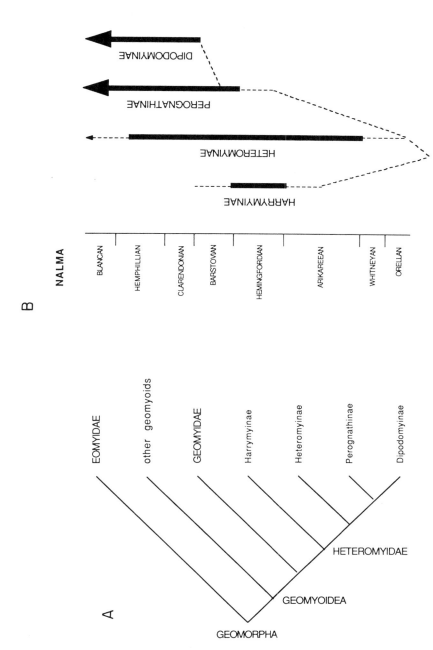

FIGURE 16.5. Phylogeny of the Heteromyidae. (A) Cladogram of heteromyid relationships. (B) Dendrogram of the Heteromyidae with age of occurrence (see Fig. 5.5 for explanation).

Heteromys and *Liomys* in the Heteromyinae, and listed *Proheteromys* as *incertae sedis* in no subfamily. Wood's (1935a) general plan for the arrangement of subfamilies of heteromyids is followed here based on the dental characters he defined. Unfortunately, little cranial material is known for the Tertiary genera referred to the Heteromyinae, and thus it is difficult to apply Wahlert's (1985, 1993) criteria for separating the subfamilies.

The most closely related of the subfamilies of heteromyids are the Perognathinae and Dipodomyinae (Wood, 1935a; Wahlert, 1985). The ancestry of the dipodomyines can be traced from within the perognathines with increased crown height of the cheek teeth, inflation of the auditory bulla, and specialization of the skeleton for saltatorial locomotion. Any of the mesodont Barstovian species of the perognathine *Cupidinimus* could have been ancestral to the dipodomyines.

The Heteromyinae and Harrymyinae cannot be considered ancestral to the remainder of the family though they are the earliest to appear in the fossil record. The derived features of the skull and dentition of *Harrymys* and heteromyines remove them from the ancestry of perognathines or heteromyines (Wood, 1935a; Wahlert, 1985, 1991).

Lindsay (1972) placed the subfamilies of heteromyids along with the subfamilies of geomyids, implying inclusion of all of the subfamilies within one family (see also Shotwell, 1967a). Based on cranial arterial circulation, Brylski (1990) included the Heteromyinae within the Geomyidae rather than the Heteromyidae. However, it is not likely that the heteromyids should be included within the Geomyidae based on numerous cranial and dental differences of the two groups (see Wood, 1935a; Wahlert, 1985).

4.3. Extrafamilial Relationships

The family of rodents most closely related to the Heteromyidae are the pocket gophers, the Geomyidae (see McLaughlin, 1984; Fahlbusch, 1985; Wahlert, 1985). Some authors derived geomyids from heteromyids in the middle Tertiary (Lindsay, 1972; Fahlbusch, 1985). If *Proheteromys* is considered ancestral to the geomyids and is within the Heteromyidae, this also implies the derivation of the Geomyidae from a heteromyid ancestor (Wahlert, 1993; Green, in press). There is no special relationship with heteromyids among the remainder of the geomyoids.

5. Problematical Taxa

5.1. "*Diprionomys*" *agrarius*

Wood (1935a) first described *Diprionomys agrarius* from the Clarendonian of Nebraska. This species has been reported from a number of Barstovian localities (Storer, 1975; Korth, 1979; Barnosky, 1986b; Voorhies, 1990a). Korth (1979) noted the difference between this species and the type species of *Diprionomys*, *D. parvus* Kellogg (1910) from the Hemphillian, and listed the species as "*D.*" *agrarius*. Martin (1984) included *agrarius* in a new genus *Oregonomys* which otherwise included species from the Hemphillian and Blancan. In all cases, this species was included in the Heteromyinae. Martin's (1984) genus is not a heteromyine because of the central union of the lophs of P_4 and the inflated auditory bullae (both characters of the Perognathinae), and therefore, it cannot include *agrarius*.

Diprionomys agrarius represents a unique genus of heteromyine. In an unpublished dissertation, Reeder (1956) erected a new genus for this species and identified several new species, all from the Barstovian and Clarendonian of the Great Plains. A formal naming and study of the available specimens referred previously to *D. agrarius* is necessary to resolve the difficulty presented by this species.

5.2. *Mookomys*

Wood (1931) named *Mookomys altifluminis* from the Hemingfordian of Montana. This species was characterized by a simple lower premolar (four-cusped), a grooved upper incisor, and low-crowned cheek teeth. Since that time, nearly all middle Tertiary heteromyids with simple premolars have been referred to the genus *Mookomys* (Wood, 1935a; Wilson, 1949c, 1960; Lindsay, 1972, 1974; Storer, 1975; Whistler, 1984). However, the cheek teeth of *Mookomys* were higher crowned than originally described (Reeder, 1956) and most of the specimens referred to this genus differ from the type species in having lower-crowned cheek teeth, differences in the morphology of the premolar, or lack a grooved upper incisor and thus are not referable to this genus. The Arikareean ?*Mookomys bodei* Wilson (1949c) is referable to the genus *Tenudomys* (Korth, 1993b); *M. formicorum* Wood (1935a) likely represented a species of *Proheteromys*; *M. subtilis* Lindsay (1972) was a species of *Heliscomys* (Chapter 15); and the remainder of the indeterminate species referred to the genus *Mookomys* are most likely referable to species of *Perognathus* (Lindsay, 1974; Storer, 1975; Whistler, 1984). At present, only one species (the type species) can be definitely referred to *Mookomys*.

5.3. *Proheteromys*

Proheteromys has been described from the Arikareean to the Barstovian (and questionably as early as the Orellan) and contains a large number of species that were highly variable in size and morphology of P_4. This genus is represented in the fossil record only by dentitions, and few of the records of this genus are of more than a few dental elements, not allowing for a thorough study of the variability of each of the species. In a recent study of several hundred isolated cheek teeth of *Proheteromys* from the Hemingfordian of South Dakota, Green (in press) was unable to determine a specific identity of the material because of the inadequate diagnoses (based on poor representation in the fossil record) of the known species.

In general, the cheek teeth of species of *Proheteromys* were low crowned (brachydont–submesodont) and with an essentially primitive heteromyine morphology of P_4. A systematic review of the species of this genus is badly needed. Questions as to the subfamilial or familial assignment cannot be fully addressed until more complete material representing larger populations are recovered and described.

5.4. *Dikkomys* and *Lignimus*

Wood (1936b) described the genus *Dikkomys* based on isolated cheek teeth from an anthill collection from the late Arikareean of Nebraska. Later, two mandibles were described from the Hemingfordian of South Dakota (Galbreath, 1948; Macdonald, 1970). The greatest collection of isolated cheek teeth of *Dikkomys* was described by Green and Bjork (1980), again from the Hemingfordian of South Dakota. Wood (1936b) referred this genus to the Geomyinae (Geomyidae). This allocation was followed by nearly all later authors. Akersten (1973) was the first to suggest that *Dikkomys* was not a geomyine.

The genus *Lignimus* was known from the Barstovian of the Great Plains and Wyoming (Storer, 1970, 1975; Barnosky, 1986a; Voorhies, 1990a). It was known only from isolated cheek teeth. Storer (1970, 1973, 1975) included this genus in the Entoptychinae (Geomyidae), but the similarity in the cheek tooth morphology of *Dikkomys* and *Lignimus* has been noted by numerous authors (Storer, 1973; Akersten, 1973; Korth, 1979; Wahlert, 1991).

A third genus with similar lower cheek teeth, *Harrymys*, was named by Munthe (1988), to which he referred a species that had previously been referred to *Dikkomys* (Black, 1961). Munthe (1988) referred *Harrymys* to the Florentiamyidae. In his detailed description of the skull of *Harrymys*, Wahlert (1991) demonstrated that this genus was clearly a heteromyid, and listed *Dikkomys* and *Lignimus* as genera that might be included in his new subfamily the Harrymyinae. Although there were some differences in the dentitions of these genera (Wahlert, 1991), the similarity of the lower cheek teeth (V-shaped hypolophid joined to the center of the metalophid) was striking and therefore all three genera are here included within the Heteromyidae (Harrymyinae) although the diagnostic cranial features of the harrymyines are not known for either *Dikkomys* or *Lignimus*. Discovery and description of cranial material of either of the latter genera may revise their current allocation.

5.5. *Schizodontomys*

Originally, Rensberger (1973b) included species of *Schizodontomys* from the Hemingfordian and possibly late Arikareean to the subfamily Pleurolicinae in the Geomyidae. Munthe (1981) questioned this allocation because of the lack of fossorial adaptations of the postcranial skeleton, and later Wahlert (1985) demonstrated that *Schizodontomys* was a heteromyid and questionably placed it in the Dipodomyinae, based on the inflation of the auditory bullae. Korth *et al.* (1990) placed this genus in the Heteromyinae according to Wahlert's (1985) diagnoses of the subfamilies of heteromyids based on cranial foramina and those of Wood (1935a) based on the morphology of the lower premolar. Most recently, Wahlert (1991) questioned whether the skull described by Korth *et al.* (1990) was referable to *Schizodontomys* and again noted the similarity of bullar inflation of *Schizodontomys* with that of dipodomyines.

Many of the cranial features of *Schizodontomys* were also present in *Harrymys*. Both had ventrally inflated bullae (not dorsally or posteriorly inflated as in heteromyines and dipodomyines). In *Harrymys* the bullae met anteroventrally and *Schizodontomys* they did not, but in the latter the separation between the bullae at the anteroventral corner was slight, and the bullae nearly touching. Other similarities of the skulls of these genera include the fusion of the masticatory and buccinator foramina and stapedial foramen present but sphenofrontal absent. However, the dentition of *Schizodontomys* had a typical geomyoid pattern on the cheek teeth and that of *Harrymys* was greatly derived.

The combination of heteromyine, dipodomyine, and harrymyine features of the skull and dentition of *Schizodontomys* makes it difficult to classify this genus within the Heteromyidae.

5.6. *Prodipodomys? mascallensis*

Downs (1956) named *Prodipodomys? mascallensis* from the Barstovian of Oregon based on a mandibular specimen with well-worn teeth. He was uncertain as to the generic allocation of this species but felt it was closer to a dipodomyine than a perognathine. Later, Reeder (1956) included this species in the heteromyine genus *Peridiomys*. Shotwell (1967a) described additional specimens of this species (which he included in the Geomyinae) with unworn cheek teeth and followed Downs in not definitely assigning this species to any currently described genus of geomyid. Lindsay (1972) referred *Prodipodomys? mascallensis* to his new genus of geomyine, *Mojavemys*. Finally, Wahlert (1991) retained the species in its original generic designation and suggested a close relationship with the Harrymyinae based on the occlusal morphology of the lower cheek teeth.

It is clear that *P.? mascallensis* was not referable to any currently known genus of heteromyid. Its cheek teeth were smaller and higher crowned than any species of *Peridiomys*. The age of this species (Barstovian) predated the otherwise first occurrence of *Prodipodomys* (Hemphillian) by several million years. It was much smaller than an of the other species of *Mojaveyms* and the cheek teeth were less lophate (cusps more distinguishable) with lower crown height. It is quite likely that *P.? mascallensis* represented an as yet unnamed genus of heteromyid.

6. Classification

Heteromyidae Gray, 1868
  Perognathinae Coues, 1875
    Perognathus Maximilian, 1839 (Hf–R)
      P. furlongi Gazin, 1930 (?Hf–B)
      P. coquorum Wood, 1935 (Cl)
      P. dunklei Hibbard, 1939 (Hp)
      P. gidleyi Hibbard, 1941 (Bl)
      P. pearlettensis Hibbard, 1941 (Bl)
      P. rexroadensis Hibbard, 1950 (Bl)
      P. minutus James, 1963 (B–Cl)
      P. maldei Zakrzewski, 1969 (Bl)
      P. henryredfieldi Jacobs, 1977 (Hp)
      P. carpenteri Dalquest, 1978 (Bl)
      P. trojectioansrum Korth, 1979 (B)
      P. stevei Martin, 1984 (Hp)
      P. brevidens Korth, 1987 (B)
    Cupidinimus Wood, 1935 (Hf–Hp)
      C. quartus (Hall, 1930) (Cl)
      C. tertius (Hall, 1930) (Cl)
      C. nebraskensis Wood, 1935* (B)
      C. halli (Wood, 1936) (B)
      C. cuyamensis (Wood, 1937) (Cl)
      C. eurekensis (Lindsay, 1972) (B)
      C. kleinfelderi (Storer, 1975) (B)
      C. saskatchewanensis (Storer, 1975) (B)
      C. bidahochiensis (Baskin, 1977) (Hp)
      C. boronensis Whistler, 1984 (Hf)
      C. lindsayi Barnosky, 1986 (B)
      C. whitlocki Barnosky, 1986 (B)
      C. avawatzensis Barnosky, 1986 (Cl)
    Trogomys Reeder, 1960 (Hf)
      T. rupinimenthae Reeder, 1960* (Hf)
    Oregonomys Martin, 1984 (Hp–Bl)

        O. sargenti (Shotwell, 1956) (Hp)
        O. magnus (Zakrzewski, 1969) (Bl)
        O. pebblespringsensis Martin, 1984* (Hp)
    Stratimus Korth, Bailey, and Hunt, 1990 (Hf)
        S. strobeli Korth, Bailey, and Hunt, 1990* (Hf)
Heteromyinae Gray, 1868
    Proheteromys Wood, 1932 (Wh–B)
        P. parvus (Troxell, 1923) (?A or ?Hf)
        P. floridanus Wood, 1932* (Hf)
        P. magnus Wood, 1932 (Hf)
        P. matthewi Wood, 1935 (Hf)
        P. formicorum (Wood, 1935) (A)
        P. nebraskensis Wood, 1937 (Wh)
        P. sulculus Wilson, 1960 (Hf)
        P. maximus James, 1963 (B)
        P. fedti Macdonald, 1963 (A)
        P. gremmelsi Macdonald, 1963 (A)
        P. ironcloudi Macdonald, 1970 (A)
    Diprionomys Kellogg, 1910 (B–Cl)
        D. parvus Kellogg, 1910* (Hp)
        D. minimus Kellogg, 1910 (Hp)
        ?*D. agrarius* Wood, 1935 (B–Cl)
    Mookomys Wood, 1931 (Hf)
        M. altifluminis Wood, 1931* (Hf)
    Peridiomys Matthew, 1924 (B)
        P. rusticus Matthew, 1924* (B)
        P. oregonensis (Gazin, 1932) (B)
        P. borealis Storer, 1970 (B)
    Schizodontomys Rensberger, 1973 (?A–Hf)
        S. harkseni (Macdonald, 1970) (Hf)
        S. greeni Rensberger, 1973* (Hf)
        S. sulcidens Rensberger, 1973 (?A)
        S. amnicolus Korth, Bailey, and Hunt, 1990 (Hf)
Dipodomyinae Coues, 1875
    Dipodomys Gray, 1841 (Bl–R)
        D. minor Gidley, 1922 (Bl)
        D. gidleyi Wood, 1935 (Bl–Pl)
        D. hibbardi Zakrzewski, 1981 (Bl)
    Prodipodomys Hibbard, 1939 (Hp–Bl)
        P. kansensis (Hibbard, 1937)* (Hp)
        P. centralis (Hibbard, 1941) (Bl)
        P. tiheni (Hibbard, 1943) (Bl)
        ?*P. mascallensis* Downs, 1956 (B)
        P. idahoensis Hibbard, 1962 (Bl)
        P. griggsorum Zakrzewski, 1970 (Hp)

  Eodipodomys Voorhies, 1975 (Cl)
   E. celtiservator Voorhies, 1975* (Cl)
 Harrymyinae Wahlert, 1991
  Harrymys Munthe, 1988 (Hf)
   H. woodi (Black, 1961) (Hf)
   H. irvini Munthe, 1988* (Hf)
  Dikkomys Wood, 1936 (Hf)
   D. matthewi Wood, 1936* (Hf)
  Lignimus Storer, 1970 (B)
   L. montis Storer, 1970* (B)
   L. transversus Barnosky, 1986 (B)

Chapter 17
Florentiamyidae

1. Characteristic Morphology	189
1.1. Skull	189
1.2. Dentition	191
1.3. Skeleton	191
2. Evolutionary Changes in the Family	192
3. Fossil Record	193
4. Phylogeny	193
4.1. Origin	193
4.2. Intrafamilial Relationships	195
4.3. Extrafamilial Relationships	196
5. Problematical Taxon: *Florentiamys agnewi*	196
6. Classification	197

1. Characteristic Morphology

1.1. Skull

Florentiamyids ranged from small to intermediate in size. Their skulls shared several features with other geomyids (parapterygoid fossa, sciuromorphy, shortened incisive foramina, auditory bulla forms back of foramen ovale). The skull had a long rostrum with the infraorbital foramen dorsal to the center of the diastema, just below middepth of the rostrum (Fig. 17.1). The surface of the upper diastema was essentially flat. Dorsally, the skull was broad. The nasal and premaxillary bones extended posteriorly on the skull to the anterior margin on the orbits in *Kirkomys*, *Hitonkala*, and *Ecclesimus*, but were shorter in *Sanctimus* and *Florentiamys*. On the most derived genera (*Florentiamys*, *Sanctimus*) there were frontal flanges above the orbits. The only markings for the temporalis muscle were low ridges on the parietal bones near their lateral extent. The auditory bullae were inflated but not enlarged. A unique feature of the skulls of all forentiamyids was a contribution of the palatine bone on the side of the skull to the anterior alar fissure (Wahlert, 1983). The palatine formed the ventral floor of this opening. The alisphenoid extended dorsally well above the glenoid fossa, nearly to the skull roof.

There were two unique features of the cranial foramina (Wahlert, 1983): the optic foramen was enlarged (measuring greater than 1 mm in diameter), and the masticatory and buccinator foramen were fused with the accessory foramen ovale on the posteroventral corner of the alisphenoid. The cranial

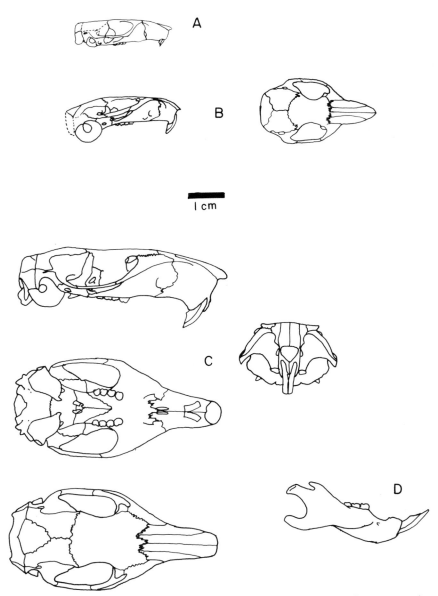

FIGURE 17.1. Skulls of Florentiamyidae. (A) Lateral view of Orellan *Ecclesimus tenuiceps*. (B) Lateral and dorsal views of Arikareean *Hitonkala andersontau*. (C) Lateral, dorsal, ventral, and anterior views of Arikareean *Sanctimus stouti*. (D) Lateral view of the mandible of Arikareean *Florentiamys kingi*. A from Korth (1989b), B from Korth (1993a), C and D from Wahlert (1983).

foramina of the medial orbital wall of *Kirkomys* were situated more anteriorly than in all other florentiamyids which retained an essentially primitive position of these foramina (Wahlert, 1984b). All genera except *Hitonkala* lacked a sphenofrontal foramen (not known for *Kirkomys*), indicating the loss of the superior branch of the stapedial artery. The posterior maxillary foramen on the palate was enclosed laterally.

The mandible was relatively deep (dorsoventrally). The coronoid process was very small and not laterally splayed as in heteromyids. The masseteric fossa had a shallow U-shaped anterior end. The ventral border of the scar was better marked (higher ridge) than the dorsal. The fossa ended anteriorly about one-third the length of the diastema. The mental foramen was ventral to the anterior terminus of the masseteric scar. The angle of the mandible was a broad process. The emargination between the condyle and angle of the mandible was deep (anteriorly extended).

1.2. Dentition

The dental formula for all florentiamyids was the same as in other geomyoids, with a single premolar and three molars on both the uppers and lowers. The microstructure of the incisor enamel was uniserial in the Arikareean *Sanctimus* and *Florentiamys* (Wahlert, 1983) but unknown in any of the other genera. The incisors were laterally compressed with gently convex or slightly flattened anterior enamel surfaces.

The cheek teeth of florentiamyids were brachydont to submesodont with bulbous cusps (Fig. 17.2). As in other geomyoids, the cusps of the teeth were arranged in two transverse rows. The upper molars had a continuous lingual cingulum that blocked the transverse valley of the teeth lingually. If stylar cusps were present on the upper molars, they were anteroposteriorly elongate. Only in the first upper molar of *Hitonkala* was the lingual cingulum interrupted by a narrow valley (Macdonald, 1963; Korth, 1992a, 1993a). The upper premolar in *Kirkomys*, *Ecclesimus*, and *Hitonkala* had a typical geomyoid pattern (isolated protocone and three-cusped metaloph) and was equal to or smaller than the first molar. In *Sanctimus* and *Florentiamys* the upper premolar was more complex and larger than any of the molars. The lower premolar was generally molariform with a complex metalophid (accessory cuspules often subequal to major cusps in size) and was larger than the molars in some species of *Florentiamys* and *Sanctimus* (Wahlert, 1983, Figs. 2, 3). Stylar cusps were well developed on the lower molars but were never as large as the principal cusps of the tooth.

1.3. Skeleton

The only postcranial skeleton reported of a florentiamyid was by Wood (1936c) who described the forelimb of *Florentiamys loomisi*. He compared

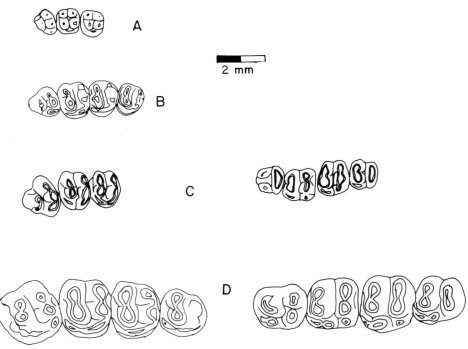

FIGURE 17.2. Dentitions of Florentiamyidae. (A) Orellan *Ecclesimus tenuiceps*, LP^4–M^2. (B) Whitneyan *Kirkomys milleri*, RP^4–M^3 (reversed). (C) Arikareean *Hitonkala macdonaldtau*, L^4–M^2 and LP$_4$–M$_3$. (D) Arikareean *Florentiamys loomisi*, RP^4–M^3 (reversed) and LP$_4$–M$_3$. A from Korth (1989b), B from Wahlert (1984b), C from Korth (1992a), D from Wahlert (1983).

the bones available to him only to heteromyids, and concluded that the limb of *Florentiamys* was more robust than any heteromyids. The limb proportions were similar to a quadruped (greater similarity to heteromyines) rather than adapted to a more saltatorial form of locomotion (reduced forelimb, enlarged hindlimb) as in perognathines and dipodomyines.

2. Evolutionary Changes in the Family

The most noticeable change within the Florentiamyidae was increased size and general robusticity of the mandible and cheek teeth. Orellan *Ecclesimus* was nearly as small as contemporary species of *Heliscomys* while the Arikareean and Hemingfordian species attained sizes larger than modern geomyids. There was also an increased complexity and enlargement of the premolar. The lower premolar is not known definitely for the earliest florentiamyids, although Korth (1989b) suggested some associations of lower dentitions originally assigned to contemporaneous heteromyid genera. This makes it difficult to trace the development of this tooth. However, the upper premolar of these genera, and of *Hitonkala*, was simple and consistent with the primitive morphology of this tooth among all geomyoids. The later and more derived florentiamyids, *Sanctimus*, *Florentiamys*, and *Fanimus*, had P^{4}s that were enlarged (larger than the molars) with a paracone subequal in size to the protocone and stylar cusps that approached the size of the major cusps.

3. Fossil Record

The entire fossil record of florentiamyids was restricted to the Great Plains of North America. The earliest record was *Ecclesimus* from the late Orellan (Korth, 1989b). The latest record was from the medial Barstovian (Voorhies, 1990a). Florentiamyids attained their greatest diversity during the Arikareean where they were represented by five genera and ten species (Fig. 17.3).

4. Phylogeny (Fig. 17.4)

4.1. Origin

As with other geomyoids, the origin of the Florentiamyidae is difficult to establish because of the lack of any Eocene rodent that was morphologically intermediate between florentiamyids and any other family of rodents that may have been ancestral. Both Wood (1936c) and Wilson (1949a) viewed *Florentiamys* as a morphological intermediate between primitive rodents and *Heliscomys* based on dental morphology, but noted the later occurrence of the

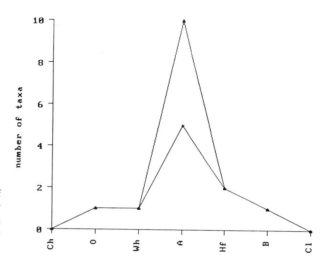

FIGURE 17.3. Occurrence of Florentiamyidae in the Tertiary of North America (▲, species; +, genera). Abbreviations as in Fig. 5.4.

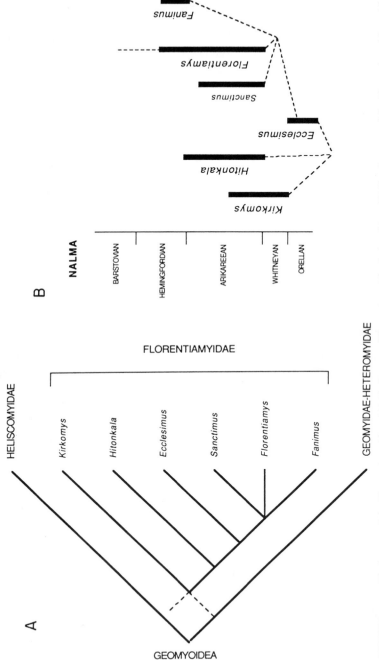

FIGURE 17.4. Phylogeny of the Florentiamyidae. (A) Cladogram of florentiamyid relationships. (B) Dendrogram of the Florentiamyidae with age of occurrence (see Fig. 5.5 for explanation).

florentiamyids in the fossil record. Rensberger (1973a) viewed *Heliscomys* as the ultimate ancestor of the florentiamyids through an intermediate form similar to *Tenudomys* Rensberger (1973b). It has been demonstrated, however, that *Tenudomys* cannot be ancestral to florentiamyids because it shared some derived characters of the skull with heteromyids and geomyids (Korth, 1993b).

With *Heliscomys* removed as a potential ancestor to later geomyoids (Korth et al., 1991), there is no other geomyoid that predates the florentiamyids in the fossil record. Again, the most likely ancestry of the florentiamyids (as with all geomyoids) may be from a sciuravid as suggested by Wilson (1949a).

4.2. Intrafamilial Relationships

Advanced dental characters unite the genera *Sanctimus*, *Florentiamys*, and *Fanimus* (complex, enlarged P^4, greater massiveness of cheek teeth). The former two genera also shared the advanced cranial characters of a bony flange above the orbits and shortened nasal bones on the skull roof. *Ecclesimus*, while retaining an essentially primitive morphology of P^4 and lacking other advanced cranial features of the *Sanctimus–Florentiamys–Fanimus* group, lacked a sphenofrontal foramen (loss of stapedial artery) as in the latter group, and was more closely related to it than other florentiamyids (Korth, 1989b). There are no features that would bar *Ecclesimus* from an ancestral position to a *Sanctimus–Florentiamys–Fanimus* clade.

Kirkomys lacked the advanced cranial features of other genera, but the more anterior position of the cranial foramina on the medial orbital wall (a derived feature), combined with its primitive dentition, appear to separate this genus out into a separate lineage from all other florentiamyids (Wahlert, 1984b). Similarly, there were no derived cranial features of *Hitonkala* that could not be considered primitive for florentiamyids, and therefore ancestral to them. However, the division of the lingual cingulum and development of stylar cusps on M^1 of *Hitonkala* are derived features, again keeping this genus from an ancestral relationship to other florentiamyids. If the presence of stylar cusps were viewed as primitive, and the continuous lingual cingulum as derived, as suggested by Wahlert (1985), then *Hitonkala* could be viewed as ancestral to other florentiamyids, at least morphologically.

Temporally, *Ecclesimus* could have easily evolved into the Arikareean and Hemingfordian *Sanctimus–Florentiamys–Fanimus* lineage. *Kirkomys* represented a unique lineage that separated from the former clade before the occurrence of *Ecclesimus* in the Orellan. The derived features of the skull of *Kirkomys* combined with its primitive dentition (see Korth, 1989b, for proposed allocation of lower dentitions) clearly separated it from all other florentiamyids.

The Arikareean and Hemingfordian *Hitonkala* had a more complex P_4, as in the lineage beginning with *Ecclesimus*, but did not show the loss of the stapedial artery already attained by Orellan times in the latter lineage. The occurrence of *Hitonkala* along with the most derived florentiamyids indicates that it was a relict of an early radiation of florentiamyids, retaining most primitive cranial and dental features of the family (Korth, 1993a).

4.3. Extrafamilial Relationships

Wood (1936c) originally named the Florentiamyinae as a subfamily of the Heteromyidae. Several subsequent authors followed this allocation (Wilson, 1949a; Macdonald, 1970). Rensberger (1973a) considered florentiamyines a subfamily of the Geomyidae, and viewed them as ancestral to entoptychine geomyids. Fahlbusch (1985) has also considered the Florentiamyinae as a subfamily of geomyids. Wahlert (1983) recognized florentiamyids as a distinct family of geomyoids and demonstrated that certain advanced cranial features of the family did not allow them to be considered ancestral to any later geomyoids.

Wahlert (1983, 1985) viewed the florentiamyids as a sister group to the heteromyid–geomyid clade. Later, with the recognition of the Heliscomyidae as a separate family, the latter was the sister group of the florentiamyid–heteromyid–geomyid clade (Korth et al., 1991, Fig. 3). The florentiamyids were united with the heteromyid–geomyid clade based on three cranial features: (1) short incisive foramina, (2) shortened nasals, and (3) laterally closed posterior maxillary foramen. Korth (1993a) demonstrated that primitively the nasals of florentiamyids were not shortened (therefore convergent with the heteromyids and geomyids). He argued that some derived cranial features of florentiamyids united them with heliscomyids (dorsal extent of the alisphenoid, enlarged optic foramen, and fusion of the buccinator, masticatory, and accessory foramen ovale), producing a florentiamyid–heliscomyid clade. The remainder of the features shared with the heteromyids and geomyids would then be considered convergent. In both scenarios, equal amounts of convergence must be accepted.

5. Problematical Taxon: *Florentiamys agnewi*

Wahlert (1983) only questionably included *F. agnewi* Macdonald (1963) in the genus *Florentiamys*. This species was known only from the type specimen, a partial mandible with P_4–M_1. It differed from other species of this genus by being much smaller and having the protostylid more posterior on M_1, joining the hypostylid and blocking the transverse valley. Korth et al. (1990) suggested that the posterior position of the protostylid on P_4 was similar to the condition in *Fanimus*, and that *F. agnewi* might be a junior synonym of *Fanimus clasoni* if the morphology of the first molar was only anomalous. However, it is impossible to determine the true generic allocation of *?F. agnewi* without a better knowledge of the range of variability of the cheek teeth. Here it is maintained as questionably in the genus *Florentiamys*.

6. Classification

Florentiamyidae Wood, 1936
    Florentiamys Wood, 1936 (A–Hf)
      F. loomisi Wood, 1936* (A)
      ?F. agnewi Macdonald, 1963 (A)
      F. tiptoni (Macdonald, 1970) (A)
      F. kennethi Wahlert, 1983 (A)
      F. kinseyi Wahlert, 1983 (A)
      Florentiamys sp. Korth, Bailey, and Hunt, 1990 (Hf)
    Hitonkala Macdonald, 1963 (A–Hf)
      H. andersontau Macdonald, 1963* (A)
      H. macdonaldtau Korth, 1992 (Hf)
    Sanctimus Macdonald, 1970 (A)
      S. stuartae Macdonald, 1970* (A)
      S. simonisi Wahlert, 1983 (A)
      S. falkenbachi Wahlert, 1983 (A)
    Kirkomys Wahlert, 1984 (Wh–A)
      K. schlaikjeri (Black, 1961) (A)
      K. milleri Wahlert, 1984* (Wh)
    Ecclesimus Korth, 1989 (O)
      E. tenuiceps (Galbreath, 1948)* (O)
    Fanimus Korth, Bailey, and Hunt, 1990 (A–Hf)
      F. clasoni (Macdonald, 1963) (A)
      F. ultimus Korth, Bailey, and Hunt, 1990* (Hf)

Chapter 18
Geomyidae

1. Characteristic Morphology	199
1.1. Skull	199
1.2. Dentition	200
1.3. Skeleton	203
2. Evolutionary Changes in the Family	203
3. Fossil Record	204
4. Phylogeny	204
4.1. Origin	204
4.2. Intrafamilial Relationships	205
4.3. Extrafamilial Relationships	208
5. Problematical Taxa	208
5.1. Pleurolicinae	208
5.2. *Tenudomys*	209
5.3. *Parapliosaccomys*	209
5.4. *Gregorymys kayi*	209
5.5. *Gregorymys larsoni*	210
6. Classification	210

1. Characteristic Morphology

1.1. Skull

All geomyids are intermediate-sized rodents. Their skulls are distinct among geomyoids but retain the basic geomyoid skull characteristics (sciuromorphous zygomasseteric structure; short incisive foramina; parapterygoid fossa present; auditory bulla forms posterior margin of foramen ovale). They also share several derived characters of the cranial foramina with heteromyids (pterygoid fossa contains entrance to sphenopterygoid canal; origin of internal pterygoid muscle extends through sphenopterygoid canal toward orbit; posterior margin of accessory foramen ovale not ossified, and lateral pterygoid flange often failing to reach auditory bulla; postglenoid foramen between squamosal bone and auditory bulla; temporal foramen absent; auditory bullae highly vesicular texture; Wahlert, 1985, p. 14). A number of features of the skulls of geomyids are the result of their fossorial habits and are shared by other fossorial rodent families such as palaeocastorine beavers and mylagaulids (low, broad neurocranium; broad, heavy zygomatic arches; strong postorbital constriction; procumbent incisors). Commonly there was a well-developed sagittal crest on the skull roof in entoptychines (Fig. 18.1). In geomyines there are two parallel parasagittal crests that nearly meet along the parietal bones and unite with well-defined occipital crests at the back of the skull.

Wahlert (1985, p. 15) listed nearly 20 unique features of the skull and cranial foramina of geomyids. The most distinctive of these features was the dorsally arched cranial diastema, reduced size of optic foramen, anterior position of the anterior alar fissure (dorsal to posterior molars), and unique "pulley" system for the temporalis muscle (alisphenoid extends dorsally to skull roof with small bony process on the squamosal). In advanced entoptychines the rostrum became elongate and approached the length of the remainder of the skull.

The mandible of geomyids is deep (dorsoventrally). The coronoid process is larger than in other geomyoids. The masseteric fossa extends anteriorly to below the premolar and is V-shaped. The diastema is deep. In geomyines the diastema is relatively short, and in advanced entoptychines the diastema is elongated. The mental foramen is just anterior to the premolar near middepth of the mandible. The angular process has a lateral flange. In geomyines the ventral border of the mandible is ventrally convex and the angle is dorsally placed, just below the condyle. In entoptychines the process was more typical of geomyoids and was directed posteriorly with a deep emargination between the angle and the condyle.

1.2. Dentition

The dental formula of geomyids is $\frac{1013}{1013}$ as in other geomyoids. The incisors are broad anteriorly and often flattened. The microstructure of the incisor is uniserial (Wahlert, 1968). In both subfamilies, the upper incisors can have one, two, or no longitudinal grooves on the enamel surface. One of the grooves occurs along the medial margin of the enamel surface, the other occurs near the center of the enamel surface.

The cheek teeth of geomyoids were at least mesodont in the most primitive species and attained total hypsodonty in several lineages. In all species that were at least mesodont there is some enamel failure on the lateral sides of the teeth. In lower-crowned species this failure appears as small chevrons. The hypsodont animals have dentine tracts that extend the entire height of the tooth. The occlusal pattern of the cheek teeth in primitive geomyids consisted of two transverse rows of three cusps (Fig. 18.2), again, as in all geomyoids. However, even in the most primitive geomyoids the transverse lophs dominated the cusps and the individual cusps were not readily recognizable as such. The overall occlusal shape of the cheek teeth of geomyids is ovate with the transverse axis being the longer.

Entoptychines had a deep transverse valley on the cheek teeth. With wear, enamel on the occlusal surface of the molars formed a U-shape (lophs

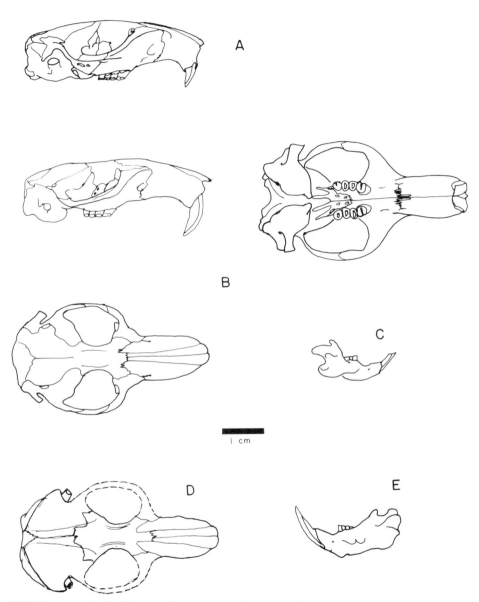

FIGURE 18.1. Skulls of Geomyidae. (A) Lateral view of Arikareean entoptychine *Gregorymys formosus*. (B) Lateral, ventral, and dorsal views of Arikareean entoptychine *Entoptychus minor* or *cavifrons*. (C) Lateral view of generalized mandible of *Entoptychus*. (D) Dorsal view of Blancan geomyine *Geomys persimilis*. (E) Lateral view of mandible of Blancan geomyine *Thomomys gidleyi*. A from Wahlert and Sousa (1988), B and C from Rensberger (1971), D from Tomida (1987), E from Zakrzewski (1969).

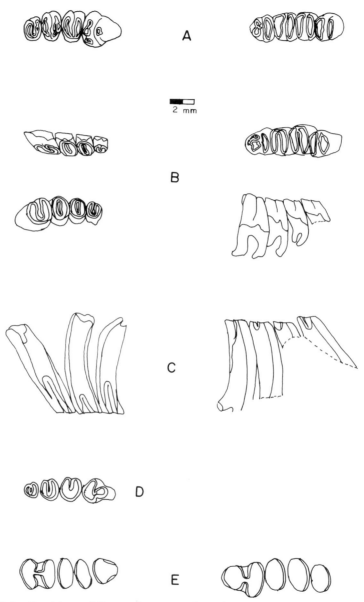

FIGURE 18.2. Dentitions of Geomyidae. (A) Early Arikareean entoptychine *Pleurolicus sulcifrons*, RP^4–M^3 and RP_4–M_3. (B) Arikareean entoptychine *Entoptychus wheelerensis*, LP^4–M^3 (with lateral view) and RP_4–M_3 (with medial view). (C) Late Arikareean entoptychine *Entoptychus individens*, lateral view of RP^4–M^2 (reversed) and medial view of LP_4–M_3. (D) Barstovian geomyine *Mojavemys alexandrae*, RP^4–M^3. (E) Blancan geomyine *Pappogeomys sansimonensis*, RP^4–M^3 and RP_4–M_3. A from Rensberger (1973b), B and C from Rensberger (1971), D from Lindsay (1972), E from Tomida (1987).

joined buccally on lowers, lingually on uppers) that had exposed dentine within. In advanced wear both the lingual and buccal ends of the lophs united, leaving a central enamel lake that was the shape of an anteroposteriorly compressed oval. The upper premolar in entoptychines was larger than the molars, and the protoloph was shorter than the metaloph (buccally). As P^4 wore down, it ultimately developed the U-Shape of the molars. The lower premolar had a narrower (buccolingually) trigonid than talonid and the cusps were usually distinguishable on the trigonid except in the most hypsodont species.

The premolars are larger than the molars as well in all geomyines. In the mesodont genera *Mojavemys* and *Pliosaccomys*, there was no enamel failure on the lateral sides of the cheek teeth, and their occlusal pattern was similar to that of mesodont entoptychines. In geomyines with more hypsodont cheek teeth (and better-developed enamel failure), the transverse valley is shallow and disappears in early stages of wear leaving only two enamel crescents (one anterior, one posterior) on the occlusal surface with the remainder of the occlusal surface as exposed dentine. Cement fills the reentrant valleys of the premolars in hypsodont geomyines.

1.3. Skeleton

The postcranial skeleton of geomyids shares many features with other fossorial rodents. The limbs are relatively shorter and the tail is short. The forelimb and manus are massive. There are well-developed muscle attachments on the major bones of the forelimb as in other digging rodents such as palaeocastorine beavers, mylagaulids, and burrowing sciurids. Rensberger (1971) demonstrated that the fossorial adaptations in entoptychines were not as well developed as in the geomyines, intermediate in development between that of the heteromyine heteromyid *Heteromys* and Recent geomyine *Geomys*.

2. Evolutionary Changes in the Family

The most obvious change in both subfamilies of geomyids was the increased hypsodonty of the cheek teeth. In entoptychines, especially the genus *Entoptychus* (Rensberger, 1971), the increased hypsodonty went from mesodonty to total hypsodonty through the stratigraphic section of the John Day Formation in Oregon (Arikareean). Similarly, in the Geomyinae, the Barstovian and Clarendonian genera *Mojavemys* and *Pliosaccomys* had mesodont cheek teeth (Wilson, 1936; Shotwell, 1967a; Lindsay, 1972), the Hemphillian genera *Parapliosaccomys* and *Pliogeomys* showed the beginnings of hypsodonty and dentine tracts on the cheek teeth (Hibbard, 1954; Shotwell, 1967a), and in the Blancan through the Pleistocene the modern genera of geomyines appeared in the fossil record and had attained complete hypsodonty of the cheek teeth.

Among the entoptychines there appeared to be the development of fossorial adaptations of the skull and skeleton from the early *Pleurolicus* through the most derived species of *Entoptychus*.

3. Fossil Record

The entire fossil record of geomyids, as with heteromyids and florentiamyids, is limited to North America. The earliest probable geomyid was the Orellan *Tenudomys basilaris* (Korth, 1989b). However, this genus cannot be referred to the Geomyidae with certainty (Korth, 1993b). The first radiation of geomyids was among the entoptychines in the Arikareean (Fig. 18.3). The Arikareean was the time of greatest diversity (in number of species) of geomyids during the Tertiary but was limited to only a few genera of entoptychines (Wood, 1936b; Rensberger, 1971, 1973b). The Entoptychinae suddenly appeared in the Arikareean and then diminished rapidly in the Hemingfordian only to become extinct by the Barstovian. One species of *Gregorymys* may have survived until the Clarendonian (Munthe, 1977).

The earliest definite geomyine was the Barstovian *Mojaveyms* (Lindsay, 1972). This makes the two subfamilies of geomyines mutually exclusive in the fossil record. Wood (1936b) considered the Arikareean to Hemingfordian *Dikkomys* a geomyine, but this genus is now considered a more likely heteromyid (Chapter 15, Section 5.4). The greatest diversity of the Geomyinae during the Tertiary was in the Blancan.

4. Phylogeny (Fig. 18.4)

4.1. Origin

As with other geomyoid families, the origin of the Geomyidae is questionable. Most authors suggest an origin from an early heteromyid (Lindsay, 1972; Fahlbusch, 1985), heliscomyid (Rensberger, 1973a,b), or florentiamyid (Rensberger, 1971). However, it has been shown that all of these families are already too specialized to have been ancestral to the geomyids (Wahlert, 1983, 1985; Korth et al., 1991). Some recent workers have considered the problematical heteromyid genus *Proheteromys* as an early ancestor of the geomyids (Wahlert, 1993; Green, in press). Because *Proheteromys* is only known from dentitions, when cranial material of this genus becomes available it will be possible to test this hypothesis. Finally, some authors have insisted that the likely origin of the geomyids was in Central America where the fossil record is not complete, and therefore any ancestral geomyids have not yet been discovered (Russell, 1968; Wahlert and Sousa, 1988).

It appears that the geomyids evolved after the geomyoid dental and cranial morphologies were well established. Therefore, the origin of the Geomyidae is most likely within some other early generalized geomyoid.

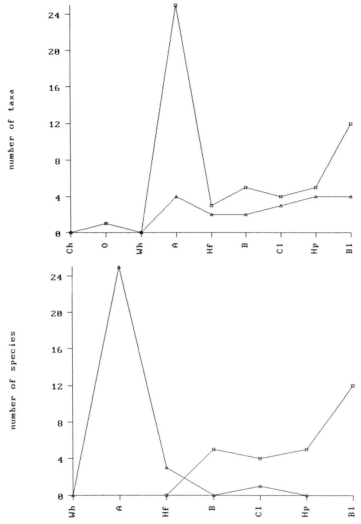

FIGURE 18.3. Occurrence of Geomyidae in the Tertiary of North America (Top: △, species; □, genera; bottom: △, Entoptychinae; □, Geomyinae). Abbreviations as in Fig. 5.4.

4.2. Intrafamilial Relationships

Early workers (Matthew, 1910) considered the entoptychines ancestral to the later Geomyinae. Wood (1936b) argued that the entoptychines were already too derived to be ancestral to the geomyines. Although the entoptychines predate geomyines and lack some of the derived cranial features of the geomyines (are more primitive), they cannot be considered ancestral to

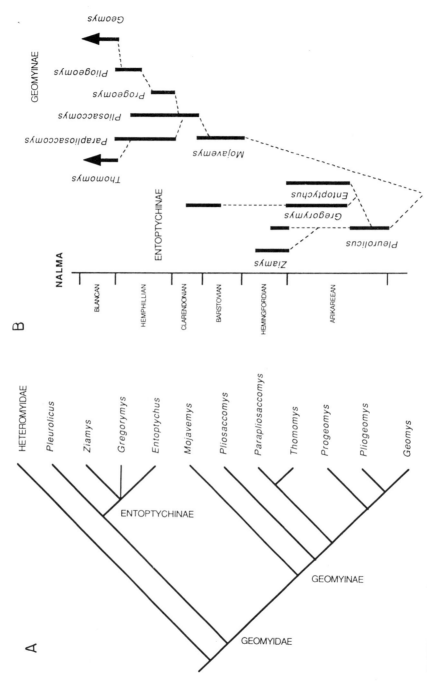

FIGURE 18.4. Phylogeny of the Geomyidae. (A) Cladogram of geomyid relationships. (B) Dendrogram of the Geomyidae with age of occurrence (see Fig. 5.5 for explanation).

Geomyinae because they possessed a number of independently derived features of the auditory bulla: anteromedial bullar processes meet at midline; mastoid region inflated and exposed on skull roof; squamosal emarginated posteriorly (Wahlert and Sousa, 1988, p. 11). The deep central transverse valley of the cheek teeth of entoptychines was also a derived feature not found in any geomyines (valley shallow, lost in early wear in geomyines; Rensberger, 1971).

Rensberger (1971) suggested two possibilities for the origin of the two subfamilies: (1) independent origins from a nonfossorial geomyoid or heteromyid (including florentiamyids) or (2) both subfamilies arose from a geomyoid stock that had already developed a semifossorial way of life. This latter hypothesized group is not yet known in the fossil record but would have had to exist in the middle Oligocene. The only possible genus of geomyoid that may have occupied this position (temporally as well as morphologically) is the problematical geomyoid genus *Tenudomys* which is only incompletely known.

Rensberger (1971) noted the greatest similarity between *Gregorymys* and *Entoptychus* and suggested that the ancestor of the *Entoptychus* radiation would have been *Gregorymys*-like. Similarly, the skull, dentition, and postcranial skeleton of *Pleurolicus* were less fossorially specialized than other entoptychines (Rensberger, 1973b), suggesting an ancestral morphotype for other entoptychines. The occurrence of *Pleurolicus* (below that of *Entoptychus*) would support this latter suggestion. Rensberger (1973b) also noted that the narrow incisor of *Tenudomys* made it more primitive than even *Pleurolicus*, and features of the cheek teeth made *Tenudomys* morphologically intermediate between the former and *Heliscomys* (which he believed to be the ancestor of all geomyoids). Again, this was supported by the biostratigraphic occurrence of *Tenudomys*, intermediate between that of *Heliscomys* and *Pleurolicus*.

Within the Entoptychinae, a morphological sequence from more primitive (nonfossorial) to derived follows *Pleurolicus–Gregorymys–Entoptychus*. This does not imply direct ancestry from one to the next, only a sequence of development of the fossorial geomyid features of the skull and skeleton, and the increased hypsodonty of the cheek teeth.

Several lineages have been identified within the Geomyinae. Shotwell (1967a) considered the Hemphillian *Parapliosaccomys* as a morphological intermediate between Clarendonian *Pliosaccomys* and Recent *Thomomys* and possibly as an ancestor of the latter as well as all other Recent genera of geomyids. Lindsay (1972, Fig. 38) placed the Barstovian *Mojavemys* as the ancestral geomyine (excepting *Dikkomys*) from which he derived *Pliosaccomys* and *Parapliosaccomys* independently. He followed Shotwell (1967a) in deriving the modern geomyines from *Parapliosaccomys*. Dalquest (1983) noted that in Shotwell's sequence of *Pliosaccomys–Parapliosaccomys–Thomomys* there was increased hypsodonty and all genera had asulcate incisors, and were geographically limited to western North America. Dalquest suggested that the lineage leading to *Geomys* separated from the *Thomomys* lineage during the Clarendonian, and the last common ancestor was *Pliosaccomys*. The lineage leading to Blancan *Geomys* was limited to central North America beginning in the early Hemphillian with *Progeomys*, through the late Hemphillian *Pliogeomys* to *Geomys*. This lineage similarly showed increase in crown height of the cheek teeth with increased dentine tracts as in the western lineage, but was distinguished by having grooved incisors on all genera.

In terms of horizon of occurrence and increased crown height of the cheek teeth (development of dentine tracts), a morphological sequence can be constructed for the Geomyinae. The Barstovian *Mojavemys* has no characters of the dentition that eliminate it from an ancestral position to later geomyines. In the Clarendonian, *Pliosaccomys* with slightly higher-crowned cheek teeth succeeds *Mojavemys*. In the original description of *Pliosaccomys*, Wilson (1936) felt that certain characters of the dentition of *Pliosaccomys* eliminated it from the ancestry of later geomyines but that it represented a stage of evolution that all later geomyines must have passed through. The next steps in the morphological sequence were the lineages that led to *Thomomys* (Shotwell, 1967a) through *Parapliosaccomys* and to *Geomys* (Dalquest, 1983) through *Progeomys* and *Pliogeomys*.

4.3. Extrafamilial Relationships

As discussed previously (Chapter 16, Section 4.3), the family that is most closely related to the Geomyidae is the Heteromyidae. These two families are more closely related than any two other families of geomyoids. No other families are derived from the Geomyidae.

5. Problematical Taxa

5.1. Pleurolicinae

Rensberger (1973b) erected the subfamily Pleurolicinae as a subfamily of the Geomyidae and included three genera: *Pleurolicus*, *Schizodontomys*, and *Tenudomys*. It has been demonstrated that *Schizodontomys* was referable to the Heteromyidae based on the skull and skeleton (Munthe, 1981; Wahlert, 1985; Korth et al., 1990; see also Chapter 16, Section 5.5). Similarly, the description of the skull of *Tenudomys* led to the conclusion that this genus lacked many of the derived features of true geomyids and it could not be referred to either the Geomyidae or Heteromyidae with certainty (Korth, 1993b). With these allocations the Pleurolicinae was reduced to the nominal genus only. Wahlert and Sousa (1988) noted the close similarity of the skull of *Pleurolicus* with that of the entoptychines *Gregorymys* and *Entoptychus* and included the former in the Entoptychinae.

Clearly, the recognition of an additional subfamily of geomyids to include only *Pleurolicus* is unnecessary, particularly considering the close relationship of *Pleurolicus* to other entoptychines.

5.2. *Tenudomys*

Rensberger (1973b) named the genus *Tenudomys* from the Arikareean of South Dakota and included it in his subfamiliy Pleurolicinae. He noted the nongeomyid features of the genus (narrow incisor, low-crowned teeth, simple P_4). Korth (1989b) named a few species of *Tenudomys* from the Orellan of Nebraska, extending its range. Finally, Korth (1993b) referred ?*Mookomys bodei* from the Arikareean of southern California to *Tenudomys* which allowed for a description of the skull. He noted a number of geomyid (deep skull, grooved palate) as well as heteromyid (narrow incisors, less pronounced postorbital constriction) features of the skull of *Tenudomys* and placed it as a sister group to the heteromyid–geomyid clade (Korth, 1993b, Fig. 2)

Tenudomys is here included in the Geomyidae with question and is presently not referable to any geomyid subfamily.

5.3. *Parapliosaccomys*

Parapliosaccomys oregonensis was named as the type species of a new genus of Hemphillian geomyine (Shotwell, 1967a). It was viewed as morphologically intermediate between the Clarendonian *Pliosaccomys* and Blancan species of *Thomomys*. Korth (1987b) included Barstovian and Clarendonian species from the Great Plains in *Parapliosaccomys* that had previously been included in *Lignimus* (Storer, 1973). The species from the Great Plains had cheek teeth that were lower crowned with a lesser development of the dentine tracts.

Akersten (1988) noted that the morphology of the lower premolar of the Great Plains species had a central enamel basin on the occlusal surface and all geomyines had a characteristic union of the lophs of the premolar at its center and no basin. Akersten argued that the Great Plains species represented a different genus than *Parapliosaccomys* and suggested that they represented a distinct subfamily of geomyids other than Geomyinae or Entoptychinae. The Great Plains species referred by Korth (1987b) to *Parapliosaccomys* are here only questionably included in this genus until a thorough review of these species is available.

5.4. *Gregorymys kayi*

Wood (1950) diagnosed *Gregorymys kayi* from the Arikareean of Montana as differing from all other species of *Gregorymys* in having higher-crowned cheek teeth with cement. He noted the similarity between G. kayi and species of Entoptychus, but included it in Gregorymys because the cheek teeth maintained roots. In a footnote, Rensberger (1971, p. 71) noted the similarity of the morphology of the cheek teeth of G. kayi to those of species of Entoptychus. He included G. kayi in Gregorymys because of its short diastema (elongated in Entoptychus).

Korth (1992a) asserted that the combination of Entoptychus and Gregorymys characters in G. kayi, as well as a distinctive morphology of P_4 (metalophid as wide as hypolophid, short anterior cingulum present), made it unique among entoptychines, and suggested that "G." kayi represented a new genus. He also suggested that specimens from the Arikareean of Wyoming referred to Entoptychus (McKenna and Love, 1972; McKenna, 1980) were also referable to this genus.

"G." kayi is known only from the type specimen (Wood, 1950) so it is difficult to establish a new genus for this species, but it appears that it does indeed represent a genus of entoptychine not yet recognized.

5.5. *Gregorymys larsoni*

Gregorymys larsoni from the late Barstovian or early Clarendonian Derby Peak fauna of Colorado (Munthe, 1977) was the only entoptychine that postdates the Hemingfordian. In general morphology of the mandible and lower cheek teeth, G. larsoni does not differ from Arikareean species of Gregorymys. Munthe (1977) proposed that a higher altitude isolation of the Derby Peak fauna may have provided a topographic barrier for the fauna that allowed a relict species of Gregorymys to persist.

6. Classification

Geomyidae Bonaparte, 1845
  Entoptychinae Miller and Gidley, 1918
    Gregorymys Wood, 1936 (A, ?B, or ?Cl)
      G. formosus (Matthew, 1907)* (A)
      G. curtus (Matthew, 1907) (A)
      G. riggsi Wood, 1936 (A)
      G. douglassi Wood, 1936 (A)
      ?G. kayi Wood, 1950 (A)
      G. larsoni Munthe, 1977 (?B or ?Cl)
      G. riograndensis Stevens, 1977 (A)
    Entoptychus Cope, 1878 (A)
      E. cavifrons Cope, 1878* (A)
      E. planifrons Cope, 1878 (A)
      E. minor Cope, 1881 (A)

      E. germannorum Wood, 1936 (A)
      E. montanensis (Hibbard and Keenmon, 1950) (A)
      E. basilaris Rensberger, 1971 (A)
      E. wheelerensis Rensberger, 1971 (A)
      E. transitorius Rensberger, 1971 (A)
      E. individens Rensberger, 1971 (A)
      E. fieldsi Nichols, 1976 (A)
      E. sheppardi Nichols, 1976 (A)
      E. grandiplanus Korth, 1992 (A)
    Pleurolicus Cope, 1878 (A–Hf)
      P. sulcifrons Cope, 1878* (A)
      P. dakotensis Wood, 1936 (A)
      P. sellardsi (Hibbard and Wilson, 1950) (?A)
      P. hemingfordensis Korth, Bailey, and Hunt, 1990 (Hf)
    Ziamys Gawne, 1975 (Hf)
      Z. tedfordi Gawne, 1975* (Hf)
      Z. hugeni Korth, Bailey, and Hunt, 1990 (Hf)
Geomyinae Bonaparte, 1845
    Pliosaccomys Wilson, 1936 (Cl–Hp)
      P. magnus (Kellogg, 1910) (Cl)
      P. dubius Wilson, 1936* (Hp)
      P. wilsoni James, 1963 (Cl)
      P. higginsensis Dalquest and Patrick, 1989 (Hp)
    Parapliosaccomys Shotwell, 1967 (B–Hp)
      P. oregonensis Shotwell, 1967* (Hp)
      ?*P. hibbardi* (Storer, 1973) (Cl)
      ?*P. annae* Korth, 1987 (B)
    Progeomys Dalquest, 1983 (Hp)
      P. sulcatus Dalquest, 1983* (Hp)
    Pliogeomys Hibbard, 1954 (Hp–Bl)
      P. buisi Hibbard, 1954* (Bl)
      P. parvus Zakrzewski, 1969 (Bl)
      P. carranzai Lindsay and Jacobs, 1985 (Hp)
    Geomys Rafinesque, 1817 (Bl–R)
    G. (Geomys) Rafinesque, 1817 (Bl–R)
      G. (G.) tobinensis Hibbard, 1944 (Bl–Pl).
      G. (G.) quinni McGrew, 1944 (Bl)
      G. (G.) garbanii White and Downs, 1961 (Bl)
      G. (G.) adamsi Hibbard, 1967 (Bl)
      G. (G.) jacobi Hibbard, 1967 (Bl)
    G. (Nerterogeomys) Gazin, 1942 (Bl–R)
      G. (N.) minor Gidley, 1922 (Bl)
      G. (N.) paenebursarius Strain, 1966 (Bl)
      G. (N.) smithi Hibbard, 1967 (Bl)
      G. (N.) anzensis Becker and White, 1981 (Bl)

Pappogeomys Merriam, 1895 (Bl–R)
   P. (Cratogeomys) Merriam, 1895 (Bl–R)
     P. (C.) bensoni Gidley, 1922 (Bl)
     P. (C.) sansimonensis Tomida, 1989 (Bl)
Thomomys Maximilian, 1839 (Bl–R)
   T. gidleyi Wilson, 1933 (Bl)
Mojavemys Lindsay, 1972 (B–Cl)
   M. alexandrae Lindsay, 1972* (B)
   M. lophatus Lindsay, 1972 (B)
   M. magnumarcus Barnosky, 1986 (B)
?Geomyidae *incertae sedis*
   Tenudomys Rensberger, 1973 (O–A)
     T. bodei (Wilson, 1949) (Wh)
     T. dakotensis (Macdonald, 1963) (A)
     T. macdonaldi Rensberger, 1973* (A)
     T. basilaris Korth, 1989 (O)

Chapter 19
Zapodidae

1. Characteristic Morphology	213
1.1. Skull	213
1.2. Dentition	214
1.3. Skeleton	214
2. Evolutionary Changes in the Family	216
3. Fossil Record	216
4. Phylogeny	217
4.1. Origin	217
4.2. Intrafamilial Relationships	219
4.3. Extrafamilial Relationships	219
5. Problematical Taxa: *Plesiosminthus* and *Schaubeumys*	220
6. Classification	221

1. Characteristic Morphology

1.1. Skull

No skulls have been described of any North American Tertiary zapodids so the following discussion is based on the general morphology of Recent species (see Fig. 2.5A). Zapodids are small rodents. The cranium is inflated and the rostrum not especially elongated. The infraorbital foramen is enlarged for the passage of a branch of the masseter muscle (hystricomorphous) and there is a distinct accessory foramen medial to the infraorbital foramen for the passage of nerves and blood vessels. The zygomatic arch is reduced to a thin strip of bone. The markings for the attachment of the temporalis muscle are low and limited to the lateral margins of the parietal bones. On the palate the incisive foramina are enlarged, extending posteriorly to the level of at least the first molar. The auditory bullae are small and not inflated.

The mandible is slender. The masseteric fossa extends anteriorly below the first molar and generally forms a V-shape. The ventral ridge of the scar is more strongly developed than the dorsal ridge. The coronoid process is pointed and sharply curved posteriorly. Similarly, the angle of the jaw terminates in a posteriorly directed point. The origin of the lower incisor is marked by a bulbous projection on the ascending ramus, ventral to the condyle.

1.2. Dentition

The dental formula for all Tertiary zapodids is $\frac{1013}{1003}$ where the upper premolar was reduced to a small, single-rooted, conical cusp. In the Recent genus *Napaeozapus* the P^4 is lost.

The incisors of zapodids are small and generally laterally compressed with narrow, gently convex anterior enamel surfaces. The upper incisor may either be smooth or have a single groove in the enamel surface. The microstructure of the incisor enamel is uniserial.

The cheek teeth are brachydont to mesodont in crown height. In sicistines and early zapodines the molars were cuspate but with well-defined lophs. In the advanced zapodines the cusps became indistinct and the occlusal pattern is dominated by lophs (Fig. 19.1).

The upper molars are nearly square in outline. M^1 is slightly larger than the posterior molars and M^3 is usually reduced posteriorly to differing degrees. In sicistines the lophs run nearly directly buccolingually connecting the corresponding buccal and lingual cusps. The mesocone is usually a distinct cusp, and a mesoloph is nearly always present. There are also well-defined anterior and posterior cingula. On the anterior cingulum of the first molar there is a distinct anterostyle at its center that is enlarged to varying degrees in different genera. The lingual cusps are connected together by an endoloph that also joins the mesocone. In zapodines, the anterior cingulum is continuous with the protocone and isolated from the mesocone, giving the lophs of the molars an oblique appearance (anterobuccal to posterolingual orientation). Any anteroposterior connections of the cusps (endoloph) are lost. In the Blancan to Recent *Zapus* the cusps are indistinguishable but the lophs of the cheek teeth are broad, filling the entire occlusal surface and divided only by extremely narrow valleys.

The lower molars are elongate (longer than wide). M_3 is reduced posteriorly to varying amounts depending on the genus. The first lower molar has a well-developed anteroconid and is narrower anteriorly (metalophid) than posteriorly (hypolophid). As in the upper molars, the sicistines and early zapodines had well-developed cusps on the lower molars. The lophs are transversely oriented with a central anteroposteriorly directed loph (ectolophid or central mure). There is a distinct mesoconid and associated mesolophid. The posterior cingulum is long and distinct from the hypolophid. In zapodines the lophs of the lower molars, as in the uppers, are oriented obliquely (posterolingually to anterobuccally) and the anteroposterior connections of the cusps are lost. In some zapodines the anteroconid is lost on M_1.

1.3. Skeleton

No postcranials have been reported for Tertiary zapodids. Skeletons of modern zapodids are small with gracile limbs and elongate tails. The hind-

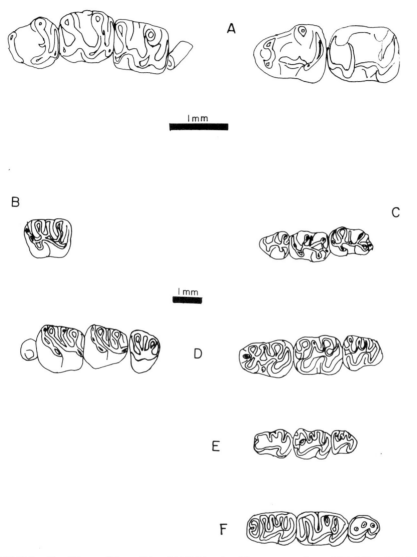

FIGURE 19.1. Dentitions of Zapodidae. (A) Bridgerian *Elymys complexus* RP^4–M^3 and LM$_1$–M$_2$ (composite). (B) Hemingfordian sicistine *Schaubeumys grangeri*, LM^1. (C) Barstovian sicistine *Schaubeumys cartomylos*, RM$_1$–M$_3$. (D) Barstovian zapodine *Megasminthus tiheni*, LP^4–M^3 (M^3 reversed) and LM$_1$–M$_3$. (E) Hemphillian zapodine *Pliozapus solus*, LM$_1$–M$_3$. (F) Blancan zapodine *Zapus rinkeri*, LM$_1$–M$_3$. A to separate scale (above). A from Emry and Korth (1989), B from Wood (1935b), C and D from Klingener (1966), E and F from Klingener (1963).

limb is enlarged relative to the forelimb for saltatorial locomotion. The fibula is always fused with the tibia.

2. Evolutionary Changes in the Family

Since the fossil record of zapodids is limited to dentitions, only changes in the teeth can be traced. It is not known whether the development of characteristics for saltatorial locomotion or hystricomorphy was present in the earliest members of the family or developed in later Tertiary genera.

Unlike many families of rodents, there was no gradual change in size through the Tertiary. Recent genera are very nearly the same size as the earliest species from the Duchesnean (Kelly, 1992). Any increase in crown height of the cheek teeth was also minimal. Zapodines have teeth that are mesodont (or submesodont) except for the earliest genus *Megasminthus* from the Barstovian which had teeth that were brachydont. Sicistines have been very conservative and modern *Sicista* has a dental morphology that varies little from that of Arikareean sicistines in crown height and occlusal pattern.

The only obvious change in the dentition was the increased lophodonty in zapodines. The earliest zapodine, *Megasminthus*, had cheek teeth that were cuspate with the lophs developed similar to those of contemporary sicistines. Hemphillian and later zapodines lost the definition of the cusps of their teeth, and the occlusal surface became filled with broader, obliquely oriented lophs.

3. Fossil Record

The earliest probable zapodid was *Elymys* from the early Bridgerian of Nevada (Emry and Korth, 1989). The dentition of *Elymys* was sciuravid-like in occlusal morphology, but the dental formula was distinctively zapodid. The next record of a zapodid in North America was *Simiacritomys* from the Duchesnean of southern California (Kelly, 1992). *Simiacritomys*, like *Elymys*, was not definitely referred to the Zapodidae, but the occlusal morphology of the cheek teeth of *Simiacritomys* more nearly approached that of early sicistines than did that of *Elymys*. By the early Arikareean, true zapodids appeared in North America (Fig. 19.2). The earliest Arikareean species of zapodids belonged to the sicistine genus *Plesiosminthus* (Macdonald, 1972; Martin, 1974). The time of greatest diversity of zapodids in North America was the Barstovian. During this time, all of the named species belonged to two genera of sicistines, *Schaubeumys* and *Macrognathomys*, and one genus of zapodine, *Megasminthus*. Geographically, this radiation of zapodids was limited to the Great Plains (Klingener, 1966; Storer, 1975; Green, 1977; Korth, 1979, 1987b). After the Clarendonian, the record of zapodids in North America was limited to the west coast and Great Basin (Hall, 1930; Wilson, 1936; Shotwell, 1956, 1970) until the Blancan where the first record of *Zapus* was

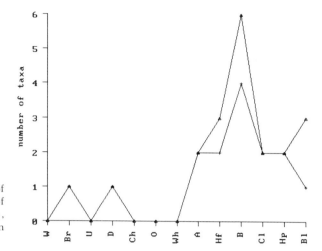

FIGURE 19.2. Occurrence of Zapodidae in the Tertiary of North America (▲, species; +, genera). Abbreviations as in Fig. 5.4.

from Kansas (Klingener, 1963). The last record of sicistines in the Tertiary of North America was the Hemphillian (Shotwell, 1970).

Zapodids were known from the Tertiary of both Europe and Asia but were never diverse. In Asia the first record of zapodids was in the late Eocene of China (Li and Ting, 1983). This material had been assigned to the sicistine genus *Plesiosminthus*. By the late Oligocene, several species of the sicistine *Parasminthus* were known (Bohlin, 1946). Species of *Parasminthus* disappeared after the early Miocene. There is no record of zapodids again in Asia until the Pleistocene (Kormos, 1930) when a species of the Recent *Sicista* appears. *Eozapus* is only known from the Holocene of Asia.

In Europe the Tertiary record of zapodids began with species of the sicistine *Plesiosminthus* which ranged from the middle to late Oligocene. The greatest diversity of zapodids was in the latest Miocene and earliest Pliocene (later than the North American radiation) where as many as four genera and seven species were known (Savage and Russell, 1983). The Recent and only surviving genus of sicistine, *Sicista*, appeared in Europe in the late Pliocene (Villafranchian) and was known from most of Europe and Asia in Pleistocene and Recent times (Nowack and Paradiso, 1983).

4. Phylogeny (Fig. 19.3)

4.1. Origin

The ancestor of zapodids (along with all other myodont rodents) is generally believed to have been among Eocene sciuravids from North America (see Chapter 6, Section 4.3), although some alternative suggestions have been made (Flynn et al., 1985, 1986). The earliest zapodid *Elymys*, although

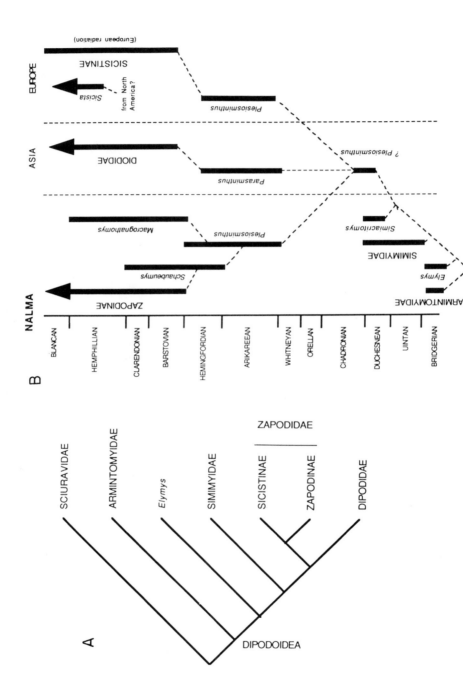

FIGURE 19.3. Phylogeny of the Zapodidae. (A) Cladogram of zapodid relationships. (B) Dendrogram of the Zapodidae with age of occurrence (see Fig. 5.5 for explanation).

questionably assigned to this family, was from the early Bridgerian and had cheek teeth that lacked a number of characteristic features of zapodids (mesolophs and mesolophids, anteroconid on M_1) and resembled sciuravids. *Elymys* was morphologically intermediate between sciuravids such as *Pauromys* that reduce the premolars and zapodids that lack at P_4 and have only a reduced P^4 (Wood, 1955). The zygomasseteric structure of the *Elymys* is not known but the distinctive hystricomorphy with an accessory nutritive foramen on the skull of zapodids (and all Dipodoidea) occurs as early as the Bridgerian and Uintan in the problematical genera *Armintomys* and *Simimys*, respectively (Chapter 22). It would not be surprising to discover that some lineage of sciuravid had already begun to develop the dipodoid hystricomorphy prior to the occurrence of *Elymys*.

4.2. Intrafamilial Relationships

Zapodines are believed to have evolved from early sicistines, notably *Plesiosminthus* (Klingener, 1963, 1966). The dental pattern of the first zapodine, *Megasminthus*, could have been derived easily from species of *Schaubeumys* that had already begun to enlarge the anterostyle on the upper molars and unite the paracone with the mesocone (rather than protocone).

Among sicistines, *Plesiosminthus* appears as the basal morphotype that could have given rise to the North American *Schaubeumys* and Asian *Parasminthus*. The Recent Eurasian *Sicista* may have been derived from the later Tertiary *Macrognathomys* from North America (Klingener, 1963; Green, 1977) which, in turn, may have been derived from the Hemingfordian *Plesiosminthus clivosus* (Shotwell, 1970).

Klingener (1966) noted that *Megasminthus* was a morphologically and chronologically ideal ancestor for *Zapus* and possibly the Hemphillian *Pliozapus*. However, a number of dental characters suggest that *Pliozapus* was not ancestral to *Zapus* but may have been more closely related to the Recent Eurasian *Eozapus* and the European Pliocene *Sminthozapus* (Sulimski, 1962, 1964; Klingener, 1966).

4.3. Extrafamilial Relationships

The family that is most closely related to the Zapodidae is the Dipodidae, the Old World jerboas. The similarity in the zygomasseteric structure, myology, dentition, and elongation of the hindlimbs has led a number of workers to include the zapodids as a subfamily of the Dipodidae (see Klingener, 1964, pp. 6–7, for review of familial classification). Nearly all authors have at least included the zapodids within the superfamily Dipodoidea.

It is likely that the earliest and most primitive dipodids from the late Miocene of Asia, *Paralactaga* and *Protalactaga* (Teilhard de Chardin and Young, 1931; Wood, 1936d; Unay, 1981; Zheng, 1982), were derived from early

Miocene species of *Parasminthus* (often referred to *Plesiosminthus* by many authors). These early dipodids had brachydont teeth and retained a P^4 as in zapodids. *Parasminthus* is distinguished from other sicistine zapodids by its more elongate upper molars that have two lingual roots (rather than one). These are features that were shared with the earliest true dipodids.

Two Eocene monotypic families of rodents from North America also appear to be referable to Dipodoidea, Simimyidae and Armintomyidae (Chapter 22). The Armintomyidae, limited to the type species *Armintomys tullbergi*, had a hystricomorphous skull, uniserial enamel of the incisors, and elongate cheek teeth as in other dipodoids (Dawson *et al.*, 1990). However, *Armintomys* retained the primitive upper dental formula (lower cheek teeth not known) and the occlusal morphology of the upper molars most closely approached that of sciuravids. *Armintomys* represented an early offshoot of dipodoid stock, not ancestral to any later dipodoids.

The Simimyidae, restricted to the genus *Simimys*, had sicistine-like cheek tooth morphology, a hystricomorphous zygomasseteric structure with an accessory nutritive foramen and reduced dental formula as in dipodoids. The upper premolar was lost in *Simimys*, leaving it with a muroid dental formula. This trait removed *Simimys* from the ancestry of later zapodids. The cheek tooth morphology of *Simimys* can be derived from that of the earlier *Elymys*, the presumed basal zapodid, and therefore was more closely related to the zapodids than was *Armintomys*.

One other problematical rodent from the Chadronian, *Nonomys*, had the muroid dental formula and the dipodoid zygomasseteric structure (Emry, 1981). This genus may have been related to the late Orellan *Diplolophus* (Chapter 23, Sections 1 and 6). The relationship of these two genera to dipodoids is uncertain, but again, may be part of an earlier Tertiary radiation.

Evidence from several fields of biology as well as paleontology suggests that the Dipodoidea are most closely related to the Muroidea (Wilson, 1949a; Wood, 1955; Klingener, 1964; Bugge, 1971; Luckett, 1985; Flynn *et al.*, 1985) and are often classified under the supergeneric grouping the Myodonta.

This dipodoid–muroid clade is most often associated with the Geomyoidea (or Geomorpha of some authors) and allocated to the suborder Myomorpha. A number of other families of rodents, fossil and Recent, have been at one time or another included in the Myomorpha (Chapter 4, Section 2). The Old World Gliridae (dormice) are myomorphous and have been questionably included in the Myomorpha along with the geomyoids and the myodonts by some authors (Wahlert, 1978; von Koenigswald, 1992) but excluded by others (Bugge, 1971; Lavocat and Parent, 1985).

5. Problematical Taxa: *Plesiosminthus* and *Schaubeumys*

The generic allocation of the small middle Tertiary sicistines from North America has long been debated. Many authors have used *Schaubeumys* Wood

(1935b), considered restricted to North America, while others have used the otherwise European genus *Plesiosminthus* Viret (1926). Korth (1980) has summarized the usage of these names in the literature and presented six diagnostic characters to separate these genera based on dentitions. He also followed Wilson (1960) by including only a single North American species *P. clivosus* in the European genus, and retaining all others in *Schaubeumys*.

Green (1992) described a large sample of "*Plesiosminthus*" *grangeri* from the Hemingfordian of South Dakota and concluded that *Plesiosminthus* and *Schaubeumys* were synonymous. However, of the six differences used to separate these genera (Wilson, 1960; Engesser, 1979; Korth, 1980) he only addressed one: the relative compression (elongation) of the lingual cusps on the first upper molar. Green (1992) measured the lingual cusps of the upper molars to detect any distinct compression. He found that in his sample the cusps were equally as wide as long although the mean measurements varied by 4 to 6% and the anteroposterior measurement was longer (as diagnosed for *Schaubeumys* by Wilson, 1960; Korth, 1980). None of the other morphologies used to distinguish these genera were addressed and no sample of any European species was measured for comparison. Even if Green (1992) were correct about the elongation of the lingual cusps of M^1, there still remains five morphological differences between *Schaubeumys* and *Plesiosminthus*. Therefore, these two genera should be recognized as such.

Green (1992) also stated that the Arikareean *Plesiosminthus geringensis* had the morphology of the lower cheek teeth as in *Schaubeumys* and upper teeth as in *Plesiosminthus*. He did not provide any criteria for this statement. In fact, the lower M_1 of *P. geringensis* (= *P. clivosus*, Korth, 1980) does match the diagnosis of *Plesiosminthus* (ectolophid obliquely connected to protoconid, mesoconid small). This indicates that there is one species of the otherwise European *Plesiosminthus* in North America.

6. Classification

Zapodidae Coues, 1875
  Sicistinae Allen, 1901
    Schaubeumys Wood, 1935 (A–Cl)
      S. grangeri Wood, 1935* (A–Hf)
      S. sabrae Black, 1958 (Hf)
      S. "sp. B" Green, 1977 (Cl)
      S. cartomylos Korth, 1987 (B)
    Plesiosminthus Viret, 1926 (A–Hf)
      P. clivosus Galbreath, 1953 (A–Hf)
    Macrognathomys Hall, 1930 (B–Hp)
      M. nanus Hall, 1930* (Cl–Hp)
      M. gemacollis Green, 1977 (B)
    Simiacritomys Kelly, 1992 (D)

        S. whistleri Kelly, 1992* (D)
Zapodinae Coues, 1875
    Megasminthus Klingener, 1966 (B)
        M. tiheni Klingener, 1966* (B)
        M. gladiofex Green, 1977 (B)
    Pliozapus Wilson, 1936 (Hp)
        P. solus Wilson, 1936* (Hp)
    Zapus Coues, 1875 (Bl–R)
        Z. rinkeri Hibbard, 1951 (Bl)
        Z. sandersi Hibbard, 1956 (Bl)
        Z. s. rexroadensis Klingener, 1963 (Bl)
Subfamily uncertain
    Elymys Emry and Korth, 1989 (Br)
        E. complexus Emry and Korth, 1989* (Br)

Chapter 20
Cricetidae—Part 1

1. Characteristic Morphology 223
   1.1. Skull 223
   1.2. Dentition 223
   1.3. Skeleton 225
2. Evolutionary Changes in the Family 225
3. Fossil Record 227
4. Phylogeny 228
   4.1. Origin 228
   4.2. Intrafamilial Relationships 228
   4.3. Extrafamilial Relationships 231
5. Problematical Taxa 232
   5.1. *Poamys* 233
   5.2. *?Copemys* 233
   5.3. *Peromyscus pliocenicus* 233
   5.4. *Pliotomodon* 233
6. Classification 234

1. Characteristic Morphology

1.1. Skull

Cricetids are small rodents. Because it is such a diverse family of rodents, many features of the skull are quite variable. The braincase is relatively inflated and the rostrum narrow (Fig. 20.1). The zygomasseteric structure is either hystricomorphous (infraorbital foramen enlarged for passage of the masseter medialis) or myomorphous (passage of masseter medialis through the infraorbital foramen and extension of the masseter lateralis onto the rostrum forming a tilted zygomatic plate). The incisive foramina vary in size from large to small depending on the subfamily. The zygoma is usually slender. The temporalis scar is limited to the lateral margins of the parietal bones except in eumyines where a distinct sagittal crest may be formed. The auditory bullae vary in size from small to greatly inflated.

The mandible is usually slender and the masseteric fossa ends anteriorly below the first molar. The coronoid process is posteriorly curved.

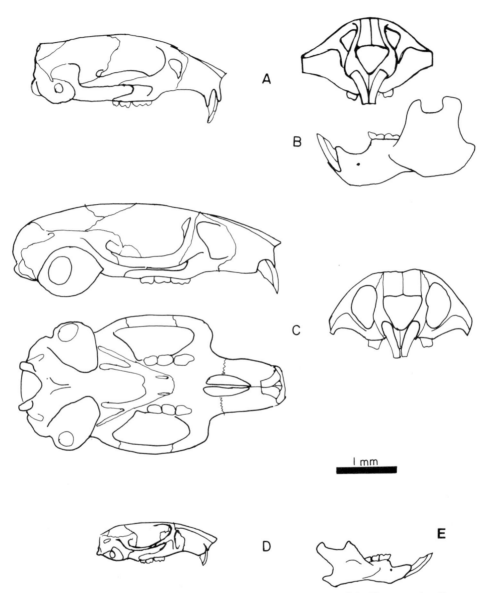

FIGURE 20.1. Skulls of Cricetidae. (A) Lateral and anterior views of Orellan eumyine *Eumys elegans*. (B) Same as A, lateral view of mandible. (C) Lateral, ventral, and anterior views of Arikareean eucricetodontine *Leidymys lockingtonianus*. (D) Lateral view of Barstovian peromyscine *Copemys longidens*. (E) Same as D, lateral view of mandible. A (lateral view) and B from Wood (1937), A (anterior view) and C from Martin (1980), D and E from Lindsay (1972).

1.2. Dentition

The dental formula for cricetids is primitive for all muroids, $\frac{1003}{1003}$. The incisors are narrow anteriorly and gently convex. The enamel surface of incisors of eucricetodontines were often ornamented by numerous ridges (Martin, 1980, Fig. 26). The primitive condition appears to have been numerous pinnate (radiating) ridges present on the earliest known cricetid, *Pappocricetodon* from China (Tong, 1992) as well as Oligocene North American species (Martin, 1980; Korth, 1981) and the genus *Paracricetodon* from the Oligocene of Europe (Hrubesch, 1957). However, Flynn et al. (1985) suggest that a smooth incisor was primitive. In later genera, the number of ridges on the incisors ranged from 1 or several to nearly 20, and they were aligned parallel to one another for the length of the incisor (Lindsay, 1978; Engesser, 1979; Martin, 1980; Martin and Corner, 1980). The ridging of the incisors is not present in all cricetids. The microstructure of the enamel of the incisors is uniserial.

The cheek teeth range from brachydont to nearly hypsodont in crown height (Fig. 20.2). Primitively, the first molar was the largest and the third molar the smallest, having been reduced posteriorly. The occlusal pattern of the molars is basically a system of cross-lophs [posterior and anterior cingula; mesoloph (-id); and major lophs] united by a single anteroposteriorly aligned loph (ectolophid on lowers, endoloph on uppers). The hypocone on the upper molars is subequal in size to that of the protocone.

M^1 is characterized by a large anterocone. It may be placed either buccally, anterior to the paracone, or more centrally. The relative size and position of the anterocone are variable among genera. M_1 similarly has an anteroconid that elongates the tooth. As with the anterocone on the upper first molar, the size and relative position of the anteroconid on M_1 are variable between genera. In sigmodontines the anterocone and anteroconid have split into two smaller cuspules, and accessory roots are added to the first molars. M_1 is narrower anteriorly, giving it an anteriorly tapered appearance. The second molar, both upper and lower, is usually rectangular in outline. The third molars taper posteriorly.

1.3. Skeleton

The skeleton of North American cricetids was relatively conservative in adaptations compared to Old World muroids (Fig. 20.3). The skeleton is basically gracile with slender limbs. The hindlimb is longer than the forelimb and the fibula is fused to the tibia. In some Old World muroids the skeleton can be specialized for saltatorial locomotion as in dipodoids and dipodomyine heteromyids whereas others are fossorially adapted as in geomyids, palaeocastorine beavers, and mylagaulids. The only variations among the North American cricetids occurred in eumyines and neotomyines that have slightly heavier skeletons.

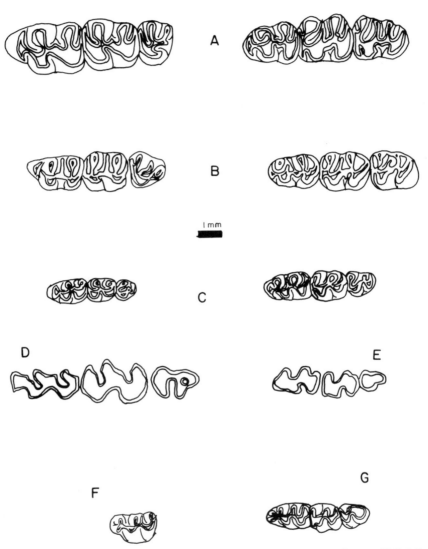

FIGURE 20.2. Dentitions of Cricetidae. (A) Orellan eumyine *Eumys elegans*, LM^1–M^3 and LM_1–M_3. (B) Arikareean eucricetodontine *Leidymys blacki*, RM^1–M^3 (reversed, composite) and LM_1–M_3. (C) Barstovian peromyscine *Copemys longidens*, LM^1–M^3 and LM_1–M_3. (D) Blancan neotomyine *Repomys maxumi*, LM^1–M^3 (composite). (E) Blancan neotomyine *Repomys panacaensis*, LM_1–M_3. (F) Clarendonian sigmodontine *Abelmoschomys simpsoni*, RM^1 (reversed). (G) Blancan sigmodontine *Calomys* (*Bensonomys*) *arizonae*, LM_1–M_3. A from Wood (1937), B from Martin (1980), C from Lindsay (1972), D and E from May (1981), F from Baskin (1986), G from Tomida (1987).

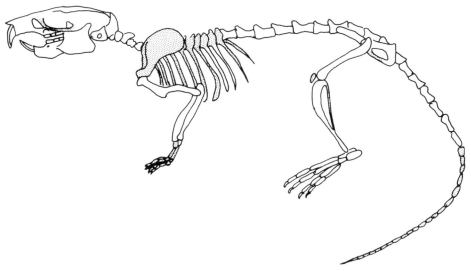

FIGURE 20.3. The skeleton of the Arikareean eucricetodont cricetid *Geringia mcgregori*. From Martin (1980). Stippled areas represent reconstructions.

2. Evolutionary Changes in the Family

Within the Cricetidae there were a number of advancements in the skull and dentition that appeared in some cases to have been attained independently in different subfamilies. The first of these changes was the development of myomorphy from hystricomorphy (Lindsay, 1977). The earliest cricetids (eucricetodontines) were hystricomorphous with only minimal development of the zygomatic plate. Later cricetids developed a zygomatic plate to varying degrees, limiting the infraorbital foramen ventrally and laterally. In North America the transition from a hystricomorphous eucricetodontine to a fully myomorphous eumyine can be traced (see Section 4.2).

Dentally, hypsodonty of the cheek teeth was attained several times. It is likely that the neotomyines were derived from brachydont to mesodont species of *Peromyscus* (May, 1981). Similarly, arvicolids which attain complete hypsodonty were derived from brachydont cricetids (Chapter 21, Section 4.1). Increased lophodonty, as in other rodent families, appears associated with increased crown height of the cheek teeth.

The alternation of the major cusps on the cheek teeth also appears to have developed. In primitive cricetids the major cusps of both upper and lower cheek teeth were directly buccolingually aligned. In North America, beginning with the Barstovian (and possibly Hemingfordian) peromyscine *Copemys*, the major cusps showed the beginnings of alternation, the lingual cusps being anterior to the buccal cusps on lower molars and the buccal cusps anterior to the lingual cusps on upper molars.

3. Fossil Record (Fig. 20.4)

The earliest record of Cricetidae in North America was a single upper molar from the Duchesnean of Saskatchewan referred to the eumyine genus *Eumys* (Storer, 1988). By Chadronian times cricetids are reported from Montana as well as Saskatchewan (Russell, 1972; Sutton and Black, 1975).

The first radiation of cricetids in North America was that of the eumyines (Fig. 20.4) that attained their greatest diversity in the Orellan (seven species). In the Arikareean the eucricetodontines (predominantly the genus *Leidymys*) diversified, but are extinct by the Barstovian.

By far the greatest number of cricetid species occurred in the North American Tertiary during the Blancan. This radiation included peromyscines, neotomyines, sigmodontines, and "microtines." The peromyscines first appeared in the Barstovian with *Copemys* which was diverse at that time (see Lindsay, 1972) and was replaced by *Peromyscus* and other genera beginning in the Hemphillian. The first neotomyines appeared in the Hemphillian (Hibbard, 1967; May, 1981). The earliest sigmodontine reported was *Abelmoschomys* from the Clarendonian of Florida. By Hemphillian times sigmodontines became very diverse.

In Asia the earliest cricetids were known and referred to a single genus, *Pappocricetodon*, from the middle to late Eocene of China (Tong, 1992). The earliest record in Europe was in the earliest Oligocene (Dawson, 1977). In Africa the first record of cricetids was in the early Miocene (Lavocat, 1978). Cricetids invaded South America after its junction with North America in the Pliocene (Reig, 1978).

The fossil record of cricetids in North America involves a number of immigration events. The first was from Asia in the late Eocene which accounted for the first occurrence of cricetids in North America. The second appears to have been in the Miocene (?Hemingfordian) with the first occurrence of peromyscines (Slaughter and Ubelaker, 1984, p. 262). After this, the number of migration events are dependent on the interpretation of the origin of different cricetid groups. Jacobs and Lindsay (1984) identified three later migrations of cricetids from Eurasia, whereas Baskin (1986) viewed the origin of these supposed immigrant groups from within North American endemic cricetids.

4. Phylogeny (Fig. 20.5)

4.1. Origin

The strongest arguments for the origin of cricetids still appear to be from the Eocene sciuravids (Chapter 6, Section 4.3) although there are some

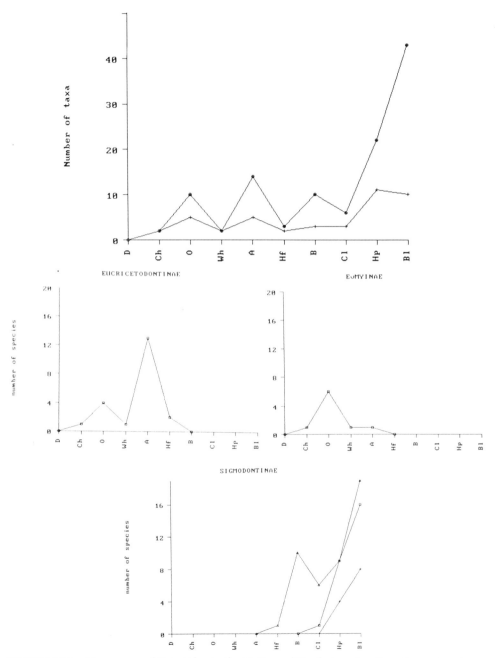

FIGURE 20.4. Occurrence of Cricetidae in the Tertiary of North America, (top: ●, species; +, genera; bottom: ▲, Peromyscini; +, Neotomyini; □, Sigmodontini). Abbreviations as in Fig. 5.4.

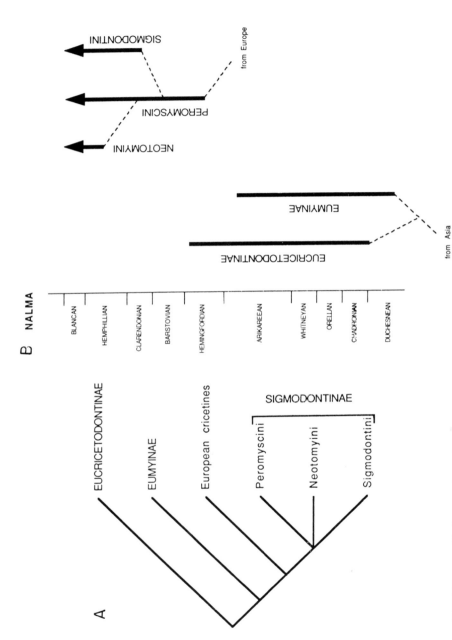

FIGURE 20.5. Phylogeny of North American Cricetidae. (A) Cladogram of cricetid relationships. (B) Dendrogram of the Cricetidae with age of occurrence (see Fig. 5.5 for explanation).

arguments against this theory. The earliest cricetids were from China but there are currently no definite sciuravids known from Asia that could have been ancestral (Dawson, 1977). The sciuravid with the most cricetid-like dentition was the North American Bridgerian *Pauromys*. The derivation of Asian cricetids from a North American sciuravid would have involved the migration of a stem cricetid-like sciuravid to Asia in the middle Eocene, and then a return migration in the latest Eocene.

4.2. Intrafamilial Relationships

The earliest and most primitive cricetids appear to have been the Eucricetodontinae. They retained the primitive hystricomorphous zygomasseteric structure (Lindsay, 1978) and brachydont cheek teeth. The earliest species had poorly developed anterocones (-conids) on the first molars and had less lophate cheek teeth than later cricetids (Tong, 1992). In North America, genera referred to the Eucricetodontinae (see Martin, 1980) gave rise to the Eumyinae. The Orellan *Eoeumys* (Martin, 1980), while stratigraphically later than the first occurrence of *Eumys* (Wood, 1980), was morphologically intermediate between eucricetodontines and eumyines. The zygomatic plate was not fully developed but shows some anterodorsal extension. Eucricetodontines were hystricomorphous while eumyines were fully myomorphous. Dentally, eumyines had more robust cheek teeth that were slightly higher crowned than eucricetodontines. *Eoeumys* was also intermediate between these two subfamilies (Korth, 1981). Thus, it appears that the exclusively North American eumyines were derived from North American members of the Eucricetodontinae. Jacobs and Lindsay (1984) argued that the North American species referred to eucricetodontines were derived from eumyines and were not eucricetodontines at all. However, this interpretation involves several evolutionary reversals in the skull (hystricomorphy to myomorphy and back to hystricomorphy) and dentition (brachydonty to mesodonty and back to brachydonty; ornamented incisors to smooth incisors back to ornamented incisors) that make such a scenario less likely.

Besides giving rise to the eumyines, the North American eucricetodonts underent a radiation in the Arikareean independent of any other cricetid groups.

Some authors have suggested that eumyines were ancestral to the later North American peromyscines (Slaughter and Ubelaker, 1984). However, the more robust skull (with a derived sagittal crest) and mandible of eumyines as well as the higher-crowned and more robust cheek teeth made them far too specialized to have been ancestral to any later cricetid.

It is generally accepted that the peromyscines immigrated to North America from Eurasia (Lindsay, 1972; Martin, 1980; Jacobs and Lindsay, 1984). However, some authors (Clark *et al.*, 1964; Engesser, 1979) suggested that the earliest North American peromyscine *Copemys* was derived from early Miocene North American eucricetodont species (notably *Leidymys*). The derivation of peromyscines from *Leidymys* was unlikely because of the relatively primitive morphology of the skull and cheek teeth of the latter and the much more specialized morphology of *Copemys* with no known intermediate forms (Martin, 1980).

The derivation of the later Tertiary peromyscines from *Copemys* is undisputed. However, the origin of the neotomyines and sigmodontines is still in question. May (1981) derived neotomyines from the Hemphillian peromyscine *Peromyscus pliocenicus* (Wilson, 1937a) which had already developed the beginnings of increased hypsodonty. On the other hand, Jacobs and Lindsay (1984) believe that neotomyines were derived from immigrants belonging to the Eurasian myospalacine cricetids.

Similarly, the neotropical sigmodontines can either be derived from North American descendants of *Copemys* (Baskin, 1986) or again represent a later Tertiary invasion of megacricetodontines from the Old World (Jacobs and Lindsay, 1984; Slaughter and Ubelaker, 1984). There is evidence in the fossil record for both of these scenarios. Studies of Recent cricetids based on parasites, soft tissue structures (glans penis), physiology, serum albumin, and other molecular characteristics have been used to support both sides of these arguments (see Baskin, 1986, for review of literature).

The classification below (Section 6) follows the hypothesis of a North American origin for both neotomyines and sigmodontines. This is the most parsimonious of the explanations and forms intermediate in morphology between the ancestral peromyscines and later neotomyines and sigmodontines are represented in the North American fossil record (May, 1981; Baskin, 1986).

4.3. Extrafamilial Relationships

Several Old World families of muroids (Muridae, Rhizomyidae, Spalacidae) all appeared during the Miocene in Asia and can be derived from cricetids (summarized in Flynn et al., 1985). These families (included as subfamilies of Muridae by Carleton and Musser, 1984) do not migrate to North America and their radiations were limited to the Old World.

The only other rodents descended from cricetids were microtines (= Arvicolidae or Microtidae of several authors). Martin (1975) suggested that the primitive "microtine" dental pattern could have been derived directly from *Copemys*. Repenning (1987), who viewed microtines as a paraphyletic group of rodents with microtine grade dental adaptations, concurred that these rodents were derived from a cricetine ancestor (perhaps in several different lineages) but noted that the continent of origin was uncertain because the first occurrence of "microtines" was nearly simultaneous in both North America and Asia (Chapter 21, Section 4.1).

5. Problematical Taxa

5.1. *Poamys*

Matthew (1924) named a new genus and species of cricetid *Poamys rivicola* from the Barstovian Lower Snake Creek beds of Nebraska based on a partial mandible with a worn M_2. Since that time the holotype has been lost from the collections of the American Museum of Natural History. No additional material referable to this species has been reported from these beds.

The morphology of the cheek tooth figured by Matthew (1924, Fig. 10) is very similar to that of the common Barstovian cricetid *Copemys* (Wood, 1936e). It is possible, and quite likely, that *Poamys* and *Copemys* are synonymous. In this case the generic name *Poamys* would have priority. However, because the holotype of *P. rivicola* was lost, and no comparison can be made with the abundant material of *Copemys*, these genera are left as distinct. The only possibility of synonymy of these genera would depend on the recovery of topotypic material of *P. rivicola* from Nebraska that would allow for such comparison.

5.2. ?*Copemys*

Tertiary species of *Copemys* were distinguished from those of *Peromyscus* by the degree of alternation of the cusps on the cheek teeth. In *Peromyscus*, the posterior arm of the protoconid is continuous with the entoconid on the lower molars and the posterior arm of the paracone is continuous with the hypocone on the upper molars; in *Copemys* these oblique connections of the molars were not complete (Lindsay, 1972).

In two species from the Hemphillian the alternation of the cusps was complete on the lower molars but not the upper molars [*C. valensis* (Shotwell, 1967b), *C. vasquezi* (Jacobs, 1977a)]. These species were clearly intermediate in morphology and time of occurrence between the earlier *Copemys* and later *Peromyscus*. Both species are listed in the classification (Section 6) as being referred to ?*Copemys* but could justifiably be included in *Peromyscus*.

5.3. *Peromyscus pliocenicus*

Wilson (1937a) named *Peromyscus pliocenicus* from the Hemphillian of California noting its larger size and greater crown height of the cheek teeth. Additional specimens of this species were later described from Oregon (Shotwell, 1967b) and California (May, 1981). May (1981) demonstrated that *P. pliocenicus* was ancestral to the later neotomyine *Repomys*.

The crown height of this species was greater than any other species of *Peromyscus*, and closely approached that of the earliest microtines. *P. pliocenicus* probably should be referred to a different genus than *Peromyscus*, representing the basal neotomyine. Here, *P. pliocenicus* is retained in *Peromyscus* with question.

5.4. *Pliotomodon*

Hoffmeister (1945) described *Pliotomodon* from the Hemphillian of California and compared it most closely with the Recent neotropical neotomyine *Neotomodon*. However, Jacobs (1977a) noted that the rodent that was most similar to *Pliotomodon* was the European cricetine *Ruscinomys hellenicus* (Freudenthal, 1970) and suggested that *Pliotomodon* was an immigrant taxon. This interpretation has been echoed by later workers (May, 1981; Jacobs and Lindsay, 1984).

Clearly, *Pliotomodon* was not part of any North American radiation of cricetids. The event of its immigration from Eurasia in the Hemphillian verifies a third immigration event in the history of cricetids in North America regardless of the interpretation of the origin of neotomyines and sigmodontines (see Section 4.2).

6. Classification

Cricetidae Rochebrune, 1883 (excluding "Microtinae")
  Eumyinae Stehlin and Schaub, 1951
    Eoeumys Martin, 1980 (O)
      E. vetus (Wood, 1937)* (O)
    Eumys Leidy, 1856 (Ch–A)
      E. elegans Leidy, 1856* (O)
      E. brachyodus Wood, 1937 (Wh–A)
      E. parvidens Wood, 1937 (O)
      E. cricetodontoides White, 1952 (O)
      E. pristinus Russell, 1972 (Ch)
    Coloradoeumys Martin, 1980 (O)
      C. galbreathi Martin, 1980* (O)
    Wilsoneumys Martin, 1980 (O)
      W. planidens (Wilson, 1949)* (O)
  Eucricetodontinae Mein and Freudenthal, 1971
    Paciculus Cope, 1879 (A)
      P. insolitus Cope, 1879* (A)
      P. montanus Black, 1961 (A)
      P. woodi (Macdonald, 1963) (A)
      P. nebraskensis Alker, 1969 (A)

Leidymys Wood, 1936 (A)
    L. nematodon (Cope, 1879)* (A)
    L. lockingtonianus (Cope, 1881) (A)
    L. parvus (Sinclair, 1905) (A)
    L. alicae (Black, 1961) (A)
    L. blacki (Macdonald, 1963) (A)
    L. cerasus Korth, 1992 (A)
Scottimus Wood, 1937 (?Ch–A)
    S. lophatus Wood, 1937* (Wh)
    S. exiguus (Wood, 1937) (O)
    S. kellamorum Black, 1961 (A)
    S. viduus Korth, 1981 (O)
    S. ambiguus Korth, 1981 (?Ch–O)
    S. longiquus Korth, 1981 (O)
Geringia Martin, 1980 (A)
    G. mcgregori (Macdonald, 1970)* (A)
    G. gloveri (Macdonald, 1970) (A)
Yatkolamys Martin and Corner, 1980 (Hf)
    Y. edwardsi Martin and Corner, 1980* (Hf)
Sigmodontinae Wagner, 1843
  Peromyscini Hershkovitz (1966)
    Copemys Wood, 1936 (Hf–Hp)
      C. loxodon (Cope, 1874)* (B)
      C. longidens (Hall, 1930) (B)
      C. dentalis (Hall, 1930) (?B–Cl)
      C. niobrarensis (Hoffmeister, 1959) (B)
      C. esmeraldensis Clark, Dawson, and Wood, 1963 (Cl)
      C. russelli (James, 1963) (B–Cl)
      C. pagei (Shotwell, 1967) (B)
      ?*C. valensis* (Shotwell, 1967) (Hp)
      C. pisinnus (Wilson, 1968) (Cl)
      C. tenuis Lindsay, 1972 (B)
      C. barstowensis Lindsay, 1972 (B)
      ?*C. vasquezi* Jacobs, 1977 (Hp)
    Poamys Matthew, 1924 (B)
      P. rivicola Matthew, 1924* (B)
    Tregomys Wilson, 1960 (B–Cl)
      T. shotwelli Wilson, 1960* (Cl)
    Peromyscus Gloger, 1841 (Hp–R)
      P. antiquus Kellogg, 1910 (Hp)
      ?*P. pliocenicus* Wilson, 1937 (Hp)
      P. kansasensis Hibbard, 1941 (Bl)
      P. cragini Hibbard, 1944 (Bl)
      P. baumgartneri Hibbard, 1954 (Bl)
      P. hagermanensis Hibbard, 1962 (Bl)

P. beckensis Dalquest, 1978 (Bl)
  P. nosher Gustafson, 1978 (Bl)
 Reithrodontomys Giglioli, 1873 (Bl–R)
  R. pratincola Hibbard, 1941 (Bl)
  R. rexroadensis Hibbard, 1952 (Bl)
  R. wetmorei Hibbard, 1952 (Bl)
  R. galushai Tomita, 1989 (Bl)
 Paronychomys Jacobs, 1977 (Hp)
  P. lemredfieldi Jacobs, 1977* (Hp)
  P. tuttlei Jacobs, 1977 (Hp)
  P. alticuspis Baskin, 1979 (Hp)
 Onychomys Baird, 1857 (Hp–R)
  O. pedroensis (Gidley, 1922) (Bl–Pl)
  O. bensoni (Gidley, 1922) (Bl)
  O. martini (Hibbard, 1937) (Hp)
  O. gidleyi (Hibbard, 1941) (Bl)
  O. fossilis (Hibbard, 1941) (Bl)
  O. larrabeei (Hibbard, 1953) (Hp)
  O. hollisteri Carleton and Eshelman, 1979 (Bl)
Neotomini Merriam, 1894
 Repomys May, 1981 (Hp–Bl)
  R. gustelyi May, 1981* (Hp)
  R. maxumi May, 1981 (Bl)
  R. panacaensis May, 1981 (Bl)
  R. arizonensis Tomita, 1989 (Bl).
 Galushamys Jacobs, 1977 (Hp)
  G. redingtonensis Jacobs, 1977* (Hp)
 Neotoma Say and Ord, 1825
  N. (Paraneotoma) Hibbard, 1967 (Hp–Bl)
   N. (P.) fossilis Gidley, 1922 (Bl)
   N. (P.) sawrockensis Hibbard, 1967 (Bl)
   N. (P.) quadriplicatus (Hibbard, 1941) (Bl)
   N. (P.) taylori Hibbard, 1967 (Bl)
   N. (P.) minutus Dalquest, 1983 (Hp)
   N. (P.) leucopetrica Zakrzewski, 1991 (Bl)
Sigmodontini Wagner, 1843
 Abelmoschomys Baskin, 1986 (Cl)
  A. simpsoni Baskin, 1986* (Cl)
 Calomys Waterhouse, 1837 (Hp–R)
  C. (Bensonomys) Gazin, 1942 (Hp–Bl)
   C. (B.) arizonae (Gidley, 1922)* (Bl)
   C. (B.) eliasi (Hibbard, 1938) (Bl)
   C. (B.) stirtoni (Hibbard, 1953) (Hp)
   C. (B.) meadensis Hibbard, 1956 (Bl)
   C. (B.) yazhi Baskin, 1978 (Hp)

        C. (B.) *gidleyi* Baskin, 1978 (Hp)
        C. (B.) *coffeyi* Dalquest, 1983 (Hp)
        C. (B.) *elachys* Lindsay and Jacobs, 1985 (Hp–Bl)
        C. (B.) *baskini* Lindsay and Jacobs, 1985 (Hp–Bl)
    Baiomys True, 1894 (Hp–R)
        B. *brachygnathus* (Gidley, 1922) (Bl)
        B. *minimus* (Gidley, 1922) (Bl)
        B. *rexroadi* Hibbard, 1941 (Bl)
        B. *kolbi* Hibbard, 1952 (Hp–Bl)
        B. *sawrockensis* Hibbard, 1953 (Bl)
        B. *aquilonius* Zakrzewski, 1969 (Bl)
    Prosigmodon Jacobs and Lindsay, 1981 (Hp–Bl)
        P. *oroscoi* Jacobs and Lindsay, 1981* (Hp–Bl)
        P. *chihuahuensis* Lindsay and Jacobs, 1985 (Hp–Bl)
    Symmetrodontomys Hibbard, 1941 (Bl)
        S. *simplicidens* Hibbard, 1941* (Bl)
    Sigmodon Say and Ord, 1825 (Bl–R)
        S. *medius medius* Gidley, 1922 (Bl)
        S. *medius hibbardi* Martin, 1979 (Bl)
    Oryzomys Baird, 1857 (Hp–R)
        O. *pliocaenicus* Hibbard, 1939 (Hp)
Subfamily uncertain
    Pliotomodon Hoffmeister, 1945 (Hp)
        P. *primitivus* Hoffmeister, 1945* (Hp)

Chapter 21
Cricetidae—Part 2 (Microtidae)

1. Characteristic Morphology	239
1.1. Skull	239
1.2. Dentition	240
1.3. Skeleton	241
2. Evolutionary Changes in the Family	241
3. Fossil Record	241
4. Phylogeny	243
4.1. Origin	243
4.2. Intrafamilial Relationships	243
4.3. Extrafamilial Relationships	244
5. Problematical Taxon: *Propliophenacomys*	245
6. Classification	245

1. Characteristic Morphology

1.1. Skull

Microtines are small to intermediate in size. The skull follows the general pattern of cricetids (and all muroids) and is fully myomorphous, the mandible sciurognathous (Fig. 21.1). The rostrum is usually shorter than in other cricetids and the skull is more robust. A postorbital process or ridge may be present on the squamosal. The zygoma is broad. A marking for the *temporalis* muscle is present on the skull on the lateral sides of the parietals but fuses into a central sagittal crest in some genera. The size of the infraorbital and incisive foramina is variable depending on genus as in other cricetids. Auditory bullae are variable in size and range from aseptate and thin-walled to septate and thick-walled, the walls being filled with spongy bone.

The mandible is similar to other muroids but modified for probalinal mastication (Repenning, 1968). There is a groove that is posterior and parallel to the dorsal edge of the masseteric fossa for the insertion of the internal masseter ("arvicoline groove"). The anterior edge of the ascending ramus originates below M_1 and rises steeply obscuring all or part of M_2 from lateral view. The area for the insertion of the diagastric muscle on the ventral border of the mandible ("chin process") is better developed than in other cricetids. The angular process is laterally deflected. The masseteric fossa is deep, anteriorly V-shaped, and ends below M_1 or anterior to it.

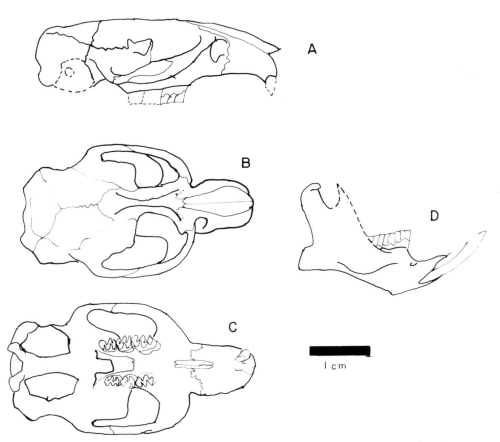

FIGURE 21.1. Skull of Blancan ondatrine *Pliopotamys minor*. (A) lateral view; (B) dorsal view; (C) ventral view. (D) lateral view of mandible. A–C from Zakrzewski (1984), D from Zakrzewski (1969).

1.2. Dentition

The dental formula of microtines is that of all muroids, consisting only of a single incisor and three molars in each quadrant. The incisors are broader anteriorly and usually asulcate with a faint groove in some species of Recent *Synaptomys* and *Micotus* (Carleton and Musser, 1984). The microstructure of the incisor enamel is uniserial as in all other muroids. In lemmings (Lemminae and Dicrostonychinae) the incisor is short (ending below the last molar) and runs lingual to the tooth row within the mandible.

The cheek teeth of all microtines are hypsodont (lateral sides of cheek teeth vertically oriented) and can be rooted or rootless. Dentine tracts occur in most genera that have greater crown height. Also in higher-crowned genera, cement is deposited between the apices of the triangles on the lateral sides of the teeth. The dentine tracts and cement both serve to anchor the teeth to the bone. Especially in rootless genera, the periodontal (Shapely) fibers which attach the teeth to the bone cannot attach to enamel, but can fuse with dentine and cement. Hence, the development of these features is increased with increased crown height of the molars. The occlusal pattern of the cheek teeth of microtines consists of a series of triangles, generally alternating, that are outlined with enamel and have exposed dentine within (Fig. 21.2). The occlusal surface of the cheek teeth is flat because of propalinal chewing that eliminates the use of any topographic features (cusps).

1.3. Skeleton

The skeleton of microtines is also similar to that of cricetids but is slightly more robust with shorter limbs. The tail is usually very short. The fibula is always fused to the tibia.

2. Evolutionary Changes in the Family

The evolutionary changes in microtines are most evident in the dentition. The greatest change within each lineage is in the occlusal pattern (number and arrangement of triangles) of the cheek teeth, notably M_1 and M^3. With increased height the cheek teeth progressively lost the roots, developed more pronounced enamel failure (dentine tracts), and had greater amounts of cement. This increase in hypsodonty occurred in several lineages (Hibbard, 1964; Guthrie, 1971; Martin, 1975; Repenning et al., 1990).

3. Fossil Record

The first record of microtines in North America (and the world) was from the early Hemphillian where two genera were known, *Goniodontomys* and *Microtoscoptes* (Wilson, 1937a; Martin, 1975). The former was only known west of the Rocky Mountains, while *Microtoscoptes* was present in faunas on both sides of the Rocky Mountains (Repenning et al., 1990). Both of these genera belonged to the subfamily Prometheomyinae (Repenning, 1987) and retained the primitive characters of microtines (rooted cheek teeth, no cement or enamel failure) and have triangles that are opposing, not alternating on the occlusal surface of the molars.

By the late Hemphillian the first arvicoline, *Promimomys*, appeared in both North America and Eurasia and represented the first worldwide dispersal event of microtines (Repenning, 1987; Repenning et al., 1990). A second genus appeared in North America, "*Propliophenacomys*" (see Section 5).

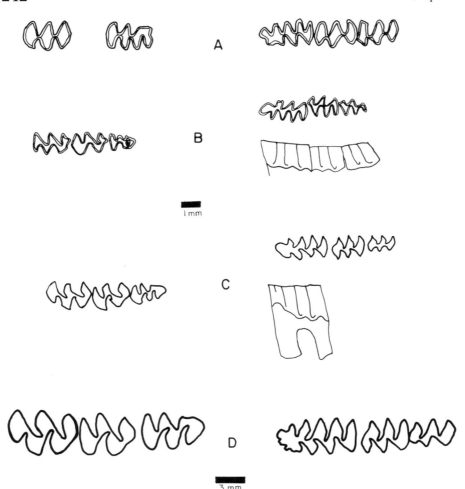

FIGURE 21.2. Dentitions of microtines. (A) Early Hemphillian prometheomyine *Goniodontomys disjunctus*, LM^1 and RM^3 (reversed) and LM_1–M_3 (first two molars reversed). (B) Late Hemphillian prometheomyine "*Propliophenacomys*" *parkeri*, RM^1–M^3 (composite, reversed) and LM_1–M_3 with lateral view. (C) Blancan arvicoline *Mimomys (Ophiomys) taylori*, LM^1–M^3 and LM_1–M_3 with lateral view of M_1 (reversed). (D) Blancan ondatrine *Pliopotamys minor*, LM^1–M^3 and RM_1–M_3 (reversed). A from Hibbard (1970b), B from Voorhies (1990b), C and D from Zakrzewski (1969). A–C to same scale (above), D to separate scale (below).

This genus evolved in North America from an earlier species of *Microtoscoptes* (Martin, 1975).

During Blancan times the microtines rapidly diversified in North America and Eurasia, and an additional six dispersal events introduced new taxa into North American faunas from Eurasia or from more northern parts of North America (Repenning, 1987; Repenning *et al.*, 1990). The number of microtines described from the Blancan of North America is greater than 30, far greater than the 4 species represented in the Hemphillian. This radiation of microtines and subsequent dispersal events continued into the Pleistocene (Repenning, 1992). By the end of the Blancan the primitive prometheomyines had diminished in number (becoming extinct in the Irvingtonian in North America) and more advanced ondatrines, lemmines, and arvicolines that were ancestral to the modern microtines and pitymyines became the dominant components of the microtine fauna (Repenning et al., 1990).

4. Phylogeny

4.1. Origin

In the past, microtines have been considered as a distinct family of muroids (= Microtidae or Arvicolidae) or as a subfamily of the Cricetidae or Muridae (Simpson, 1945; Chaline et al., 1977; Carleton and Musser, 1984). More recently, microtines have been considered a polyphyletic group that consisted of five distinct lineages derived from different brachydont cricetid ancestors, and are part of the family Cricetidae (Repenning, 1987; Repenning et al., 1990). Repenning (1987) divided these lineages into five subfamilies, one that originated in North America and four that originated in Eurasia.

Martin (1975) considered a generalized *Copemys*-type cricetid as the ancestor of the North American microtine radiation. The oldest ondatrine was *Ischymomys* that occurred in the latest Miocene of Russia (Topachevsky et al., 1978). The possible cricetid ancestors of the Arvicolinae were known from Eurasia and may be contemporaneous with the first occurrence of that subfamily, *Promimomys* (Repenning, 1987). The ancestors of the remaining two Eurasian lineages, Lemminae and Dicrostonychinae, cannot be determined with any certainty. The emergence of these latter subfamilies was subsequent to those of the other lineages (Repenning, 1987).

4.2. Intrafamilial Relationships

Some authors have argued that the evolution of the North American microtines was endemic with a limited number of immigration events from Eurasia (Martin, 1975, 1979). However, recent work involving both the European and North American microtine faunas has indicated that there were as many as eight migration events into North America during the Tertiary and Pleistocene (Repenning, 1980, 1984, 1987; Repenning et al., 1990). The amount of evolution of microtines that occurred in North America appears limited and usually at the species level. Within each of the four subfamilies of microtids present in the Tertiary of North America there were recognizable lineages.

The earliest lineage (subfamily) of microtines in North America were the

Prometheomyinae which may have evolved from North American cricetids. Within this group some evolution on the generic level is recognizable. Although contemporaneous, it is likely that early Hemphillian *Goniodontomys* was ancestral to *Microtoscoptes* which, in turn, gave rise to the late Hemphillian "*Propliophenacomys*" (Repenning et al., 1990). "*Propliophenacomys*" has also been suggested as the ancestor for two Blancan prometheomyines, *Pliophenacomys* and *Pliolemmus* (Martin, 1975; Repenning et al., 1990). Zakrzewski (1984) considered *Pliophenacomys* as the likely ancestor of *Guildayomys*. Hibbard and Dalquest (1973) also suggested that *Pliophenacomys* was ancestral to the ondatrine *Proneofiber*, but Martin (1979) argued that *Pliophenacomys* was too specialized to have been ancestral to *Proneofiber*.

Little evolution of the Arvicolinae above the species level occurred in North America. Martin (1975, 1979) believed that the Hemphillian *Promimomys* (the earliest arvicoline) was directly ancestral to a later subgenus of *Mimomys*, *Ogmodontomys*, which gave rise to two other North American subgenera of the genus, *Cosomys* and *Ophiomys*. Hibbard and Zakrzewski (1967) also believed that the latter two taxa were derived from *Ogmodontomys*. However, both *M. (Cosomys)* and *M. (Ophiomys)* were immigrants into North America, *Mimomys* having evolved in Eurasia, and *M. (Ogmodontomys)* was more likely derived from *M. (Cosomys)* (Repenning, 1987; Repenning et al., 1990). The only other suggested ancestral relationship among North American microtines above the species level is that of *Promimomys* being ancestral to the Blancan *Nebraskomys* (Hibbard, 1970a; Martin, 1975) and *Nebraskomys* as ancestral to the Pleistocene genus *Atopamys* (Hibbard, 1972; Martin, 1975).

Although the Ondatrinae in North America were derived from Eurasian ancestors, *Pliopotamys* certainly gave rise to *Ondatra* during Blancan times in North America (Hibbard and Zakrzewski, 1967; Eshelman, 1975; Martin, 1979).

The only branch of the Lemminae from the Tertiary of North America is the Synaptomyini (bog lemmings). The earliest synaptomyine, *Plioctomys*, first occurred in Europe (Suchov, 1976) but immigrated to North America during the Blancan where it was the base of a lineage of synaptomyines that evolved through *Mictomys* and ultimately to *Synaptomys* in the Pleistocene (Repenning and Grady, 1988).

4.3. Extrafamilial Relationships

Microtines are clearly related to other muroids because of their dental formula, myomorphous zygomasseteric structure, uniserial incisor enamel, fused tibia and fibula, and many soft tissue structures (Carleton and Musser, 1984). Microtines, whether considered as a monophyletic group of rodents or as several separate lineages that attain a microtine grade of dentition, were definitely derived from cricetid ancestors that had brachydont cheek teeth (Martin, 1975; Repenning, 1987).

No other groups of muroids can be derived from microtines, as they are clearly too specialized.

5. Problematical Taxon: *Propliophenacomys*

Martin (1975) erected the genus Propliophenacomys and included two species, P. uptegrovensis (type species) and P. parkeri, both from the Hemphillian of Nebraska. Later, Voorhies (1984) demonstrated that the type and only specimen of P. uptegrovensis was a Recent intrusive into the fauna and was actually a specimen of the Recent microtine Phenacomys intermedius. Hence, the genus Propliophenacomys was placed in synonymy with Phenacomys. However, the second species described by Martin (1975) had dental characters distinct from Phenacomys and did indeed represent a distinct genus of prometheomyine (Voorhies, 1984)

More recently, Voorhies (1990b) listed Martin's second species as ?Cseria parkeri. The genus Cseria is otherwise European and included as a subgenus of the arvicoline Mimomys (Repenning and Grady, 1988). The occurrence of Mimomys (Cseria) in Europe was also younger than that of parkeri occurring in the late Ruscinian (equivalent to Blancan II). If parkeri were considered a prometheomyine, derived from and ancestral to later members of that subfamily (Martin, 1975; Repenning et al., 1990), it is not likely that it is referable to an arvicoline such as Cseria.

"Propliophenacomys" parkeri represented a distinct microtine that was intermediate between Microtoscoptes and Pliophenacomys. Only the formal naming of a new genus to replace Propliophenacomys is necessary to resolve the difficulties of using this generic assignment.

6. Classification

Cricetidae Rochebrune, 1883
  Lemminae Gray, 1825
    Synaptomyini von Koenigswald and Martin, 1984
      Plioctomys Suchov, 1976
        P. rinkeri Hibbard, 1956 (Bl)
        P. mimomiformis Suchov, 1976* (Bl)
      Mictomys True, 1894
        M. vetus (Wilson, 1933) (Bl)
        M. landesi (Hibbard, 1941) (Bl)
  Prometheomyinae Kretzoi, 1955
    Ellobiini Gill, 1872
      Microtoscoptes Schaub, 1934 (Hp)

        M. hibbardi (Martin, 1975) (Hp)
     Goniodontomys Wilson, 1937 (Hp)
        G. disjunctus Wilson, 1937* (Hp)
   Pliophenacomyini Repenning, Fejfar, and Heinrich, 1990
     Pliophenacomys Hibbard, 1938 (Bl)
        P. primaevus Hibbard, 1938* (Bl)
        P. finneyi Hibbard and Zakrzewski, 1972 (Bl)
        P. osborni Martin, 1972 (Bl)
        P. dixonensis Zakrzewski, 1984 (Bl)
        P. wilsoni Lindsay and Jacobs, 1985 (Bl)
     Gildayomys Zakrzewski, 1984 (Bl)
        G. hibbardi Zakrzewski, 1984* (Bl)
     Unnamed genus
        "*Propliophenacomys*" *parkeri* Martin, 1975 (Hp)
     Pliolemmus Hibbard, 1938 (Bl)
        P. antiquus Hibbard, 1938* (Bl)
Arvicolinae Gray, 1821
  Arvicolini Gray, 1821
     Promimomys Kretzoi, 1955 (Hp)
        P. mimus (Shotwell, 1956) (Hp)
     Mimomys Major, 1902 (Bl)
        M. monahani Martin, 1972 (Bl)
     Mimomys (*Cosomys*) Wilson, 1932 (Bl)
        M. (*C.*) *primus* Wilson, 1932* (Bl)
        M. (*C.*) *sawrockensis* (Hibbard, 1957) (Bl)
     Mimomys (*Ogmodontomys*) Hibbard, 1941 (Bl)
        M. (*O.*) *poaphagus poaphagus* Hibbard, 1941* (Bl)
        M. (*O.*) *transitionalis* Zakrzewski, 1967 (Bl)
     Mimomys (*Ophiomys*) Hibbard and Zakrzewski, 1967 (Bl)
        M. (*O.*) *parvus* (Wilson, 1933) (Bl)
        M. (*O.*) *meadensis* (Hibbard, 1956) *(Bl)*
        M. (*O.*) *taylori* (Hibbard, 1959)* (Bl)
        M. (*O.*) *magilli* Hibbard, 1972 (Bl)
        M. (*O.*) *fricki* Hibbard, 1972 (Bl)
        M. (*O.*) *mcknighti* Gustafson, 1978 (Bl)
     Nebraskomys Hibbard, 1957 (Bl)
        N. mcgrewi Hibbard, 1957* (Bl)
        N. rexroadensis Hibbard, 1970 (Bl)
     Hibbardomys Zakrzewski, 1984 (Bl)
        H. marthae Zakrzewski, 1984* (Bl)
        H. skinneri Zakrzewski, 1984 (Bl)
        H. fayae Zakrzewski, 1984 (Bl)
        H. voorhiesi Zakrzewski, 1984 (Bl)
     Allophaiomys Kormos, 1932 (Bl–Pl)
        A. pliocaenicus (Kormos, 1932) (Bl–Pl)

    Phenacomys Merriam, 1889 (Bl–R)
      P. gryci Repenning, 1987 (Bl)
Ondatrinae Kretzoi, 1955
  Ondatrini Kretzoi, 1955
    Pliopotamys Hibbard, 1938 (Bl)
      P. minor (Wilson, 1933) (Bl)
      P. meadensis Hibbard, 1938* (Bl)
    Ondatra Link, 1795 (Bl–R)
      O. idahoensis Wilson, 1933 (Bl)

Chapter 22
Monotypic Families

1. Simimyidae	249
1.1. Known Morphology	249
1.2. Occurrence	251
1.3. Relationships	251
1.4. Recognized Species	252
2. Armintomyidae	252
2.1. Known Morphology	252
2.2. Occurrence	252
2.3. Relationships	252
2.4. Recognized Species	253
3. Laredomyidae	253
3.1. Known Morphology	253
3.2. Occurrence	253
3.3. Relationships	253
3.4. Recognized Species	254

1. Simimyidae

1.1. Known Morphology (Fig. 22.1C,D)

Simimys was a small rodent. Only the rostrum of the skull is known for Simimys. The infraorbital foramen was enlarged and it appears that the masseter had extended through it onto the rostrum, thus making it clearly hystricomorphous. Anterior and slightly ventral to the infraorbital foramen is a separate, smaller neurovascular foramen as in dipodoid rodents. Ventral to the neurovascular foramen is an additional minute foramen. The ventral surface of the anterior zygomatic arch is apparently unmodified. The rostrum is relatively heavy (dorsoventrally thick) for the size of the animal. There is no indication of tapering anteriorly. The incisive foramina are small.

    The dental formula of Simimys was $\frac{1003}{1003}$, the same as in muroid rodents. The cheek teeth were brachydont and cuspate with well developed crosslophs and longer (anteroposteriorly) than wide (buccolingually). The third molars were the smallest of the cheek teeth. M^1 was extended anteriorly in the parastylar area by the protoloph connecting to the parastyle, running obliquely across the tooth. The metaloph paralleled the protoloph, usually uniting with the paracone. The protoconule was well developed. There was a mesoloph present. The posterior two molars were similar to M^1 except for the

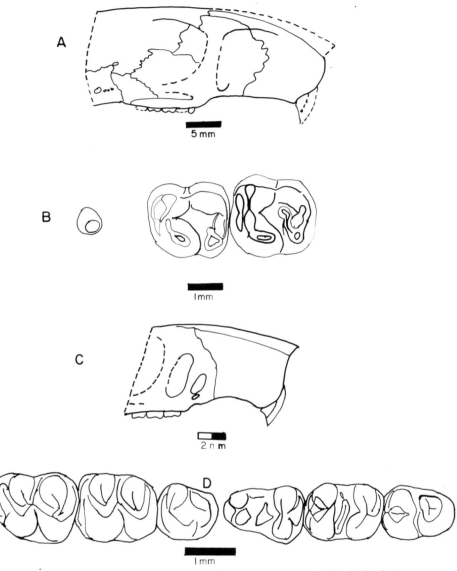

FIGURE 22.1. *Armintomys* and *Simimys*. (A) Reconstruction of the skull of *Armintomys tullbergi*. (B) LP^3, M^1–M^2 of *A. tullbergi*. (C) Reconstruction of the skull of *Simimys simplex*. (D) Composite dentition of *S. simplex*, LM^1–M^3 and LM$_1$–M$_3$. All drawn to different scale (below each panel). A and B from Dawson et al. (1990), C from Emry (1981), D from Lillegraven and Wilson (1975).

anterior extension and oblique protoloph. M^3 had a greatly reduced posterior half.

M_1 was the longest of the lower molars. It was narrower (buccolingually) anteriorly than posteriorly. There was a small, central anteroconid present. The trigonid cusps were united by the metalophulid II, the talonid cusps by the hypolophid. There were also a complete ectolophid and mesolophid. As in the upper molars, M_3 was reduced posteriorly.

1.2. Occurrence

Specimens of *Simimys simplex* were limited to Uintan and Duchesnean localities in southern California (Wilson, 1935a, 1949d) and possibly the early Chadronian of Texas (Wood, 1974).

1.3. Relationships

Wilson (1935a,b) originally named two species of *Simimys* from the Uintan of southern California, *S. simplex* and *S. vetus*. Later, he (Wilson, 1949d) named a third species from the Duchesnean, *S. (?) murinus*. More recently, based on much more fossil material from southern California, Lillegraven and Wilson (1975) included all three species in the type species *S. simplex*.

Because of the dipodoid-like zygomasseteric structure of the skull and the muroid-like dental formula of *Simimys*, it has been assigned to different families of rodents throughout its history of study.

In the original description of *Simimys*, Wilson (1935a,b) referred it to the Cricetidae. Lindsay (1968, 1977) followed this familial allocation. However, a number of authors have included *Simimys* in the Zapodidae (Wood, 1937, 1974; Schaub, 1958; Lillegraven and Wilson, 1975; Emry and Korth, 1989). Other authors have been more general about the affinities of *Simimys*. Emry (1981) included *Simimys* in the Muroidea *incertae sedis*. Stehlin and Schaub (1951) included it within the Dipodoidea. Still other authors have classified *Simimys* as a primitive myomorph that predated the differentiation of the muroids and dipodoids or as Myomorpha *incertae sedis* (Simpson, 1945; Wilson, 1949d; Dawson, 1966). Wood (1980) erected the Simimyidae to accommodate this unique genus.

The dipodoid-like zygomasseteric structure and similarity of the cheek tooth morphology to early zapodids make *Simimys* more likely referable to the Dipodoidea rather than the Muroidea or Cricetidae. The reduction in dental formula to that of muroids appears to be a convergent character rather than a shared derived one because such a reduction in the dental formula occurs several times within the Rodentia; for example, the enigmatic North American rodents *Diplolophus* and *Nonomys* (Chapter 23, Sections 1 and 6), the African hystricomorph family Myophiomyidae (Lavocat, 1973), and the Eocene Asian ctenodactyloid *Ivanantonia* (Shevyreva, 1989). However, the Simimyidae cannot be ancestral to any later dipodoids because of specializations of the dentition and skull (see Wilson, 1935a, 1949d). The Simimyidae represent an aberrant side branch of an early dipodoid radiation (Fig. 19.3).

1.4. Recognized Species

Simimys simplex (Wilson, 1935) (U–D, ?Ch)
S. landeri Kelly, 1992 (D)

2. Armintomyidae

2.1. Known Morphology (Fig. 22.1A,B)

The skull of *Armintomys* was fully hystricomorphous with the infraorbital foramen widened and opening anteriorly. There are clear muscle markings on the rostrum for the attachment of the masseter that has passed through the infraorbital foramen. The posterior half of the skull is unknown. The cranial foramina of the medial orbital wall do not differ from those of primitive sciuravids or ischyromyids except for the presence of multiple (three) interorbital foramina just anterior to the optic foramen.

The microstructure of the upper incisors of *Armintomys* was ". . . pauciserial that is almost uniserial" (Dawson et al., 1990, p. 140). There is a shallow groove along the center of the upper incisors. The upper dental formula for *Armintomys* was 1-0-2-3. P^3 is a small peg. P^4 is not known, but was smaller than the molars based on its alveolus. M^1 and M^2 were anteroposteriorly elongated, similar to the condition in dipodoid rodents. These teeth were brachydont and cuspate with weak development of the transverse lophs. The hypocone was large, nearly equal to the protocone in size. The conules were absent or greatly reduced, and there was no anteroposterior connection between the protoloph and metaloph.

2.2. Occurrence

Armintomys is represented by a single partial skull from the earliest Bridgerian of Wyoming (Dawson et al., 1990).

2.3. Relationships

Armintomys was the earliest known rodent to possess either hystricomorphy or uniserial enamel. The hystricomorphy, uniserial enamel of the incisors, and occlusal proportions of the upper molars of *Armintomys* clearly ally it with the Dipodoidea. The upper molars of *Armintomys* had a sciuravid-like occlusal pattern suggesting a sciuravid origin for this genus. The combination of the derived dipodoid features and primitive sciuravid-like dentition and dental formula of *Armintomys* led Dawson *et al.* (1990) to establish a distinct family for this genus. It appears that it was part of an early radiation of dipodoid rodents in North America but *Armintomys* itself was not directly ancestral to any later dipodoids (Fig. 19.3).

2.4. Recognized Species

Armintomys tullbergi Dawson, Krishtalka, and Stucky, 1990 (Br)

3. Laredomyidae

3.1. Known Morphology (Fig. 22.2A)

The cheek teeth of *Laredomys* are quite small and brachydont (all measurements less than 1 mm; Wilson and Westgate, 1991). The teeth are wider (buccolingually) than long (anteroposteriorly) and are highly lophate in that the individual cusps or conules cannot be distinguished. Most of the cheek teeth assigned to *Laredomys* have three complete transverse lophs that are nearly parallel, and a fourth loph that ends at the center of the basin of the tooth or intersects with another loph. One presumed lower molar has a short anteriorly directed loph running from the metalophid(?) to the anterior cingulum(?). Because *Laredomys* is so poorly known, and none of the referred specimens were found in association with a mandible or maxilla, the recognition of these teeth as upper or lower is uncertain.

3.2. Occurrence

Laredomys is known only from five isolated cheek teeth from Uintan sediments near the town of Laredo, Texas (Wilson and Westgate, 1991).

3.3. Relationships

The morphology of the cheek teeth of *Laredomys* with its combination of small size, brachydonty, and high degree of lophodonty is unique among North American rodents, particularly from the Eocene. The degree of lophodonty in the cheek teeth of *Laredomys* was attained by Oligocene and later Old World glirids, but the pattern of the lophs in the latter is different than

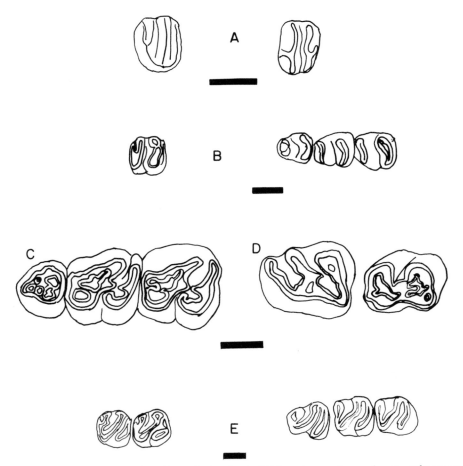

FIGURE 22.2. Dentitions of problematical rodents. (A) *Laredomys riograndensis*, right upper molar (left) and left lower molar. (B) *Griphomys alecer*, RM^1 or M^2 and LP_4–M_2. (C) *Pipestoneomys bisulcatus*, RM^1–M^3. (D) *Pipestoneomys pattersoni*, RP_4 and RM_1. (E) *Zetamys nebraskensis*, RP^4–M^1 and LP_4–M_2. Bar = 1 mm. A from Wilson and Westgate (1991), B from Wilson (1940), C and D from Alf (1962), E from Martin (1974).

that of *Laredomys*. *Laredomys* cannot, at present, be recognized as closely related to any known North or South American rodent, fossil or Recent.

3.4. Recognized Species

Laredomys riograndensis Wilson and Westgate, 1991 (U)

Chapter 23

Rodents Not Referable to Recognized Families

1. *Diplolophus*	256
1.1. Record	256
1.2. Known Morphology	256
1.3. History of Investigation	256
2. *Griphomys*	256
2.1. Record	256
2.2. Known Morphology	257
2.3. History of Investigation	258
3. *Guanajuatomys*	258
3.1. Record	258
3.2. Known Morphology	258
3.3. History of Investigation	260
4. *Jimomys*	260
4.1. Record	260
4.2. Known Morphology	260
4.3. History of Investigation	261
5. *Marfilomys*	262
5.1. Record	262
5.2. Known Morphology	262
5.3. History of Investigation	262
6. *Nonomys*	263
6.1. Record	263
6.2. Known Morphology	263
6.3. History of Investigation	264
7. *Pipestoneomys*	264
7.1. Record	264
7.2. Known Morphology	264
7.3. History of Investigation	265
8. *Texomys*	265
8.1. Record	265
8.2. Known Morphology	265
8.3. History of Investigation	266
9. *Zetamys*	266
9.1. Record	266
9.2. Known Morphology	266
9.3. History of Investigation	267

1. *Diplolophus*

1.1. Record

Late Orellan of the Great Plains. All known specimens are partial maxillae or mandibles with cheek teeth and some isolated cheek teeth.

1.2. Known Morphology (Fig. 23.1B,D)

The dental formula is $\frac{1003}{1003}$, characteristic of muroid rodents. The cheek teeth are made of six bulbous cusps arranged to two transverse rows, characteristic of geomyoid rodents. M^1 has a well-developed anterocone. The masseteric scar on the mandible extends anterior to P_4 and is a shelflike structure as in heteromyids and some eomyids.

1.3. History of Investigation

Troxell (1923b) named the genus *Diplolophus* and included two species, *D. insolens* (the type species) and *D. parvus*. He failed to refer this genus to any family of rodents. Hay (1930) listed *Diplolophus* as a sciurid. Wood (1935a) demonstrated *D. parvus* was referable to the heteromyid genus *Proheteromys* and was much younger than Troxell had originally indicated. Wood (1937) discussed the familial affinities of the genus, but retained it in "Rodentia, *incertae sedis*." He noted that the greatest similarity in the morphology of the cheek teeth was with the geomyoid rodents, but that the dental formula was not, suggesting that the three cheek teeth would have to be the last premolar and the first two molars, having lost the M_3.

The only review of *D. insolens* was by Barbour and Stout (1939), in which they synonymized *Gidleumys adspectans* Wood (1936b), but failed to refer the genus to any family. In his classification of all mammals, Simpson (1945) included *Diplolophus* in "?Geomyidae *incertae sedis*" along with *Griphomys* Wilson (1940), another enigmatic rodent. In his description of the upper dentition of *Nonomys*, Emry (1981) noted the similarity in its dental formula with that of *Diplolophus* but stated that any morphological similarities between the cheek teeth of these genera were only superficial (see Section 6).

2. *Griphomys*

2.1. Record

Griphomys was reported from the Uintan and Duchesnean of southern California (Wilson, 1940; Lindsay, 1968; Lillegraven, 1977) and the early

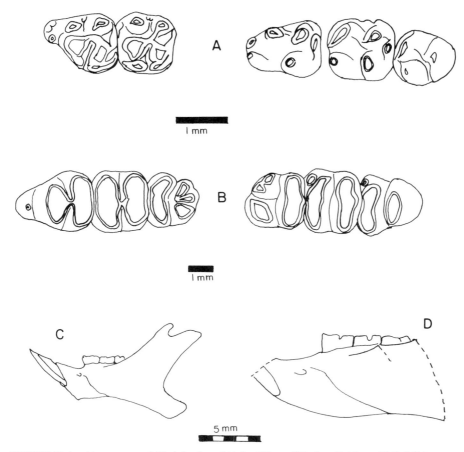

FIGURE 23.1. *Nonomys* and *Diplolophus*. (A) dentition of *N. simplicidens*, RM^1–M^2 (reversed) and LM$_1$–M$_3$. (B) Dentition of *D. insolens*, RM^1–M^3 and LM$_1$–M$_3$. (C) Lateral view of mandible of *N. simplicidens*. (D) Lateral view of mandible of *D. insolens*. All not to same scale (scale below each panel). A (uppers) and C from Emry (1981), A (lowers) from Emry and Dawson (1972), B and D from Barbour and Stout (1939).

Chadronian of Montana (Sutton and Black, 1975). Known specimens include partial maxillae and mandibles along with numerous isolated cheek teeth.

2.2. Known Morphology (Fig. 22.2B)

A partial maxilla reported by Wilson (1940) demonstrates that *Griphomys* is sciuromorphous. The mandible is relatively deep (dorsoventrally) and sciurognathous. The masseteric scar on the mandible extends anteriorly to below P$_4$.

The dental formula is $\frac{1013}{1013}$. The lower incisor has an anterior enamel surface that is nearly flat. The microstructure of the enamel of the incisor has not been studied. Cheek teeth are brachydont. The occlusal morphology of all cheek teeth is dominated by two transverse lophs that unite the major cusps and are divided by a central transverse valley and an anterior cingulum that runs parallel to the other lophs. The cusps are reduced and nearly indistinguishable from the lophs that contain them. The lophs on the lower molars are weakly convex posteriorly, and anteriorly on the upper molars. The trigonid on P_4 is narrower (buccolingually) than that of the molars. Any mesoconids present are only rudimentary and minute in size.

2.3. History of Investigation

When Wilson (1940, p. 93) first named *Griphomys alecer*, he listed the family as "Incertae Sedis, or Geomyoidea(?)." He related it to the geomyoids based on a number of characters: (1) sciuromorphy; (2) anterior extension of the masseteric scar on the mandible; (3) dental formula; (4) cheek teeth basically four-cusped and bilophate with only a vestigial mesoconid on the lower molars. Simpson (1945, p. 81) listed *Griphomys* along with *Diplolophus* as "?Geomyidae *incertae sedis*." Lindsay (1968) classified *Griphomys* questionably within the Geomyidae. Wood (1974, 1980) noted the similarity between the cheek teeth of *Griphomys* and the problematical Chadronian *Meliakrouniomys* (see Chapter 14, Section 5.3) and classified both as eomyids. Sutton and Black (1975, p. 310) did not include *Griphomys* in any family, listing it as "Family indet." Lillegraven (1977) in naming a new species, *G. toltecus* also from the Uintan of southern California, followed Lindsay and included the genus as a questionable geomyid.

Wilson's (1940) criteria for including *Griphomys* in the Geomyoidea appear valid, and have not been questioned by any later investigators. Unfortunately, the microstructure of the enamel has not been studied. If it were known, it would be possible to establish whether this genus was an eomyid or not (Wahlert and von Koenigswald, 1985). If *Griphomys* was not an eomyid, it represented the earliest occurrence of any of the other families of geomyoids.

3. *Guanajuatomys*

3.1. Record

Late Eocene or early Oligocene (?Uintan–?Chadronian) of central Mexico (Guanajuato). Known only from a single mandible with complete dentition.

3.2. Known Morphology (Fig. 23.2A)

The mandible is dorsoventrally deep with a strongly convex downward ventral border. The masseteric fossa is weakly defined and ends anteriorly

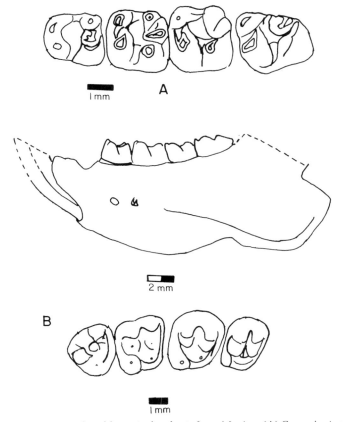

FIGURE 23.2. Dentitions of problematical rodents from Mexico. (A) *Guanajuatomys hibbardi*, occlusal view of LP$_4$–M$_3$ and lateral view of mandible. (B) *Marfilomys aewoodi*, occlusal view of RP⁴–M³. A from Black and Stephens (1973), B from Ferrusquia-Villafranca (1989).

below M$_2$. There are two mental foramina at middepth of the mandible below P$_4$–M$_1$. The diastema is very short and shallow. The mandible was originally described as hystricognathous by Black and Stephens (1973), but has been considered by others as sciurognathous (Korth, 1984; Wilson, 1986).

The dental formula is $\frac{????}{1013}$. The incisor is nearly flat anteriorly. No study of the microstructure of the enamel has been made. The cheek teeth are cuspate and the enamel in the basins is smooth. P$_4$ is the smallest of the cheek teeth with a narrower trigonid. There is no ectolophid and the hypoconulid is enlarged (nearly equal in size to the major cusps) and elongated anteriorly, extending into the talonid basin. M$_1$ and M$_2$ are subequal in size and squared in outline. M$_3$ is slightly expanded posteriorly.

3.3. History of Investigation

Guanajuatomys hibbardi was first described by Black and Stephens (1973) but they did not assign it to any family. They described the mandible as hystricognathous but noted that the dental morphology was such that it could not be considered as ancestral to any of the New World hystricognaths (= Caviomorpha). They also noticed the similarity of the cheek tooth morphology to that of early ctenodactyloids from Asia. No subsequent authors have included *Guanajuatomys* within any family of rodents.

Wood (1975) included *Guanajuatomys* in his suborder Franimorpha based on the hystricognathy of the mandible. However, more recent workers have failed to see the hystricognathy of the mandible and considered it sciurognathous (Korth, 1984; Wilson, 1986).

The lower dentition of *G. hibbardi* more closely approaches that of the Uintan *Protoptychus* (Turnbull, 1991) than any other North American rodent. The cheek teeth of *G. hibbardi* are lower crowned, P_4 is less reduced, and M_3 is more posteriorly elongate than those of *P. hatcheri*, but otherwise they are quite similar. The mandible of *Guanajuatomys* and *Protoptychus* are markedly different. *Protoptychus* has a slender mandible with a nearly flat ventral border and a narrow, delicate angle. In *Guanajuatomys* the jaw is robust with a strangely curved ventral border and a broad angle.

The characters cited by Black and Stephens (1973) as shared by *Guanajuatomys* and early ctenodactyloids from Asia appear to be primitive for rodents (see Dawson *et al.*, 1984). A relationship with protoptychids appears the most likely. However, not enough material of *Guanajuatomys* is known to defnitely assign it to the Protoptychidae.

4. *Jimomys*

4.1. Record

Late Arikareean of Oregon (John Day) and earliest Barstovian of Nebraska. The type species *J. labaughi* is known from nine partial mandibles with lower dentitions, and *J. lulli* is known from a single partial mandible with M_1–M_2.

4.2. Known Morphology (Fig. 23.3A)

The mandible is slender. The diastema is relatively long and deep (extends well ventral to the tooth row). There is a single mental foramen and the masseteric fossa is weakly defined but extends anterior to the cheek teeth.

The dental formula is $\frac{????}{1013}$. The incisor is convex anteriorly and has enamel with uniserial microstructure. The cheek teeth are mesodont in crown height. The lower premolar is the largest of the cheek teeth, being nearly twice the length of the posterior molars. P_4 has three transverse lophs that are gently concave anteriorly; a small enamel lake appears in the center of the anterior

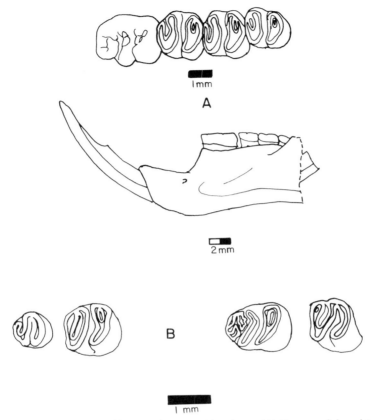

FIGURE 23.3. Dentition of problematical geomyoid rodents. (A) *Jimomys labaughi*, occlusal view of LP_4–M_3 and lateral view of mandible. (B) *Texomys ritchiei*, RM^2 and M^3 and LP_4 and LM_1 or M_2. A from Wahlert (1976), B from Slaughter (1981).

loph on moderately worn specimens. The posterior molars are roughly the same size. There is a deep central, transverse valley on the molars which separates the metalophid from the hypolophid. The anterior cingulum attaches to the metalophid at the protoconid and the posterior cingulum attaches to the hypolophid at the hypoconid. The narrow valley between the cingula and the major lophs leads to the formation of small enamel lakes near the lingual end of the lophs on moderately worn teeth. P_4 is two-rooted, M_1 and M_2 are four-rooted, and M_3 has two anterior and one posterior root.

4.3. History of Investigation

Wahlert (1976) first described *Jimomys*. He recognized two species, *J. labaughi* (type species) from the early Barstovian of Nebraska, and transferred

Florentiamys lulli Wood (1936c) from the Arikareean of Oregon to the genus. Wahlert (1976) referred a single mandible from Texas to *J. labaughi* that was later included in the genus *Texomys* (Slaughter, 1981). Wahlert (1976) included *Jimomys* in the Geomyoidea *incertae sedis* noting that the structure of the mandible was advanced over that of eomyids and more similar to that in geomyids and heteromyids. He suggested that *Jimomys* represented a separate lineage of geomyoid that had been distinct from other geomyoids since at least the Arikareean. He did not name a new family to include *Jimomys*.

Slaughter (1981) described another geomyoid from the Texas Coastal Plain, *Texomys*, that was similar to *Jimomys* in the morphology of the cheek teeth. He did not assign *Jimomys* and his new genus to a family, but followed Wahlert (1976) by including these genera in Geomyoidea *incertae sedis*.

5. *Marfilomys*

5.1. Record

Late Eocene or early Oligocene (?Uintan–?Chadronian) of central Mexico (Guanajuato). Known only from a partial skull with complete dentition.

5.2. Known Morphology (Fig. 23.2B)

The known skull morphology of *Marfilomys* is similar to that of primitive rodents (notably reithroparamyine ischyromyids, sciuravids, and cocomyids). The rostrum is relatively short and the incisive foramina small, and the zygomasseteric structure is protrogomorphous.

The upper dental formula for *Marfilomys* is that of primitive rodents, retaining a peglike P^3. The incisors are laterally compressed and gently convex anteriorly. No study of the microstructure of the enamel has been reported. The upper cheek teeth are bunodont and brachydont. P^3 is a small peg. P^4 is molariform and nearly as large as the molars. The cheek teeth are characerized by large bulbous cusps, an enlarged metaconule (subequal in size to the metacone), and a large hypocone situated more lingual than the protocone. The transverse lophs are weak and the anterior and posterior cingula are well developed.

5.3. History of Investigation

Ferrusquia-Villafranca (1989) described and named *Marfilomys aewoodi* from the Paleogene of central Mexico. Although it is similar in size and from the same area and horizon as *Guanajuatomys*, Ferrusquia-Villafranca insisted that his material did not belong to the latter genus because of several dental differences of the cheek teeth (higher crowned, more transversely elongate).

However, because M. aewoodi is only represented by a skull and upper dentition and G. hibbardi is only represented by a mandible with lower dentition, no direct comparison was possible. Ferrusquia-Villafranca (1989) provided a thorough comparison of the skull and dentition of Marfilomys with 27 different families of Paleogene rodents. He found that it had the greatest similarity to 7 families including reithroparamyines, cylindrodontids, and protoptychids from North America, octodontids from South America, and Eocene ctenodactyloids from Asia. He concluded that the similarity to the Asian families was the result of convergence and suggested that the similarities with reithroparamyines and octodontids more nearly reflected the true relationships of the genus. He was unable to assign Marfilomys to any family but suggested that it may well represent a distinct family of its own.

It is still possible that Marfilomys represents the skull and upper cheek teeth of Guanajuatomys but, of course, it will be necessary to find associated upper and lower dentitions to establish this. The similarities with the Asian ctenodactyloids are more likely retained primitive characters rather than the result of convergence. The differences between the cheek teeth of Marfilomys and the early Oligocene Plattypitamys from South America are far greater than the similarities cited by Ferrusquia-Villafranca (1989, pp. 105–106). The cheek teeth of the South American genus are higher crowned, more lophate with metaconules absent, and P^4 less molariform than in Marfilomys. The simlarities appear to be primitive features of the cheek teeth (e.g., large hypocones).

As concluded by Ferrusquia-Villafranca (1989), there is no currently known family of rodents to which Marfilomys can be assigned. A better knowledge of the Paleogene history of rodents in Central America may be necessary to determine the true affinities of Marfilomys.

6. *Nonomys*

6.1. Record

Early Chadronian of Wyoming and Texas. All known specimens are isolated cheek teeth or maxillae and mandibles.

6.2. Known Morphology (Fig. 23.1A,C)

The dental formula is $\frac{1003}{1003}$, characteristic of muroid rodents. The zygomasseteric structure is hystricomorphous with an accessory foramen for the transmission of nerves and blood vessels, as in dipodoid rodents. The upper cheek teeth are brachydont and have four major cusps and two smaller stylar cusps as in geomyoid rodents, but retain low, short lophs that unite these cusps across the central basin of the teeth. M^1 has an enlarged anterocone (parastylar area). The lower cheek teeth are similarly bunodont and cuspate with low lophs connecting the buccal cusps. The stylar area is not as well developed as in the upper teeth. M_1 is the largest of the teeth.

6.3. History of Investigation

Nanomys simplicidens was first described by Emry and Dawson (1972) from the Chadronian of Wyoming (later renamed *Nonomys*, Emry and Dawson, 1973). They referred their new genus to the Cricetidae. Martin (1980) erected a new subfamily within the Cricetidae, the Nonomyinae which contained *Nonomys* and another species *Subsumus candelariae* Wood (1974) from the early Chadronian of Texas. In his description of the upper dentition and zygomasseteric structure of *Nonomys*, Emry (1981) synonymized *S. candelariae* with *N. simplicidens* and demonstrated that the dipodoid-like hystricomorphy of *Nonomys* removed it from the Cricetidae. Emry (1981) noted that the zygomasseteric structure and dental formula of *Nonomys* most closely resembled those of the Uintan *Simimys* from California, though the morphology of their respective cheek teeth were quite different. He suggested that *Nonomys* was part of an early radiation of myodont rodents and should be referred to "Muroidea, Family *incertae sedis*." He briefly compared *Nonomys* with *Diplolophus* noting the simlarity in dental formula, but felt that the anteriorly running connections of the cusps in the cheek teeth of *Nonomys* that were lacking in *Diplolophus* were significant enough to refute any possible relationship between these genera. However, the simlarity between the upper cheek teeth of *Nonomys* and *Diplolophus* is striking, and the connections of the cusps in the cheek teeth of the former, older genus may not be as significant as asserted by Emry (1981), and a *Diplolophus*-like morphology of the cheek teeth could have been derived from a *Nonomys*-like dentition by the late Orellan. This proposed relationship could be easily supported or refuted when the zygomasseteric structure of *Diplolophus* becomes known.

7. *Pipestoneomys*

7.1. Record

Chadronian of Montana and Nebraska and possibly Duchesnean of Saskatchewan. All known specimens are partial maxillae or mandibles with cheek teeth or isolated cheek teeth.

7.2. Known Morphology (Fig. 22.2C,D)

Cheek teeth are mesodont and lophate. The dental formula appears to be $\frac{1013}{1013}$. Upper cheek teeth are characterized by enlarged anterocone on P^4; an oblique loph connecting the paracone to the hypocone and its separation from the protocone–parastylar loph by a deep, oblique valley; and irregular lophule development in the central basin of the cheek teeth. Lower cheek teeth develop a characteristic pair of isolated enamel lakes with wear.

7.3. History of Investigation

Pipestoneomys bisulcatus was named by Donohoe (1956) based on dental elements from the middle Chadronian Pipestone Springs fauna of Montana. He placed this rodent in the familiy Aplodontidae. Alf (1962) named a second species, *P. pattersoni* from the Chadronian of Nebraska, and questionably referred the genus to the family Castoridae. Later, Black (1965) noted that a number of dental characters of the Castoridae were lacking in *Pipestoneomys* and that the genus was probably not referable to either the aplodontids or castorids, but retained the genus questionably in the latter pending discovery of additional, more complete material. All records of *Pipestoneomys* have been from the Chadronian of North America (see also Ostrander, 1985).

Storer (1987) reported several isolated molars of the eomyid *Yoderimys* from the Duchesnean of Saskatchewan. These teeth, however, are distinct from those of later *Yoderimys*, and more closely approach those of *Pipestoneomys*. These specimens appear to be the cheek teeth of a primitive (more brachydont, less lophate) species of *Pipestoneomys* rather than an eomyid, thus extending the range of this genus back into the Duchesnean.

8. *Texomys*

8.1. Record

Late Hemingfordian to Barstovian of the Texas Coastal Plain and the early Hemingfordian of Panama. All species represented almost entirely by isolated cheek teeth with some mandibular fragments.

8.2. Known Morphology (Fig. 23.3B)

The mandible of *Texomys* is known only from small fragments. The only feature observable is the diastema which is shallow and relatively short and the mental foramen which is situated below the posterior half of the diastema.

The cheek teeth of *Texomys* are mesodont. The dental formula appears to be $\frac{1013}{1013}$, although no complete tooth rows have been reported. The premolars are the largest cheek teeth. Both the upper and lower premolars have three large cross-lophs. On P_4 they are gently concave anteriorly. The anterior loph is closed anteriorly by a short anterior cingulum that encloses a small basin. The lophs are oriented obliquely (posterolingually) on the tooth. The lower molars have a deep central transverse valley and two major lophs. All of the transverse lophs on the molars have numerous complications of the enamel. The lophs on the molars are obliquely oriented. The anterior cingulum joins the metalophid at the protoconid and extends nearly the entire width of the tooth. The posterior cingulum joins the hypolophid at its center. All lower cheek teeth are two-rooted. The upper molars are nearly identical to the lowers but can be differentiated by possessing three roots (one lingual, two buccal). The upper premolar also has three major cross-lophs that are obliquely oriented with a distinct anterocone. Both M_3 and M^3 are smaller than the anterior molars.

8.3. History of Investigation

Slaughter (1981) named two species of *Texomys*, *T. ritchiei* (type species) from the early Barstovian and *T. stewarti* from the early Hemingfordian. He also cited a distinct population from the late Hemingfordian that he referred to *Texomys* sp. He noted the similarity of the cheek teeth of *Texomys* with those of *Jimomys* from the Barstovian of Nebraska (Wahlert, 1976) and referred the genus to Geomyoidea *incertae sedis*.

Texomys is distinct from *Jimomys* based on the characters of the cheek teeth cited by Slaughter (1981, p. 113) as well as the depth and length of the diastema (deeper and longer in *Jimomys*) and the number of roots of the lower molars (two in *Texomys* and three or four in *Jimomys*). It is likely that both *Texomys* and *Jimomys* represented a distinct radiation of geomyoids as suggested by Wahlert (1976) and were probably referable to a distinct family closely related to the Heteromyidae and Geomyidae.

9. *Zetamys*

9.1. Record

Early Arikareean of Nebraska and possibly Whitneyan of South Dakota. All known specimens are of dental elements including partial mandibles and maxillae.

9.2. Known Morphology (Fig. 22.2E)

The dental formula is $\frac{1013}{1013}$. Cheek teeth are mesodont and lophate. All cheek teeth attain a Z-pattern of lophs composed of anterior and posterior transverse lophs and an obliquely oriented loph (running in a posterolingual to anterobuccal direction). A remnant of a mesolophid (reduced to a small cuspule) is variably present on the lower cheek teeth. Incisor, mandibular, and cranial morphology are not known for *Zetamys*.

9.3. History of Investigation

Martin (1974) first described *Zetamys* from the early Arikareean of Nebraska and did not refer it to any family of rodents, listing it under the heading "Superfamily indet." He thoroughly compared *Zetamys* with other rodents with similar lophate teeth; the early Oligocene phiomyid *Gaudeamus* from Egypt, the late Oligocene/early Miocene eomyid *Rhodanomys* from Europe, and the Recent Central American capromyid *Plagiodontia*. However, Martin was unable to demonstrate any type of special relationship between these genera and *Zetamys*.

Besides the type species, *Z. nebraskensis*, Martin (1974) cited a second unnamed species from an anthill collection from South Dakota. He noted the more primitive morphology of the South Dakota species (lower crowned, less lophate cheek teeth) which corresponded to its slightly older age (Whitneyan).

No other authors have discussed the possible relationships of *Zetamys*, and no additional material of this genus has been recovered since its original description. Therefore, no further conclusions can be made about its systematics beyond those suggested by Martin (1974).

III

Dynamics of the North American Rodent Fauna through Time

Chapter 24
Changes in the Rodent Fauna

1. Tertiary	271
1.1. Clarkforkian	271
1.2. Wasatchian	271
1.3. Bridgerian	272
1.4. Uintan	272
1.5. Duchesnean	272
1.6. Chadronian	273
1.7. Orellan	273
1.8. Whitneyan	274
1.9. Arikareean	274
1.10. Hemingfordian	275
1.11. Barstovian	275
1.12. Clarendonian	276
1.13. Hemphillian	276
1.14. Blancan	276
2. Pleistocene	277
2.1. Irvingtonian	277
2.2. Rancholabrean	278

1. Tertiary

1.1. Clarkforkian

Rodents first appeared in the Clarkforkian of Montana (Jepsen, 1937). The entire rodent fauna from North America during the Clarkforkian was comprised of members of the Ischyromyidae but showed an amazing diversity with as many as three subfamilies having been already differentiated (Korth, 1984). This implies either the origin and a very rapid radiation in the late Clarkforkian, or the origin of rodents on another continent (most likely Asia) and immigration of the stem ischyromyid to North America.

1.2. Wasatchian

The first occurrence of three additional families of rodents occurred in the Wasatchian, the Sciuravidae, Eutypomyidae, and Cylindrodontidae (Korth, 1984). However, the number of ischyromyids described makes up 76% of the known species of Wasatchian rodents.

Of the three new families, only one, the Eutypomyidae, appears to be derivable from an ischyromyid stock (Dawson, 1966; Korth, 1984). This suggests a separate origin and immigration of the cylindrodonts and sciuravids.

1.3. Bridgerian

Again in the Bridgerian, the North American rodent fauna was dominated by ischyromyid species (50% of known species) including the appearance and radiation of the large-sized manitshines. The Sciuravidae were also at their zenith at this time, making up nearly a third of the rodent fauna.

The first occurrence of two problematical dipodoid rodents, *Armintomys* (Dawson et al., 1990) and *Elymys* (Emry and Korth, 1989), marked an early radiation of dipodoids in North America in the Bridgerian. If *Elymys* was correctly assigned to the Zapodidae, then this was the first occurrence of a modern family of rodents in the North American record.

1.4. Uintan

During the Uintan the ischyromyids maintained their dominance of the rodent fauna but their percentage of the known species was declining (49%). The number of families of rodents recorded from the Uintan increased to eight with the first appearance of the Protoptychidae (Turnbull, 1991), Aplodontidae (Black, 1971), Eomyidae (Storer, 1987), and Simimyidae (Wood, 1980). All of these families may have been derived from North American endemic species of ischyromyids or sciuravids except perhaps the protoptychids. The presence of *Simimys* was another part of the early dipodoid radiation in North America that began in the Bridgerian.

The problematical geomyoid *Griphomys* also appeared in the Uintan (Wilson, 1940), again demonstrating the much increased diversity of rodents at this time.

1.5. Duchesnean

Fewer species of rodents have been described from the Duchesnean than any of the previous ages except the Clarkforkian. This is believed to reflect a paucity in the fossil record rather than a true decline in the number of rodent species.

The greatest number of species reported from the Duchesnean belonged to the Ischyromyidae, but they only represent one-third of the rodent fauna. The paramyine ischyromyids, which had outnumbered all other subfamilies of ischyromyids until this time, had greatly declined and the first occurrence of ischyromyines is recorded (Black, 1971). The second most common family was the Eomyidae which constituted about 21% of the known species of Duchesnean rodents (Storer, 1987).

Two families of early rodents had their last occurrence in the Duchesnean, the Sciuravidae (Dawson, in Black and Sutton, 1984) and Protoptychidae (Chapter 8, Section 3). The first record of the Heliscomyidae also occurred in the Duchesnean (Storer, 1988), continuing the geomyoid radiation started with *Griphomys* in the Uintan.

The first cricetid from North America was recorded from the Duchesnean of Saskatchewan (Storer, 1988). The Cricetidae, while most likely derived from a sciuravid ancestor (Chapter 6, Section 4.3, and Chapter 20, Section 4.1) which are exclusively North American, were immigrant taxa from Asia where the family first appeared near the middle Eocene (Tong, 1992).

1.6. Chadronian

By Chadronian times the composition of the rodent fauna had drastically changed from that of the earlier Eocene times. The species of Ischyromyidae only made up 21% of the known rodents, and belonged almost entirely to the specialized Ischyromyinae (Wood, 1980). The dominant family in the Chadronian was the Eomyidae which had reached its peak of diversity (Fahlbusch, 1973, 1979), constituting over 42% of the known rodent fauna. The Chadronian was also the most likely time of emigration of eomyids to Europe (Fahlbusch, 1979). The Cylindrodontidae were also at their zenith during the Chadronian (Wood, 1974) but became extinct by the end of the interval.

Two Recent families of rodents had their first appearance in the fossil record during the Chadronian: the Sciuridae (Black, 1965) and Castoridae (Emry, 1972a). The Sciuridae may have been derived from North American Eocene reithroparamyine ischyromyids (Meng, 1990; Korth and Emry, 1991), and the same may have been true for the Castoridae though it was not as evident (Korth, 1984).

Also during this time, there were a few problematical genera of rodents that appeared in the fossil record. *Pipestoneomys* from Montana and Nebraska was of uncertain familial assignment and known only from dental elements (Donohoe, 1956; Alf, 1962). Two genera from northern Mexico, *Guanajuatomys* and *Marfilomys*, were also of uncertain affinities but may have had some relationship with South American caviomorph rodents (Black and Stephens, 1973; Ferrusquia-Villafranca, 1989). The probable dipodoid rodent *Nonomys* also occurred in the Chadronian (Emry and Dawson, 1972; Emry, 1981).

1.7. Orellan

The Eomyidae, the most diverse family of rodents during the Orellan, made up one-fourth of the known rodent species but had reduced their numbers almost in half from their greater abundance in the Chadronian (Korth, 1989b).

The Cricetidae had rapidly diversified since their first occurrence and represented 21% of the known species. This diversity of cricetids was almost entirely restricted to the subfamily Eumyinae (Martin, 1980). The Heliscomyidae were at their most abundant, though only represented by five species (Korth et al., 1991). Prosciurine aplodontids were similarly at their greatest diversity during the Orellan (Korth, 1989a). The earliest florentiamyid was also reported from the Orellan (Korth, 1989b).

The problematical rodents *Diplolophus* and *Tenudomys* were known from the late Orellan (Korth, 1989b). The affinities of *Diplolophus* are uncertain, but it may be closely related to the Chadronian probable dipodoid *Nonomys* (Chapter 23, Section 1). *Tenudomys* may represent the first occurrence of the Geomyidae (Korth, 1993b).

1.8. Whitneyan

The number of rodent species reported from the Whitneyan (14) is much lower than during any other provincial age except the Clarkforkian. Similar to the Duchesnean, this low number is viewed as a reflection of the rarity of small-mammal-bearing deposits of this age rather than a true decline in the number and diversity of rodents. This is evidenced by the fact that a number of rodent families that appear in the fossil record in the Orellan and again in the Arikareean are absent in the Whitneyan (e.g., Eutypomyidae, Eomyidae, and Heliscomyidae).

Prosciurine aplodontids continued to be diverse during the Whitneyan (Korth, 1989a). It was also the time of the last occurrence of the Ischyromyidae, being represented by only a single species of *Ischyromys* (Heaton, 1993).

The Heteromyidae first appeared in the Whitneyan of the Great Plains. The single known species was *Proheteromys nebraskensis* Wood (1937).

1.9. Arikareean

There was a dramatic change in the rodent fauna in the Arikareean as compared with the Chadronian through Whitneyan times. There were explosive radiations of a number of rodent groups: allomyine and meniscomyine aplodontids (Rensberger, 1983), palaeocastorine castorids (Martin, 1987), entoptychine geomyids (Rensberger, 1971, 1973b), florentiamyids (Wahlert, 1983), and eucricetodont cricetids (Martin, 1980). All of these rodents were at their highest levels of diversity during the Arikareean and diminished rapidly afterwards. Of these groups, the aplodontids appear to have been the only immigrant taxa that are derived from earlier rodents from Eurasia (Wang,

1987). The remainder of these rapidly radiating groups were likely derived from earlier North American forms.

The Arikareean was the time of the first record of the Mylagaulidae which was represented by as many as five species and three genera (Rensberger, 1979; Korth, 1992a). Marmotine sciurids also first appeared during the Arikareean (Black, 1963). At this time there was a reinvasion of eomyids from Europe (Fahlbusch, 1973, 1979; Korth, 1992a).

The unusual rodent *Zetamys* was known only from the Arikareean of Nebraska. Its affinities are unclear but may be related to earlier north African phiomyids (Martin, 1974).

1.10. Hemingfordian

The composition of the North American rodent fauna once again underwent a drastic change in the Hemingfordian. This change was not so much in the number of families of rodents represented as in the relative diversities of the families present. The rodent groups with the dramatic increases in diversity in the Arikareean were reduced to only a few species or became extinct.

The Heteromyidae were the only family that had a marked increase in diversity. This diversity included the introduction of the Perognathinae (Korth and Reynolds, 1991). Within the Sciuridae, the marmotines were becoming much more diverse, a trend that continued until the end of the Tertiary (Black, 1963), and there was the first occurrence of petauristines in North America (Skwara, 1986; Pratt and Morgan, 1989). The palaeocastorine beavers disappeared and were replaced by castoroidine castoroids (Stirton, 1935).

The Hemingfordian was a time of transition for the Cricetidae. During this time there was the last record of the archaic eucricetodonts (Martin and Corner, 1980; Martin and Green, 1984) and the first record of peromyscines (Slaughter and Ubelaker, 1984). This did not appear to be a case of replacement of the eucricetodonts by the peromyscines. Rather, the former may have been extinct long before the appearance of the latter. Geographically these occurrences were not coincident; the last eucricetodonts were from the northern Great Plains and the first peromyscines from Texas. The peromyscines were clearly immigrants from Eurasia and not evolved from North American eumyines or eucricetodonts (Martin, 1980; Jacobs and Lindsay, 1984).

1.11. Barstovian

During the Barstovian there was a shift to a rodent fauna that more closely reflected that of Recent times including the last occurrence of some older rodent families; Eutypomyidae, Florentiamyidae, and Heliscomyidae (Voorhies, 1990a; Chapter 15, Section 3).

The remainder of the changes in the Barstovian rodent fauna over that of the Hemingfordian occurred within the families of rodents present at the subfamily level rather than the introduction of new families. Geomyine geomyids first appeared and replaced the earlier entoptychines (Lindsay, 1972). The peromyscine cricetids began to diversify, but were limited to only two genera, *Copemys* and *Tregomys* (Lindsay, 1972; Voorhies, 1990a). Heteromyids were at their greatest diversity of the Tertiary, and the perognathines became the dominant subfamily (Lindsay, 1972; Storer, 1975; Korth, 1979). Zapodids, although never very diverse in North America, had their greatest diversity during the Barstovian (Green, 1977), and the first zapodines evolved (Klingener, 1966).

Two problematical geomyoids were also reported from the early Barstovian, *Jimomys* (Wahlert, 1976) and *Texomys* (Slaughter, 1981).

1.12. Clarendonian

There was little change in the rodent fauna during the Clarendonian. The first probable sigmodontine cricetid was reported from Florida (Baskin, 1986) and the first dipodomyine heteromyid was reported from Nebraska (Voorhies, 1975). All other families showed no marked change in relative diversity or introduction of new subfamily groups.

1.13. Hemphillian

The Hemphillian was, again, a time of transition for the North American rodent fauna. The Mylagaulidae and Eomyidae became extinct although the latter persisted in Europe until the Pleistocene (Fahlbusch, 1979). The diversity of cricetids increased more than threefold with the introduction of "microtines," neotomyines and greater diversity of sigmodontines and peromyscines.

A number of immigrant taxa characterized the Hemphillian. Along with the first microtines (which may represent more than a single invasion; Repenning, 1987), there was the first occurrence of *Castor* (Tedford et al., 1987).

1.14. Blancan

Only seven families of rodents were represented in the fossil record of the Blancan. This number was less than in any other Tertiary horizon since the Wasatchian. Three of the families were at their greatest diversity.

Cricetids, especially the microtines, were the rodents with the greatest number of species in the Blancan. The number of sigmodontines, neotomyines, and peromyscines doubled from that of the Hemphillian. Microtines increased from only 4 species in the Hemphillian to over 30 in the Blancan (Repenning, 1987). In the case of microtines, it appears that much of the increase in diversity was the result of immigration rather than speciation events of indigenous North American species.

Sciurids were also at their greatest diversity of the Tertiary during Blancan times. This diversity was entirely limited to the marmotine ground squirrels. Castorids were at their greatest diversity since the Arikareean. However, the castorid radiation was mainly within the castoroidines, whereas the earlier radiation was of palaeocastorines. This same pattern was repeated in the geomyids that were also at their greatest diversity since the Arikareean. Similarly, the geomyid diversity was within the Geomyinae rather than the Arikareean entoptychines.

Heteromyid species were predominantly those of dipodomyines and to a lesser extent the perognathines. The Zapodidae were represented by three species of the Recent *Zapus* (Klingener, 1963).

Near the end of the Blancan, hystricomorphous–hystricognathous rodents (Caviomorpha) were introduced from South America into North America (Wilson, 1935c; Savage and Russell, 1983).

2. Pleistocene

2.1. Irvingtonian

Nine families of rodents were represented in the fossil record of the Irvingtonian. The Sciuridae remained dominated by the marmotines, including the first record of *Cynomys* (prairie dog). However, the Sciurini first appeared after having been absent from the fossil record since the Arikareean. Similarly, petauristines reappeared in the fossil record after their absence since the Clarendonian. However, the lack of tree squirrels and petauristines in the fossil record of the later Tertiary may have resulted from the poor preservation potential of their woodland habitats (Kurten and Anderson, 1980).

The Geomyidae were not as diverse as in the Blancan and were limited to the derived genera of geomyines. Heteormyids consisted of species of the Recent genera *Dipodomys* and *Perognathus*, but also included species of the predominantly Blancan *Prodipodomys* (Hibbard, 1962).

The Castoridae were still dominated by the castoroidines, especially the largest genus, *Castoroides*, although three species of *Castor* were also known from Irvingtonian horizons (Kurten and Anderson, 1980).

The Cricetidae were the dominant rodents in the Irvingtonian. Sigmodontine, peromyscine, and neotomyine cricetids all flourished at this time. There were three dispersal events among the microtines resulting in immigrant species into North America beginning with the first appearance of

Microtus in North America at the beginning of the Irvingtonian (Repenning *et al.*, 1990). As in the Blancan, over 30 species of microtines have been described from the Irvingtonian (Kurten and Anderson, 1980; Savage and Russell, 1983).

The Zapodidae were surprisingly diverse in the Irvingtonian, being represented by four genera and six species including a sicistine (Klingener, 1963; Martin, 1989). This matched the greatest diversity of the zapodids in North America during the entire Tertiary.

Two families of immigrant caviomorph rodents from South America were reported from the Irvingtonian, the porcupines, Erethizontidae, and capybaras, Hydrochoeridae (Simpson, 1928; White, 1970).

2.2. Rancholabrean

Over 80% of the reported fossil rodent species from the Rancholabrean survived until Recent times (Savage and Russell, 1983). The Aplodontidae lacked a fossil record in the Blancan but were represented by the Recent species *Aplodontia rufa* in the Rancholabrean (Kurten and Anderson, 1980).

The majority of the rodent families in the Rancholabrean increased in diversity from their known records in the Irvingtonian. A few families (Castoridae and Zapodidae) reduced in variety. The Castoridae were limited to species of *Castor* and *Castoroides*, the latter becoming extinct by the end of the Rancholabrean and the former reduced in diversity to a single species. Zapodids declined to the two Recent genera, *Zapus* and *Napaeozapus* (Klingener, 1963; Guilday *et al.*, 1964).

The Heteromyidae were dominated by two genera, the dipodomyine *Dipodomys* and the perognathine *Perognathus* (Savage and Russell, 1983), but there was the first record of a heteromyine, *Liomys*, from the Rancholabrean of Mexico (Jakway, 1958). This was the first record of a heteromyine since the Hemphillian. Among the cricetids, the genera that made up the majority of the diversity of the family were *Peromyscus* and *Microtus*.

The same two families of South American caviomorphs were represented in the Rancholabrean as in the Irvingtonian (Hay, 1926; Simpson, 1928; White, 1970; Voorhies, 1985).

Chapter 25
Tertiary Rodent Faunas of Other Continents

1. Europe	279
1.1. Eocene	279
1.2. Oligocene	281
1.3. Miocene	281
1.4. Pliocene	282
2. Asia	282
2.1. Eocene	282
2.2. Oligocene	282
2.3. Miocene	282
2.4. Pliocene	283
3. Africa	283
3.1. Eocene	283
3.2. Oligocene	283
3.3. Miocene	283
3.4. Pliocene	284
4. South America	284
4.1. Oligocene	284
4.2. Miocene	284
4.3. Pliocene	285

1. Europe

1.1. Eocene

As in North America, the earliest rodents from Europe (Sparnacian) belonged to primitive Ischyromyidae (Godinot, 1981) (Table 25.1). These ischyromyids were likely immigrants. By the end of the early Eocene (Cuisian) there were also representatives of the hystricomorphous–sciuromorphous family Theridomyidae and myomorphous Gliridae (Savage and Russell, 1983). The glirids were derived from a reithroparamyine ancestor (*Microparamys*) in Europe (Hartenberger, 1971) but the origin of theridomyids is uncertain (Hartenberger, 1969). This proportion of rodents, numerous ischyromyids and rare glirids and theridomyids, was maintained until the later Eocene (Robiacian) when theridomyids began to rapidly diversify, glirids increased in number, and ischyromyids were greatly reduced in diversity.

By the end of the Eocene, theridomyids were at their peak of diversity

TABLE 25.1. Occurrence of Rodents in North America[a]

	Ck	W	Br	U	D	Ch	O	Wh	A	Hf	B	Cl	Hp	Bl
Ischyromyidae	5	23	20	24	8	10	3	1						
Sciuravidae		5	13	7										
Cylindrodontidae		1	4	6	3	10								
Protoptychidae			?1	2	1									
Aplodontidae				1	1	3	7	6	22	1	3	2	1	
Mylagaulidae									5	6	4	4	3	
Sciuridae						1	4	2	5	8	10	6	11	16
Eutypomyidae		1	1	2	4	2	2		1	1	1			
Castoridae						1	2	1	13	1	7	4	5	8
Eomyidae				3	5	20	12		3	2	4	1	3	
Heliscomyidae					1	1	5		1	1	1			
Heteromyidae							1		6	14	21	11	9	12
Florentiamyidae							1	1	10	2	1			
Geomyidae							?1		25	3	5	4	5	12
Zapodidae			?1		1				2	3	6	2	2	3
Cricetidae						2	10	2	15	3	9	6	21	42
Arvicolidae													4	31
Armintomyidae				1										
Loredomyidae				1										
Simimyidae				1	2									
Diplolophus						1								
Griphomys				1	2									
Guanajuatomys						1								
Jimomys									1		1			
Marfilomys						1								
Nonomys						1								
Pipestoneomys						2								
Texomys										1	1			
Zetamys									1					

[a]Numbers represent number of species described from each land-mammal age. Question mark indicates species tentatively referred to this family.
Abbreviations: Ck, Clarkforkian; W, Wasatchian; Br, Bridgerian; U, Uintan; D, Duchesnean; Ch, Chadronian; O, Orellan; Wh, Whitneyan; A, Arikareean; Hf, Hemingfordian; B, Barstovian; Cl, Clarendonian; Hp, Hemphillian; Bl, Blancan.

(Hartenberger, 1973), the Ischyromyidae had their last appearance, and glirids showed an increased diversity.

1.2. Oligocene

The early Oligocene rodent fauna of Europe, like that of the latest Eocene, was dominated by species of theridomyids. However, at this time there was the first appearance of a number of rodent families: Eomyidae, Aplodontidae, Sciuridae, Castoridae, and Cricetidae; possibly all immigrant taxa (Hartenberger, 1989). Of these families, the Eomyidae arrived from North America (Fahlbusch, 1973, 1979) and the Cricetidae were likely immigrants from Asia (Tong, 1992). The other families were more difficult to trace. The sciurids and aplodontids were known from earlier horizons in North America and may have immigrated from there. The origin of castorids is uncertain, initially appearing in Europe, Asia, and North America almost simultaneously (Martin, 1987). The sciurids present in the European Oligocene included the first occurrence of petauristines (Heissig, 1979), predating the arrival of sciurines in Europe.

In the middle Oligocene, theridomyids remained the most diverse group, but the Cricetidae rapidly increased in diversity (Vianey-Liaud, 1979). The aplodontids had their greatest diversity in the early Oligocene but diminished in number in the middle and late Oligocene (Schmidt-Kittler and Vianey-Liaud, 1979). The middle Oligocene marked the first occurrence of zapodids in Europe (Schaub, 1930).

The late Oligocene rodent fauna of Europe was similar to that of the middle but theridomyids began to be reduced in number of species and cricetids were the most diverse family of rodents. There were also lesser increases in the diversity of nearly all other rodent families present.

1.3. Miocene

The early Miocene (Agenian) rodent fauna differed only very little from that of the latest Oligocene. The number of species of glirids was nearly three times that in the Oligocene (Savage and Russell, 1983). This was also the last occurrence of theridomyids in Europe. The slightly younger Orleanian fauna was dominated by species of cricetids, glirids, and sciurids. Eomyids were also at their greatest diversity at this time, although they persisted in Europe until the Pleistocene (Fahlbusch, 1979).

The relative proportions of the rodent families remained essentially the same throughout the remainder of the Miocene with the introduction of Muridae (Schaub, 1938) and probable Spalacidae (Van de Weerd and Daams, 1979) from Asia in the later Miocene (Vallesian) and the first appearance of Hystricidae in the latest Miocene (Turolian).

1.4. Pliocene

The rodent fauna in the Pliocene of Europe was dominated by species of cricetids and murids (Savage and Russell, 1983). Sciurids and glirids were numerous but not as diverse as the muroid rodents. All other families of rodents continued at approximately the same level of diversity as they had in the Miocene throughout the Pliocene.

2. Asia

2.1. Eocene

Unlike Europe and North America, the Eocene faunas of Asia were dominated by ctenodactyloid rodents (Hussain *et al.*, 1978; Dawson *et al.*, 1984). There were only rare occurrences of ischyromyids (Li, 1963; Dawson, 1968b; Shevyreva, 1984; Qi, 1987). In the middle to late Eocene, the very first cricetid appeared (Tong, 1992), and there is evidence of an early zapodid (Li and Ting, 1983). The probable sciuravid or eomyid *Zelomys* also was described from the late Eocene (Wang and Li, 1990). The ischyromyids, *Zelomys*, and cricetids all may have been immigrants from North America where the most likely ancestors of all of these rodents may be found.

2.2. Oligocene

The Oligocene of Asia was similarly dominated by species of ctenodactyloids (Wood, 1977). However, a number of other rodent families invaded Asia: Cylindrodontidae (Wood, 1974), Zapodidae (Bohlin, 1946), Aplodontidae (Rensberger and Li, 1986; Wang, 1987), and Eomyidae (Wang and Emry, 1991). The problematical Tsaganomyidae were restricted to the Oligocene of Asia (Wood, 1974). The first occurrence of castorids was also in the early Oligocene of Asia, nearly simultaneous with their first occurrence in both Europe and North America (Martin, 1987).

The first spalacids (tachyoryctoidines) appear to have evolved in Asia at this time (Flynn *et al.*, 1985). Similarly, by the latest Oligocene or earliest Miocene the first dipodoid *Protalactaga* also had evolved in Asia (Wood, 1936d; Argyropulo, 1939).

2.3. Miocene

During the Miocene the ctenodactyloids lost their dominance in the Asian rodent fauna, and are lacking in beds of late Miocene age and younger. Murids first evolved in the middle Miocene of Asia (Jacobs, 1977b) as did rhizomyids (Flynn, 1982). The African hystricomorphous–hystricognathous Thryonomyidae were known from the middle Miocene of Asia (Hinton, 1933). Hystricids first appeared in Asia at this time as well.

The late Miocene saw the greatest diversity of zapodids in Asia (Savage and Russell, 1983), postdating the Barstovian radiation of zapodids in North America (Chapter 19, Section 3).

2.4. Pliocene

The Pliocene rodent fauna of Asia, as in North America and Europe, was dominated by muroid rodents. The European and Asian rodent faunas differed from that of North America in the presence of murids that are lacking in North America. The cricetid and arvicolid radiations in the Pliocene of Europe and Asia were echoed in North America (Repenning, 1987).

The remainder of the rodent fauna from the Pliocene of Asia was similar to that of the late Miocene, containing members of the Rhizomyidae, Castoridae, Hystricidae, and Dipodidae.

3. Africa

3.1. Eocene

Only two rodent species have been described from the late Eocene of Africa, a phiomyid and an anomalurid (Jaeger et al., 1985). The Phiomyidae were hystricomorphous–hystricognathous and the Anomaluridae were hystricomorphous–sciurognathous. The origin of both families is uncertain but they may have evolved in Africa.

3.2. Oligocene

The entire rodent fauna of the Oligocene of Africa was limited to species of phiomyoid rodents from Egypt, included in one family, Phiomyidae, by Wood (1968) but separated into three families by Lavocat (1973).

3.3. Miocene

The phiomyoid rodents were greatly diversified by the early Miocene. At this time there was also the first record of the protrogomorphous–hystricognathous bathyergoids and the hystricomorphous–sciurognathous pedetids (Lavocat, 1973). The bathyergoids appear to have evolved in Africa from a phiomyoid ancestor, but the ancestry of the Pedetidae is uncertain. The hystricomorphous–hystricognathous Thryonomyidae evolved in Africa and dispersed into Asia in the early Miocene. Along with this interchange with Eurasia, Sciuridae and Cricetidae invaded Africa for the first time (Lavocat, 1973).

By the late Miocene, phiomyoids had waned in number and disappeared from the fossil record of Africa. Ctenodactylids, originally from Asia, invaded Africa just as they were becoming extinct in Asia (Lavocat, 1961). The Gliridae of Europe also spread into Africa by the late Miocene (Ternanian).

3.4. Pliocene

The only major change in the African rodent fauna in the Pliocene was the introduction of murids from Eurasia. They first appeared in the early Pliocene (Slaughter and James, 1979) and rapidly radiated to the most numerous type of rodents on the continent (Lavocat, 1978; Savage and Russell, 1983). The remainder of the African Pliocene rodent fauna was similar to that of the Miocene as well as the modern fauna including: Thryonomyidae, Hystricidae, Bathyergidae, Ctenodactylidae, Anomaluridae, Pedetidae, Cricetidae, and Sciuridae.

4. South America

4.1. Oligocene

The fossil record of rodents in South America was limited to the radiation of endemic forms throughout the Tertiary until the union of the North and South American continents near the end of the Tertiary. The origin of the South American hystricomorphous–hystricognathous rodents (= Caviomorpha) is in question. Some authors believe that they originated from North American rodents that rafted to South America, while others argue that an immigration from Africa is more likely (see George, 1993, for summary of proposed theories). Regardless of their origin, rodents first appeared in South America in the early Oligocene (Deseadan) and were already diversified into at least six recognizable families: Octodontidae, Echimyidae, Chinchillidae, Dasyproctidae, Eocardiidae, and Erethizontidae (Wood, 1949; Wood and Patterson, 1959).

An additional family, the Cephalomyidae, appeared later in the Oligocene (Colhuehuapian), and the erethizontids became the most diverse group. No other changes occurred throughout the Oligocene.

4.2. Miocene

The early Miocene fauna was merely a continuation of the Oligocene fauna with only a minor change in proportions: the Erethizontidae diminished in variety while the other families increased (Ameghino, 1887). It was the last occurrence of the Eocardiidae. However, in the middle Miocene (Friasian) two new families evolved, the Caviidae and Dinomyidae (Fields, 1957; Hirschfeld and Marshall, 1976).

By the late Miocene (Huayquerian), two additional caviomorph families appeared in the fossil record, the Abrocomidae and Hydrochoeridae (Savage and Russell, 1983).

4.3. Pliocene

The early Pliocene (Montehermosan) marked the first record of an invasion of rodents from North America. The only family of North American rodents to reach South America at this time was the Cricetidae, as part of the sigmodontine radiation (Reig, 1978). Another caviomorph family had its first occurrence, the Myocastoridae (Rovereto, 1914). The remainder of the fauna was relatively constant with fluctuations in the relative number of species of each of the indigenous families. The late Pliocene fauna (Chapadmalalan) was only a continuation of the earlier Pliocene fauna with an increase in the number of sigmodontine cricetids (Marshall et al., 1983).

References

Akersten, W. A., 1973, Evolution of geomyine rodents with rooted cheek teeth, unpublished Ph.D. dissertation, University of Michigan.

Akersten, W. A., 1988, Affinities of "*Pliosaccomys*" and "*Parapliosaccomys*" from the Great Plains, (abstract), *J. Vertebr. Paleontol.* (Suppl.) **3**:8A.

Alf, R., 1962, A new species of the rodent *Pipestoneomys* from the Oligocene of Nebraska, *Breviora* **172**:1–7.

Alston, E. R., 1876, On the classification of the order Glires, *Proc. Zool. Soc. London* **1876**:61–98.

Ameghino, F., 1887, Enumeración sistemática de los especies de mamíferos fósiles coleccionados por Carlos Ameghino en los terrenos eocenos de la Patagonia austral y depositados en el Museo de la Plata, *Bol. Mus. La Plata* **1**:1–26.

Argyropulo, A. I., 1939, Sciuromorpha and Dipodidae (Glires, Mammalia) in the Tertiary of Kazakhstan, *C. R. Acad. Sci. URSS* **25**:171–175.

Barbour, E. H., and Stout, T. M., 1939, The White River Oligocene rodent *Diplolophus*, *Bull. Univ. Nebr. State Mus.* **2**:29–36.

Barnosky, A. D., 1986a, Arikareean, Hemingfordian, and Barstovian mammals from the Miocene Colter Formation, Jackson Hole, Teton County, Wyoming, *Bull. Carnegie Mus. Nat. Hist.* **26**:1–69.

Barnosky, A. D., 1986b, New species of the Miocene rodent *Cupidinimus* (Heteromyidae) and some evolutionary relationships within the genus, *J. Vertebr. Paleontol.* **6**:46–64.

Baskin, J. A., 1980, Evolutionary reversal in *Mylagaulus* (Mammalia, Rodentia) from the late Miocene of Florida, *Am. Mid. Nat.* **104**:155–162.

Baskin, J. A., 1986, The late Miocene radiation of Neotropical sigmodontine rodents in North America, in: *Vertebrates, Phylogeny, and Philosophy* (K. M. Flanagan and J. A. Lillegraven, eds.), pp. 287–304, *Contrib. Geol. Univ. Wyo. Spec. Pap.* 3.

Berggren, W. A., Kent, D. V., Flynn, J. J., and Van Couvering, J. A., 1985, Cenozoic geochronology, *Bull. Geol. Soc. Am.* **96**:1407–1418.

Black, C. C., 1961, Rodents and lagomorphs from the Miocene Fort Logan and Deep River Formations of Montana, *Postilla* **48**:1–20.

Black, C. C., 1963, A review of the North American Tertiary Sciuridae, *Bull. Mus. Comp. Zool. Harvard Univ.* **130**:109–248.

Black, C. C., 1965, Fossil mammals from Montana. Part 2. Rodents from the early Oligocene Pipestone Springs local fauna, *Ann. Carnegie Mus.* **38**:1–48.

Black, C. C., 1968, The Oligocene rodent *Ischyromys* and discussion of the family Ischyromyidae, *Ann. Carnegie Mus.* **39**:273–305.

Black, C. C., 1969, The fossil rodent genera *Horatiomys* and *Palustrimus*—Juvenile geomyoid rodents, *J. Mammal.* **50**:815–817.
Black, C. C., 1971, Paleontology and geology of the Badwater Creek area, central Wyoming. Part 7. Rodents of the family Ischyromyidae, *Ann. Carnegie Mus.* **43**:179–217.
Black, C. C., 1974, Paleontology and geology of the Badwater Creek area, central Wyoming. Part 9. Additions to the cylindrodont rodents from the late Eocene, *Ann. Carnegie Mus.* **45**:151–160.
Black, C. C., and Kowalski, K., 1974. The Pliocene and Pleistocene Sciuridae (Mammalia, Rodentia) from Poland, *Acta Zool. Cracov.* **19**:461–485.
Black, C. C., and Stephens, J. J., III, 1973, Rodents from the Paleogene at Guanajuato, Mexico, *Occas. Pap. Mus. Tex. Tech Univ.* **14**:1–10.
Black, C. C., and Sutton, J. A., 1984, Paleocene and Eocene rodents of North America, in: *Papers in Vertebrate Paleontology Honoring Robert Warren Wilson* (R. M. Mengel, ed.), pp. 67–84, Carnegie Mus. Nat. Hist. Spec. Publ. 9.
Black, C. C., and Wood, A. E., 1956, Variation and tooth replacement in a Miocene mylagaulid rodent, *J. Paleontol.* **30**:672–684.
Bleedfeld, A. R., and McKenna, M. C., 1985, Skeletal integrity of *Mimolagus rodens* (Lagomorpha, Mammalia), *Am. Mus. Novit.* **2806**:1–5.
Bohlin, B., 1946, The fossil mammals from the Tertiary deposit of Taben-buluk, western Kansu. Part II: Simplicidentata, Carnivora, Artiodactyla, Perissodactyla, and Primates, *Palaeontol. Sin. New Ser. 8b* **123**:1–259.
Borisoglebskaya, M. B., 1967, New genus of beaver from the Oligocene of Kazakhstan, *Byull. Mosk. Ova. Ispyt. Prir. Otd. Biol.* **72**:129–135 (in Russian).
Bown, T. M., 1982, Geology, paleontology, and correlation of Eocene volcaniclastic rocks, southeast Absaroka Range, Hot Springs County, Wyoming, *U.S. Geol. Surv. Prof. Pap.* **1201–A**:1–75.
Brandt, J. F., 1855, Beiträge zur náhern Kenntnis der Sáugethiere Russlands, *Mem. Acad. Imp. Sci. St. Petersburg Ser. 6* **9**:1–375.
Brylski, P., 1990, Development and evolution of the carotid circulation in geomyoid rodents in relationship to their craniomorphology, *J. Morphol.* **204**:33–45.
Bugge, J., 1971, The cephalic arterial system in mole-rats (Spalacidae) bamboo rats (Rhizomyidae), jumping mice and jerboas (Dipodoidea) and dormice (Glirioidea) with special reference to the systematic classification of rodents, *Acta Anat.* **79**: 165–180.
Bugge, J., 1974, The cephalic arterial system in insectivores, primates, rodents and lagomorphs, with special reference to the systematic classification, *Acta Anat.* **87**(Suppl. 62):1–160.
Burke, J. J., 1934, New Duchesne River rodents and a preliminary survey of the Adjidaumidae, *Ann. Carnegie Mus.* **23**:391–398.
Burke, J. J., 1936, *Ardynomys* and *Desmatolagus* in North American Oligocene, *Ann. Carnegie Mus.* **25**:135–154.
Carleton, M. D., and Musser, G. G., 1984, Muroid rodents, in: *Orders and Families of Recent Mammals of the World* (S. Anderson and J. K. Jones, Jr., eds.) pp. 289–379, Wiley, New York.
Carroll, R. L., 1987, *Vertebrate Paleontology and Evolution*, Freeman, San Francisco.
Cassiliano, M., 1980, Stratigraphy and vertebrate paleontology of the Horse Creek–Trial Creek area, Laramie County Wyoming, *Contrib. Geol. Univ. Wyo.* **19**:25–68.
Chaline, J., and Mein, P., 1979, *Les Rongeurs et l'Evolution*, Doin, Paris.
Chaline, J., Mein, P., and Petter, F., 1977, Les grandes lignes d'une classification évolutive des Muroidea, *Mammalia* **41**:245–252.
Cifelli, R. L., Ibui, A. K., Jacobs, L. L., and Thorington, R. W., Jr., 1986, A giant tree squirrel from the late Miocene of Kenya, *J. Mammal.* **67**:274–283.
Clark, J. B., Dawson, M. R., and Wood, A. E., 1964, Fossil mammals from the Lower Pliocene of Fish Lake Valley, Nevada, *Bull. Mus. Comp. Zool.* **131**:27–63.
Cope, E. D., 1874, On a new mastodon and rodent, *Proc. Acad. Nat. Sci. Philadelphia* **1874**: 221–223.
Cope, E. D., 1884, The Vertebrata of the Tertiary formations of the West, *Rep. U. S. Geol. Surv. Terr.* **3**:1–1009.

Cuvier, G., 1817, *La réne animal distribué d'aprés son organisation*, Vol. 1, Paris.
Dalquest, W. W., 1983, Mammals of the Coffee Ranch local fauna, Hemphillian of Texas, *Pearce-Sellards Ser. Tex. Mem. Mus.* **38**:1–41.
Dalquest, W. W., and Patrick D. B., 1989, Small mammals from the early and medial Hemphillian of Texas, with descriptions of a new bat and gopher, *J. Vertebr. Paleontol.* **9**:78–88.
Dashzeveg, D., 1990, New trends in adaptive radiation of early Tertiary rodents (Rodentia, Mammalia), *Acta Zool. Cracov.* **33**:37–44.
Dashzeveg, D., and Russell, D. E., 1988, Palaeocene and Eocene Mixodontia (Mammalia, Glires) of Mongolia and China, *Palaeontology* **31**:129–164.
Dawson, M. R., 1961, The skull of *Sciuravus nitidus*, a middle Eocene rodent, *Postilla* **53**:1–13.
Dawson, M. R., 1962, A sciuravid rodent from the middle Eocene of Wyoming, *Am. Mus. Novit.* **2075**:1–5.
Dawson, M. R., 1966, Additional late Eocene rodents (Mammalia) from the Uinta Basin, Utah, *Ann. Carnegie Mus.* **38**:98–144.
Dawson, M. R., 1967, Lagomorph history and the stratigraphic record, in: *Essays in Paleontology and Stratigraphy, R. C. Moore Commemorative Volume* (C. Teichert and E. L. Yockelson, eds.), *Dept. Geol. Univ. Kans. Spec. Publ.* **2**:287–316.
Dawson, M. R., 1968a, Middle Eocene rodents (Mammalia) from northeastern Utah, *Ann. Carnegie Mus.* **39**:327–370.
Dawson, M. R., 1968b, Oligocene rodents (Mammalia) from the East Mesa, Inner Mongolia, *Am. Mus. Novit.* **2324**:1–12.
Dawson, M. R., 1977, Late Eocene rodent radiations: North America, Europe and Asia, *Geobios Mem. Spec.* **1**:195–209.
Dawson, M. R., and Krishtalka, L., 1984, Fossil history of the families of Recent mammals, in: *Orders and Families of Recent Mammals of the World* (S. Anderson and J. K. Jones, Jr., eds.), pp. 11–57, Wiley, New York.
Dawson, M. R., Li, C. K., and Qi, T., 1984, Eocene ctenodactyloid rodents (Mammalia) of eastern and central Asia, in: *Papers in Vertebrate Paleontology Honoring Robert Warren Wilson* (R. M. Mengel, ed.), pp. 138–150, *Carnegie Mus. Nat. Hist. Spec. Publ.* 9.
Dawson, M. R., Krishtalka, L., and Stucky, R. K., 1990, Revision of the Wind River faunas, early Eocene of central Wyoming. Part 9. The oldest known hystricomorphous rodent (Mammalia: Rodentia), *Ann. Carnegie Mus.* **59**:135–147.
de Blainville, H. M. D., 1816, Prodrome d'une nouvelle distribution systématique du régne animal, *Bull. Soc. Philomath. Paris Ser.* 3, **3**:105–124.
de Bruijn, H., and Unay, E., 1989, Petauristinae (Mammalia, Rodentia) from the Oligocene of Spain, Belgium, and Turkish Thrace, in: *Papers on Fossil Rodents in Honor of Albert Elmer Wood* (C. C. Black and M. R. Dawson, eds.), *Sci. Ser. Nat. Hist. Mus. Los Angeles Co.* **33**:139–146.
de Jong, W. W., 1985, Superordinal affinities of Rodentia studied by sequence analysis of eye lens protein, in: *Evolutionary Relationships among Rodents: A Multidisciplinary Analysis* (W. P. Luckett and J.-L. Hartenberger, eds.), pp. 211–226, Plenum Press, New York.
Dehm, R., 1950, Die Nagetiere aus dem Mittel-Miocan (Burdigalium) von Sintershof-West bei Eichstatt in Bayern, *Neues Jahrb. Mineral. Geol. Palaeontol. Abh. Abt.* **91**:321–428.
Donohoe, J. C., 1956, New aplodontid rodent from Montana Oligocene, *J. Mammal.* **37**:264–268.
Douglass, E., 1902, Fossil Mammalia of the White River beds of Montana, *Trans. Am. Philos. Soc.* **20**:230–279.
Downs, T., 1956, The Mascall fauna from the Miocene of Oregon, *Univ. Calif. Publ. Geol. Sci.* **31**:199–354.
Ellerman, J. R., 1940, *The Families and Genera of Living Rodents*, Vols. I and II, British Museums (Natural History), London.
Emry, R. J., 1972a, A new species of *Agnotocastor* (Rodentia, Castoridae) from the early Oligocene of Wyoming, *Am. Mus. Novit.* **2485**:1–7.
Emry, R. J., 1972b, A new heteromyid rodent from the early Oligocene of Natrona County, Wyoming, *Proc. Biol. Soc. Washington* **85**:175–190.

Emry, R. J., 1981, New material of the Oligocene muroid rodent Nonomys, and its bearing on muroid origins, Am. Mus. Novit. **2712**:1–14.

Emry, R. J., and Dawson, M. R., 1972, A unique cricetid (Rodentia, Mammalia) from the early Oligocene of Natrona County, Wyoming, Am. Mus. Novit. **2508**:1–14.

Emry, R. J., and Dawson, M. R., 1973, Nonomys, a new name for the cricetid (Rodentia, Mammalia) genus Nanomys Emry and Dawson, J. Paleontol. **47**:1003.

Emry, R. J., and Korth, W. W., 1989, Rodents of the Bridgerian (middle Eocene) Elderberry Canyon local fauna of eastern Nevada, Smithsonian Contrib. Paleobiol. **67**:1–14.

Emry, R. J., and Korth, W. W., 1993, Evolution in Yoderimyinae (Eomyidae: Rodentia), with new material from the White River Formation (Chadronian) at Flagstaff Rim, Wyoming, J. Paleontol. **67**:1047–1057.

Emry, R. J., and Thorington, R. W., 1982, Descriptive and comparative osteology of the oldest fossil squirrel, Protosciurus (Rodentia: Sciuridae), Smithsonian Contrib. Paleobiol. **47**:1–35.

Engesser, B., 1979, Relationships of some insectivores and rodents from the Miocene of North America and Europe, Bull. Carnegie Mus. Nat. Hist. **14**:1–68.

Engesser, B., 1990, Die Eomyidae (Rodentia, Mammalia) der Molasse der Schweiz und Savoyens, Schweiz. Palaeontol. Abh. **112**:1–144.

Eshelman, R. E., 1975, Geology and paleontology of the early Pleistocene (late Blancan) White Rock fauna from northcentral Kansas, Univ. Mich. Pap. Paleontol. **13**:1–60.

Fagan, S. R., 1960, Osteology of Mylagaulus laevis, a fossorial rodent from the Upper Miocene of Colorado, Univ. Kans. Paleontol. Contrib. Vert. **9**:1–32.

Fahlbusch, V., 1970, Populationsverschievungen bei tertiären Nagetieren, eine Studie an oligozänen und miozänen Eomyidae Europas, Bayer. Akad. Wiss. Math. Naturwiss. Kl. Abh. **145**:1–136.

Fahlbusch, V., 1973, Die stammesgeschichtlichen Beziehungen zwischen den Eomyiden (Mammalia, Rodenta) Nordamerikas und Europas, Mitt. Bayer. Staatssamml. Palaeontol. Hist. Geol. **13**:141–175.

Fahlbusch, V., 1979, Eomyidae—Geschichte einer Säugetierfamilie, Palaeontol. Z. **53**:88–97.

Fahlbusch, V., 1985, Origin and evolutionary relationships among geomyoids, in: Evolutionary Relationships among Rodents: A Multidisciplinary Analysis (W. P. Luckett and J.-L. Hartenberger, eds.), pp. 617–630, Plenum Press, New York.

Ferrusquia-Villafranca, I., 1989, A new rodent genus from central Mexico and its bearing on the origin of the Caviomorpha, in: Papers on Fossil Rodents in Honor of Albert Elmer Wood (C. C. Black and M. R. Dawson, eds.), Sci. Ser. Nat. Hist. Mus. Los Angeles Co. **33**:91–118.

Fields, R. W., 1957, Hystricomorph rodents from the late Miocene of Colombia, South America, Univ. Calif. Publ. Geol. Sci. **32**:273–403.

Fischer de Waldheim, G. F., 1809, Sur l'Elasmotherium et le Trogontherium, deux animaux fossiles et inconnus de la Russie, Mem. Soc. Nat. Moscou **2**:250–268.

Flanagan, K. M., 1986, Early Eocene rodents from the San Jose Formation, San Juan Basin, New Mexico, in: Vertebrates, Phylogeny, and Philosophy (K. M. Flanagan and J. A. Lillegraven, eds.), Univ. Wyoming Contrib. Geol., Spec. Pap. **3**:197–220.

Flynn, L. J., 1982, Systematic revision of Siwalik Rhizomyidae (Rodentia), Geobios **15**:327–389.

Flynn, L. J., Jacobs, L. L., and Lindsay, E. H., 1985, Problems in muroid phylogeny: Relationship to other rodents and origin of major groups, in: Evolutionary Relationships among Rodents: A Multidisciplinary Analysis (W. P. Luckett and J.-L. Hartenberger, eds.), pp. 589–616, Plenum Press, New York.

Flynn, L. J., Jacobs, L. L., and Cheema, I. U., 1986, Baluchimyinae, a new ctenodactyloid rodent subfamily from the Miocene of Baluchistan, Am. Mus. Novit. **2841**:1–58.

Flynn, L. J., Russell, D. E., and Dashzeveg, D., 1987, New Glires (Mammalia) from the early Eocene of the People's Republic of Mongolia. Part II. Incisor morphology and enamel microstructure, Proc. K. Ned. Akad. Wet. Ser. B **90**:143–154.

Forsyth Major, C. J., 1893, On some Miocene squirrels, Proc. Zool. Soc. London **1893**:179–215.

Freudenthal, M., 1970, A new Ruscinomys (Mammalia, Rodentia) from the late Tertiary (Pikermian) of Samos, Greece, Am. Mus. Novit. **2402**:1–10.

Friant, M., 1932, Contribution á l'étude de la differentiation des dents jugales chez les mammiféres, *Publ. Mus Hist. Nat.* **1**:1–132.
Galbreath, E. C., 1948, An additional specimen of the rodent *Dikkomys* from the Miocene of Nebraska, *Trans. Kans. Acad. Sci.* **51**:316–317.
Galbreath, E. C., 1953. A contribution to the Tertiary geology and paleontology of northeastern Colorado, *Univ. Kans. Paleontol. Contrib. Vertebr.* **4**:1–120.
Galbreath, E. C., 1984, On *Mesogaulus paniensis* (Rodentia) from Hemingfordian (middle Miocene) deposits of northeastern Colorado, in: *Papers in Vertebrate Paleontology Honoring Robert Warren Wilson* (R. M. Mengel, ed.), pp. 85–89, *Carnegie Mus. Nat. Hist. Spec. Publ.* 9.
Gazin, C. L., 1961, New sciuravid rodents from the Lower Eocene Knight Formation of western Wyoming, *Proc. Biol. Soc. Washington* **74**:193–194.
Gazin, C. L., 1962, A further study of the Lower Eocene mammalian faunas of southwestern Wyoming, *Smithsonian Misc. Collect.* **144**:1–98.
Geoffroy Saint-Hilaire, E., 1833, *Revue Encyclopedique* **59**:95.
George, W., 1993, The strange rodents of Africa and South America, in: *The Africa–South America Connection* (W. George and R. Lavocat, eds.), pp. 119–141, Oxford University Press (Clarendon), London.
Gidley, J. W., 1907, A new horned rodent from the Miocene of Kansas, *Proc. U. S. Nat. Mus.* **32**:627–636.
Gidley, J. W., 1912, The lagomorphs an independent order, *Science* **36**:285–586.
Gingerich, P. D., 1976, Cranial anatomy and evolution of early Tertiary Plesiadapidae (Mammalia, Primates), *Univ. Mich. Pap. Paleontol.* **15**:1–141.
Godinot, M., 1981, Les mammiféres de Rians (Eocéne Inférieur, Provence), *Palaeovertebrata* **10**:43–126.
Green, M., 1977, Neogene Zapodidae (Mammalia: Rodentia) from South Dakota, *J. Paleontol.* **51**:996–1015.
Green, M., 1992, Comments on North American fossil Zapodidae (Rodentia: Mammalia) with reference to *Megasminthus*, *Plesiosminthus*, and *Schaubeumys*, *Occas. Pap. Mus. Nat. Hist. Univ. Kansas* **148**:1–11.
Green, M., in press, The status of the genera *Proheteromys* Wood 1932 and *Mookomys* Wood 1931, *Occas. Pap. Mus. Nat. Hist. Univ. Kansas*.
Green, M., and Bjork, P. R., 1980, On the genus *Dikkomys* (Geomyoidea, Mammalia), in: *Palaeovertebra, Mémoire Jubilare en Hommage á René Lavocat* (J. Michaux, ed.), pp. 343–353.
Guilday, J. E., Martin, P. S., and McCrady, A. D., 1964, New Paris No. 4: A Pleistocene cave deposit in Bedford County, Pennsylvania, *Bull. Nat. Speleol. Soc.* **26**:121–194.
Guthrie, R. D., 1971, Factors regulating the evolution of microtine tooth complexity, *Z. Saeugetierkd.* **36**:37–54.
Hall, E. R., 1930, Rodents and lagomorphs from the later Tertiary of Fish Lake Valley, Nevada, *Univ. Calif. Publ. Geol. Sci.* **19**:295–312.
Harris, J. M., and Wood, A. E., 1969, A new genus of eomyid rodent from the Oligocene Ash Spring local fauna of Trans-Pecos, Texas, *Pearce-Sellards Ser. Tex. Mem. Mus.* **14**:1–7.
Hartenberger, J.-L., 1969, Les Pseudosciuridae (Mammalia, Rodentia) de l'Eocéne moyen de Bouxwiller Lissieu, Egerkingen, *Palaeovertebrata* **3**:27–61.
Hartenberger, J.-L., 1971, Contribution á l'étude des geners *Gliravus* et *Microparamys* (Rodentia) de l'Eocéne d'Europe, *Palaeovertebrata* **4**:97–135.
Hartenberger, J.-L., 1973, Étude systématique des Theridomyoidea (Rodentia) de l'Eocéne supérieur, *Mem. Soc. Geol. Fr.* **117**:1–76.
Hartenberger, J.-L., 1980, Donnée et hypothéses sur la radiation initiale des rongeurs, *Palaeovert., Mem. Jubil. en homage á René Lavocat* (J. Michaux, ed.), pp. 285–301.
Hartenberger, J.-L., 1985, The order Rodentia: Major questions on their evolutionary origin, relationships and suprafamilial systematics, in: *Evolutionary Relationships among Rodents: A Multidisciplinary Analysis* (W. P. Luckett and J.-L. Hartenberger, eds.), pp. 1–34, Plenum Press, New York.

Hartenberger, J.-L., 1989, Summary of Paleogene rodents in Europe, in: *Papers on Fossil Rodents in Honor of Albert Elmer Wood* (C. C. Black and M. R. Dawson, eds.), Sci. Ser. Nat. Hist. Mus. Los Angeles Co. **33**:119–128.

Hay, O. P., 1899, Notes on the nomenclature of some North America fossil vertebrates, *Science* **10**:253–254.

Hay, O. P., 1902, Bibliography and catalogue of the fossil Vertebrata of North America, *Bull. U. S. Geol. Surv.* **179**:1–868.

Hay, O. P., 1926, A collection of Pleistocene vertebrates from southwestern Texas, *Proc. U. S. Natl. Mus.* **68**:1–18.

Hay, O. P., 1930, Second bibliography and catalog of the fossil vertebrates of North America, *Carnegie Inst. Washington Publ.* **390**:1–1074.

Heaton, T. H., 1993, The Oligocene rodent *Ischyromys* of the Great Plains; replacement mistaken for anagenesis, *J. Paleontol.* **67**:297–308.

Heissig, K., 1979, Die fruehesten Flughoernchen und primitive Ailuravinae (Rodentia, Mammalia) aus dem sueddeutschen Oligozaen, *Mitt. Bayer. Staatssamml. Palaeontol. Hist. Geol.* **19**:139–169.

Hibbard, C. W., 1949, Techniques of collecting microvertebrate fossils, *Univ. Mich. Contrib. Mus. Paleontol.* **8**:7–19.

Hibbard, C. W., 1954, A new Pliocene vertebrate fauna from Oklahoma, *Pap. Mich. Acad. Sci.* **39**:339–359.

Hibbard, C. W., 1962, Two new rodents from the early Pleistocene of Idaho, *J. Mammal.* **43**:482–485.

Hibbard, C. W., 1964, A contribution to the Saw Rock Canyon local fauna of Kansas, *Pap. Mich. Acad. Sci.* **49**:115–127.

Hibbard, C. W., 1967, New rodents from the late Cenozoic of Kansas, *Pap. Mich. Acad. Sci. Arts Lett.* **52**:115–131.

Hibbard, C. W., 1970a, A new microtine rodent from the Upper Pliocene of Kansas, *Univ. Mich. Contrib. Mus. Paleontol.* **23**:99–103.

Hibbard, C. W., 1970b, The Pliocene rodent *Microtoscoptes disjunctus* (Wilson) from Idaho and Wyoming, *Univ. Mich. Contrib. Mus. Paleontol.* **23**:95–98.

Hibbard, C. W., 1972, Class Mammalia, in: *Early Pleistocene pre-glacial and glacial rocks and faunas of north-central Nebraska* (M. G. Skinner and C. W. Hibbard, eds.), *Bull. Am. Mus. Nat. Hist.* **148**:77–125.

Hibbard, C. W., and Dalquest, W. W., 1973, *Proneofiber*, a new genus of vole (Cricetidae: Rodentia) from the Pleistocene Seymour Formation of Texas, and its evolutionary and stratigraphic significance, *Quat. Res. (N.Y.)* **3**:269–274.

Hibbard, C. W., and Schultz, C. B., 1948, A new sciurid of Blancan age from Kansas and Nebraska, *Bull. Univ. Nebr. State Mus.* **3**:19–29.

Hibbard, C. W., and Zakrzewski, R. J., 1967, Phyletic trends in the late Cenozoic microtine *Ophiomys* gen. nov., from Idaho, *Univ. Mich. Contrib. Mus. Paleontol.* **21**:255–271.

Hinton, M. A. C., 1926, *Monograph of Voles and Lemmings*, Vol. 1, British Museum (Natural History).

Hinton, M. A. C., 1933, Diagnoses of new genera and species of rodents from Indian Tertiary deposits, *Ann. Mag. Nat. Hist.* Ser. 10, **12**:620–622.

Hirschfeld, S. E., and Marshall, L. G., 1976, Revised faunal list of the La Venta fauna (Friasian–Miocene) of Colombia, South America, *J. Paleontol.* **50**:433–436.

Hoffmeister, D. F., 1945, Cricetine rodents of the middle Pliocene Mulholand fauna, California, *J. Mammal.* **26**:186–190.

Howe, J. A., 1966, The Oligocene rodent *Ischyromys* in Nebraska, *J. Paleontol.* **40**:1200–1210.

Hrubesch, K., 1957, *Paracricetodon dehmi*, n. sp., ein neuer Nager aud dem Oligozän Mitteleuropas, *Neues Jahrb. Geol. Palaeontol. Abh.* **105**:250–271.

Hussain, S. T., de Bruijn, H., and Leinders, J. M., 1978, Middle Eocene rodents from the Kala Chitta Range (Punjab, Pakistan), *Proc. K. Ned. Akad. Wet.* **81**:74–112.

Ivy, L. D., 1990, Systematics of late Paleocene and early Eocene Rodentia (Mammalia) from the Clarks Fork Basin, Wyoming, *Univ. Mich. Contrib. Mus. Paleontol.* **28**:21–70.

Jacobs, L. L., 1977a, Rodents of the Hemphillian age Redington local fauna, San Pedro Valley, Arizona, *J. Paleontol.* **51**:505–519.

Jacobs, L. L., 1977b, A new genus of murid rodent from the Miocene of Pakistan and comments on the origin of Muridae, *PaleoBios* **25**:1–11.

Jacobs, L. L., and Lindsay, E. H., 1984, Holarctic radiation of Neogene muroid rodents and the origin of South American cricetids, *J. Vertebr. Paleontol.* **4**:265–272.

Jaeger, J.-J., 1976, Les rongeurs du Miocéne de Beni Mellal, *Palaeovertebrata* **7**:91–100.

Jaeger, J.-J., Denys, C., and Coiffait, B., 1985, New Phiomorpha and Anomaluridae from the late Eocene of north-west Africa: Phylogenetic implications, in: *Evolutionary Relationships among Rodents: A Multidisciplinary Analysis* (W. P. Luckett and J.-L. Hartenberger, eds.), pp. 567–588, Plenum Press, New York.

Jakway, G. E., 1958, Pleistocene lagomorpha and Rodentia from the San Josecito Cave, Nuevo Leon, Mexico, *Trans. Kans. Acad. Sci.* **61**:313–327.

James, G. T., 1963, Paleontology and nonmarine stratigraphy of the Cuyama Valley badlands, California. Part 1. Geology, faunal interpretations, and systematic descriptions of Chiroptera, Insectivora and Rodentia, *Univ. Calif. Publ. Geol. Sci.* **45**:1–154.

Jepsen, G. L., 1937, A Paleocene rodent, *Paramys atavus*, *Proc. Am. Philos. Soc.* **78**:291–301.

Kellogg, L., 1910, Rodent fauna of the late Tertiary beds at Virgin Valley and Thousand Creek, Nevada, *Univ. Calif. Publ. Geol. Sci.* **5**:421–437.

Kelly, T. S., 1992, New Uintan and Duchesnean (middle and late Eocene) rodents from the Sespe Formation, Simi Valley, California, *South Calif. Acad. Sci. Bull.* **91**:97–120.

Kielan-Jaworowska, Z., and Trofimov, B. A., 1980, Cranial morphology of the Cretaceous eutherian mammal *Barunlestes*, *Acta Palaeontol. Pol.* **25**:167–186.

Klingener, D., 1963, Dental evolution of *Zapus*, *J. Mammal.* **44**:248–260.

Klingener, D., 1964, The comparative myology of four dipodoid rodents (genera *Zapus*, *Napaeozapus*, *Sicista*, and *Jaculus*), *Misc. Publ. Mus. Zool. Univ. Mich.* **124**:1–100.

Klingener, D., 1966, Dipodoid rodents from the Valentine Formation of Nebraska, *Occas. Pap. Mus. Zool. Univ. Mich.* **644**:1–9.

Klingener, D., 1968, Rodents of the Mio-Pliocene Norden Bridge local fauna, Nebraska, *Am. Midl. Nat.* **80**:65–74.

Kormos, T., 1930, Diagnosen neur Säugetiere aus der Oberpliozänen Fauna des Somlyoberges bei Püspökfürdö, *Ann. Hist.-Nat. Mus. Nation. Hung.* **27**:237–246.

Korth, W. W., 1979, Geomyoid rodents from the Valentine Formation of Knox County, Nebraska, *Ann. Carnegie. Mus.* **48**:287–310.

Korth, W. W., 1980, Cricetid and zapodid rodents from the Valentine Formation of Knox County, Nebraska, *Ann. Carnegie Mus.* **49**:307–322.

Korth, W. W., 1981, New Oligocene rodents from western North America, *Ann. Carnegie Mus.* **50**:289–318.

Korth, W. W., 1984, Earliest Tertiary evolution and radiation of rodents in North America, *Bull. Carnegie Mus. Nat. Hist.* **24**:1–71.

Korth, W. W., 1987a, Sciurid rodents (Mammalia) from the Chadronian and Orellan (Oligocene) of Nebraska, *J. Paleontol.* **61**:1247–1255.

Korth, W. W., 1987b, New rodents (Mammalia) from the late Barstovian (Miocene) Valentine Formation, Nebraska, *J. Paleontol.* **61**:1058–1064.

Korth, W. W., 1988a, The rodent *Mytonomys* from the Uintan and Duchesnean (Eocene) of Utah, and the content of the Ailuravinae (Ischyromyidae, Rodentia), *J. Vertebr. Paleontol.* **8**:290–294.

Korth, W. W., 1988b, *Paramys compressidens* Peterson and the systematic relationships of the species of *Paramys* (Paramyinae, Ischyromyidae), *J. Paleontol.* **62**:468–471.

Korth, W. W., 1988c, A new species of beaver (Rodentia, Castoridae) from the middle Oligocene (Orellan) of Nebraska, *J. Paleontol.* **62**:965–967.

Korth, W. W., 1989a, Aplodontid rodents (Mammalia) from the Oligocene (Orellan and Whitneyan) Brule Formation, Nebraska, *J. Vertebr. Paleontol.* **9**:400–414.

Korth, W. W., 1989b, Geomyoid rodents (Mammalia) from the Orellan (middle Oligocene) of Nebraska, in: *Papers on Fossil Rodents in Honor of Albert Elmer Wood* (C. C. Black and M. R. Dawson, eds.), *Sci. Ser. Nat. Hist. Mus. Los Angeles Co.* **33**:31–46.

Korth, W. W., 1992a, Fossil small mammals from the Harrison Formation (late Arikareean: earliest Miocene), Cherry County, Nebraska, *Ann. Carnegie Mus.* **61**:69–131.

Korth, W. W., 1992b, Cylindrodonts (Cylindrodontidae, Rodentia) and a new genus of eomyid, *Paranamatomys*, (Eomyidae, Rodentia) from the Chadronian of Sioux County, Nebraska, *Trans. Neb. Acad. Sci.* **19**:75–82.

Korth, W. W., 1993a, The skull of *Hitonkala* (Florentiamyidae, Rodentia) and relationships within Geomyoidea, *J. Mammal.* **74**:168–174.

Korth, W. W., 1993b, Review of the Oligocene (Orellan and Arikareean) genus *Tenudomys* Rensberger (Geomyoidea: Rodentia), *J. Vertebr. Paleontol.* **13**:335–341.

Korth, W. W., and Bailey, B. E., 1992, Additional specimens of *Leptodontomys douglassi* (Eomyidae, Rodentia) from the Arikareeann (late Oligocene) of Nebraska, *J. Mammal.* **73**:651–662.

Korth, W. W., and Emry, R. J., 1991, The skull of *Cedromus* and a review of the Cedromurinae (Rodentia, Sciuridae), *J. Paleontol.* **65**:984–994.

Korth, W. W., and Reynolds, R. E., 1991, A new heteromyid from the Hemingfordian of California and the first occurrence of sulcate incisors in heteromyids, (abstract), *J. Vertebr. Paleontol.* **11**(Suppl.):41A.

Korth, W. W., and Tabrum, A. R., in press, Fossil mammals of the Orellan Cook Ranch local fauna, southwest Montana. Part 1: Eomyid and heliscomyid rodents, *Ann. Carnegie Mus.*

Korth, W. W., Bailey, B. E., and Hunt, R. M., Jr., 1990, Geomyoid rodents from the early Hemingfordian (Miocene) of Nebraska, *Ann. Carnegie Mus.* **59**:25–47.

Korth, W. W., Wahlert, J. H., and Emry, R. J., 1991, A new species of *Heliscomys* and recognition of the family Heliscomyidae (Geomyoidea: Rodentia), *J. Vertebr. Paleontol.* **11**:247–256.

Korvenkontio, V. A., 1934, Mikroskopische Untersuchungen an Nagerincisiven, unter Hinweis auf die Schmelzstruktur der Backenzäne, *Ann. Zool. Soc. Zool. Botan. Fenn. Vanamo* pp. 1–274.

Kowalski, K., 1974, Middle Oligocene rodents from Mongolia, *Palaeontol. Pol.* **30**:148–178.

Kurten, B., and Anderson, E., 1980, *Pleistocene Mammals of North America*, Columbia University Press, New York.

Landry, S. O., 1957, The interrelationships of the New and Old World hystricomorph rodents, *Univ. Calif. Publ. Zool.* **56**:1–118.

Lavocat, R., 1951, Révision de la faune des mammifères Oligocénes d'Auvergne et du Valay, Editions "Sciences et Avenir", Paris.

Lavocat, R., 1961, Le gisement de vertébrés fossiles de Beni Mellal, *Notes Mem. Serv. Geol. Maroc.* **155**:1–144.

Lavocat, R., 1969, La systématique des rongeurs hystricomorphes et la dérive des continents, *C. R. Acad. Sci.* **269**:1496–1497.

Lavocat, R., 1973, Les rongeurs du Miocene d'Afrique oriental; 1, Miocene inferieur, *Ec. Pract. Hautes Etudes Inst. Montpellier Mem. Trav.* **1**:1–284.

Lavocat, R., 1978, Rodentia and Lagomorpha, in: *Evolution of African Mammals* (V. J. Maglio and H. B. S. Cooke, eds.), pp. 68–89, Harvard University Press, Cambridge, Mass.

Lavocat, R., and Parent, J.-P., 1985, Phylogenetic analysis of middle ear features in fossil and living rodents, in: *Evolutionary Relationships among Rodents: A Multidisciplinary Analysis* (W. P. Luckett and J.-L. Hartenberger, eds.), pp. 333–354, Plenum Press, New York.

Leidy, J., 1856, Notices of remains of extinct Mammalia discovered by Dr. F. V. Hayden in Nebraska Territory, *Proc. Acad. Nat. Sci. Philadelphia* **8**:88–90.

Leidy, J., 1858, Notice of remains of extinct Vertebrata, from the valley of the Niobrara River, collected during the exploring expedition of 1857, in Nebraska, under the command of Lieut.

G. K. Warren, U. S. Top. Eng., by Dr. F. V. Hayden, *Proc. Acad. Nat. Sci. Philadelphia* **1858:** 20–29.
Leidy, J., 1869, The extinct mammalian fauna of Dakota and Nebraska, including an account of some allied forms from other localities, together with a synopsis of the mammalian remains of North America, *J. Acad. Nat. Sci. Philadelphia* **2:**1–472.
Li, C. K., 1963, Paramyids and sciuravids from North China, *Vertebr. PalAsiat.* **7:**151–160.
Li, C. K., 1977, Paleocene eurymyloids (Anagalida, Mammalia) of Qianshan, Anhui, *Vertebr. PalAsiat.* **15:**103–118.
Li, C. K., and Ting, S. Y., 1983, The Paleogene mammals of China, *Bull. Carnegie Mus. Nat. His.* **21:**1–93.
Li, C. K., and Ting, S. Y., 1985, Possible phylogenetic relationship of Asiatic eurymylids and rodents, with comments on mimotonids, in: *Evolutionary Relationships among Rodents: A Multidisciplinary Analysis* (W. P. Luckett and J.-L. Hartenberger, eds.), pp. 35–58, Plenum Press, New York.
Li, C. K., and Ting, S. Y., 1993, New cranial and postcranial evidence for the affinities of eurymylids (Rodentia) and mimotonids (Lagomorpha), in: *Mammal Phylogeny—Placentals* (F. S. Szalay, M. J. Novacek, and M. C. McKenna, eds.), pp. 151–158, Springer-Verlag, Berlin.
Li, C. K., and Yan, D. F., 1979, The systematic position of eurymylids (Mammalia) and the origin of Rodentia, *Abstr., 12th Annu. Conf. and 3rd Natl. Congr. Paleontol. Soc. China,* pp. 155–156.
Li, C. K., Chiu, C. S., Yan, D. F., and Hsieh, S. H., 1979, Notes on some early Eocene mammalian fossils of Hengtung, Hunan, *Vertebr. PalAsiat.* **17:**71–82.
Li, C. K., Wilson, R. W., Dawson, M. R., and Krishtalka, L., 1987, The origin of rodents and lagomorphs, *Curr. Mammal.* **1:**97–108.
Lillegraven, J. A., 1969, Latest Cretaceous mammals of upper part of Edmonton Formation of Alberta, Canada, and review of marsupial–placental dichotomy in mammalian evolution, *Univ. Kans. Paleontol. Contrib. Vertebr.* **12:**1–122.
Lillegraven, J. A., 1977, Small rodents (Mammalia) from Eocene deposits of San Diego County, California, *Bull. Am. Mus. Nat. Hist.* **158:**221–262.
Lillegraven, J. A., and Wilson, R. W., 1975, Analysis of *Simimys simplex*, an Eocene rodent (?Zapodidae), *J. Paleontol.* **49:**856–874.
Lindsay, E. H., 1968, Rodents from the Hartman Ranch local fauna, California, *PaleoBios* **6:**1–22.
Lindsay, E. H., 1972, Small mammal fossils from the Barstow Formation, California, *Univ. Calif. Publ. Geol. Sci.* **93:**1–104.
Lindsay, E. H., 1974, The Hemingfordian mammal fauna of the Vedder locality, Branch Canyon Formation, Santa Barbara County, California. Part II: Rodentia (Eomyidae and Heteromyidae), *PaleoBios* **16:**1–20.
Lindsay, E. H., 1977, *Simimys* and origin of the Cricetidae (Rodentia: Muroidea), *Geobios* **10:**597–623.
Lindsay, E. H., 1978, *Eucricetodon asiaticus* (Matthew and Granger), an Oligocene rodent (Cricetidae) from Mongolia, *J. Paleontol.* **52:**590–595.
Linnaeus, C., 1758, *Systema naturae per regna tria naturae, secundum classes, ordines, genera, species cum characteribus, differentiis, synonymis, locis,* Vol. 1, Laurentii Salvii, Stockholm.
Luckett, W. P., 1985, Superordinal and intraordinal affinities of rodents: Developmental evidence from the dentition and placentation, in: *Evolutionary Relationships among Rodents: A Multidisciplinary Analysis* (W. P. Luckett and J.-L. Hartenberger, eds.), pp. 227–276, Plenum Press, New York.
Luckett, W. P., and Hartenberger, J.-L., 1985, Evolutionary relationships among rodents: Comments and conclusions, in: *Evolutionary Relationships among Rodents: A Multidisciplinary Analysis* (W. P. Luckett and J.-L. Hartenberger, eds.), pp. 685–712, Plenum Press, New York.
Lychev, G. F., 1978, A new species of beaver of the genus *Agnotocastor* from the early Oligocene of Kazakhsan, *Paleontol. Zhr.* **1978:**128–130 (in Russian).
Macdonald, J. R., 1963, The Miocene faunas from the Wounded Knee area of western South Dakota, *Bull. Am. Mus. Nat. Hist.* **125:**139–238.

Macdonald, J. R., 1970, Review of the Miocene Wounded Knee faunas of southwestern South Dakota, *Bull. Los Angeles Co. Mus. Nat. Hist. Sci.* **8**:1–82.

Macdonald, L. J., 1972, Monroe Creek (early Miocene) microfossils from the Wounded Knee area, South Dakota, *S. D. Geol. Surv. Rep. Invest.* **105**:1–43.

McGrew, P. O., 1941a, The Aplodontoidea, *Field Mus. Nat. Hist. Geol. Ser.* **9**:1–30.

McGrew, P. O., 1941b, Heteromyids from the Miocene and Lower Oligocene, *Field Mus. Nat. Hist. Geol. Ser.* **8**:55–57.

McKenna, M. C., 1961, A note on the origin of rodents, *Am. Mus. Novit.* **2037**:1–5.

McKenna, M. C., 1962, Collecting small fossils by washing and screening, *Curator* **5**:221–235.

McKenna, M. C., 1975, Toward a phylogenetic classification of the Mammalia, in: *Phylogeny of the Primates* (W. P. Luckett and F. S. Szalay, eds.), pp. 21–46, Plenum Press, New York.

McKenna, M. C., 1980, Remaining evidence of Oligocene sedimentary rocks previously present across the Bighorn Mountains, Wyoming, *Univ. Mich. Pap. Paleontol.* **37**:131–164.

McKenna, M. C., 1982, Lagomorph interrelationships, *Geobios Mem. Spec.* **6**:213–233.

McKenna, M. C., and Love, J. D., 1972, High-level strata containing early Miocene mammals on the Bighorn Mountains, Wyoming, *Am. Mus. Novit.* **2400**:1–31.

McLaughlin, C. A., 1984, Protrogomorph, sciuromorph, castorimorph, myomorph (geomyoid, anomaluroid, pedetoid and ctenodactyloid) rodents, in: *Orders and Families of Recent Mammals of the World* (S. Anderson and J. K. Jones, Jr., eds.), pp. 267–288, Wiley, New York.

Marsh, O. C., 1872, Preliminary description of new Tertiary mammals, *Am. J. Sci.* **4**:202–224.

Marshall, L. G., Hoffstetter, R., and Pascual, R., 1983, Mammals and stratigraphy: Geochronology of the continental mammal-bearing Tertiary of South America, *Palaeovertebr. Mem. Extraord.* pp. 1–93.

Martin, J. E., 1984, A survey of Tertiary species of *Perognathus* (Perognathinae) and a description of a new genus of Heteromyinae, in: *Papers in Vertebrate Paleontology Honoring Robert Warren Wilson* (R. M. Mengel, ed.), pp. 90–121, *Carnegie Mus. Nat. Hist. Spec. Publ.* 9.

Martin, J. E., and Green, M., 1984, Insectivora, Sciuridae, and Cricetidae from the early Miocene Rosebud Formation in South Dakota, in: *Papers in Vertebrate Paleontology Honoring Robert Warren Wilson* (R. M. Mengel, ed.), pp. 28–40, *Carnegie Mus. Nat. Hist. Spec. Publ.* 9.

Martin, L. D., 1974, New rodents from the Lower Miocene Gering Formation of western Nebraska, *Occas. Pap. Mus. Nat. Hist. Univ. Kans.* **32**:1–12.

Martin, L. D., 1975, Microtine rodents from the Ogallala Pliocene of Nebraska and the early evolution of the Microtinae in North America, *Univ. Mich. Pap. Paleontol.* **12**:101–110.

Martin, L. D., 1979, The biostratigraphy of arvicoline rodents in North America, *Trans. Nebr. Acad. Sci.* **7**:91–100.

Martin, L. D., 1980, The early evolution of the Cricetidae in North America, *Univ. Kans. Paleontol. Contrib.* **102**:1–42.

Martin, L. D., 1987, Beavers from the Harrison Formation (early Miocene) with a revision of *Euhapsis*, *Dakoterra* **3**:73–91.

Martin, L. D., and Bennett, D. K., 1977, The burrows of the Miocene beaver *Palaeocastor*, western Nebraska, U.S.A., *Palaeogeogr. Palaeoclimatol. Palaeoecol.* **22**:173–193.

Martin, L. D., and Corner, R. G., 1980, A new genus of cricetid rodent from the Hemingfordian (Miocene) of Nebraska, *Univ. Kans. Paleontol. Contrib.* **103**:1–6.

Martin, R. A., 1989, Early Pleistocene zapodid rodents from the Java Local Fauna of northcentral South Dakota, *J. Vertebr. Paleontol.* **9**:101–109.

Matthew, W. D., 1902, A horned rodent from the Colorado Miocene, with a revision of the Mylagauli, beavers, and hares of the American Tertiary, *Bull. Am. Mus. Nat. Hist.* **16**:291–310.

Matthew, W. D., 1905, Notice of two new genera of mammals from the Oligocene of South Dakota, *Bull. Am. Mus. Nat. Hist.* **21**:21–26.

Matthew, W. D., 1910, On the osteology and relationships of *Paramys*, and the affinities of the Ischyromyidae, *Bull. Am. Mus. Nat. Hist.* **28**:43–72.

Matthew, W. D., 1918a, A revision of the Lower Eocene Wasatch and Wind River faunas, *Bull. Am. Mus. Nat. Hist.* **38**:429–483.

Matthew, W. D., 1918b, Contributions to the Snake Creek fauna, with notes upon the Pleistocene of western Nebraska, *Bull. Am. Mus. Nat. Hist.* **38**:197–199.

Matthew, W. D., 1924, Third contribution to the Snake Creek fauna, *Bull. Am. Mus. Nat. Hist.* **50**:59–210.

Matthew, W. D., and Granger, W., 1923, New Bathyergidae from the Oligocene of Mongolia, *Am. Mus. Novit.* **101**:1–5.

Matthew, W. D., and Granger, W., 1925, Fauna and correlation of the Gashato Formation of Mongolia, *Am. Mus. Novit.* **189**:1–12.

Matthew, W. D., Granger, W., and Simpson, G. G., 1929, Additions to the fauna of the Gashato Formation of Mongolia, *Am. Mus. Novit.* **376**:1–12.

May, S. R., 1981, *Repomys* (Mammalia: Rodentia gen. nov.) from the late Neogene of California and Nevada, *J. Vertebr. Paleontol.* **1**:219–230.

Meng, J., 1990, The auditory region of *Reithroparamys delicatissimus* (Mammalia, Rodentia) and its systematic implications, *Am. Mus. Novit.* **2972**:1–35.

Michaux, J., 1968, Les Paramyidae (Rodentia) de l'Eocéne inférieur du bassin de Paris, *Palaeovertebrata* **1**:135–193.

Miller, G. S., and Gidley, J. W., 1918, Synopsis of the supergeneric groups of rodents, *J. Wash. Acad. Sci.* **8**:431–448.

Miller, G. S., and Gidley, J. W., 1920, A new fossil rodent from the Oligocene of South Dakota, *J. Mammal.* **1**:73–74.

Misonne, X., 1957, Mammiféres oligocéne de Hoogbutsel et de Hoeleden. 1. Rongeurs et Ongulés, *Bull. Mus. Nat. Hist. Belge.* **33(51)**:1–16.

Munthe, J., 1977, A new species of *Gregorymys* (Rodentia, Geomyidae) from the Miocene of Colorado, *PaleoBios* **26**:1–12.

Munthe, J., 1980, Rodents of the Miocene Daud Khel Local Fauna, Mianwali District, Pakistan; Part 1, Sciuridae, Gliridae, Ctenodactylidae, and Rhizoomyidae, *Contrib. Biol. Geol.* **34**:1–36.

Munthe, J., 1988, Miocene mammals of the Split Rock area, Granite Mountains Basin, central Wyoming, *Univ. Calif. Publ. Geol. Sci.* **126**:1–136.

Munthe, L. K., 1981, Skeletal morphology and function of the Miocene rodent *Schizodontomys harkseni*, *PaleoBios* **35**:1–33.

Nichols, R., 1976, Early Miocene mammals from the Lemhi Valley of Idaho, *Tebiwa* **18**:9–47.

Novacek, M. J., 1977, Aspects of the problem of variation, origin and evolution of the eutherian auditory bulla, *Mammal Rev.* **7**:131–149.

Novacek, M. J., 1980, Cranioskeletal features in tupaiids and selected Eutheria as phylogenetic evidence, in: *Comparative Biology and Evolutionary Relationships of Tree Shrews* (W. P. Luckett, ed.), pp. 35–93, Plenum Press, New York.

Novacek, M. J., 1985, Cranial evidence for rodent affinities, in: *Evolutionary Relationships among Rodents: A Multidisciplinary Analysis* (W. P. Luckett and J.-L. Hartenberger, eds.), pp. 59–82, Plenum Press, New York.

Novacek, M. J., 1986a, The primitive eutherian dental formula, *J. Vertebr. Paleontol.* **6**:191–196.

Novacek, M. J., 1986b, The skull of leptictid insectivorans and the higher-level classification of eutherian mammals, *Bull. Am. Mus. Nat. Hist.* **183**:1–112.

Nowack, R. M., and Paradiso, J. L., 1983, *Walker's Mammals of the World*, 4th ed., Vol. I and II, The Johns Hopkins University Press, Baltimore.

Olson, E. C., 1940, Cranial foramina of North American beavers, *J. Paleontol.* **14**:495–501.

Osborn, H. F., 1902, American Eocene primates, and the supposed rodent family Mixodectidae, *Bull. Am. Mus. Nat. Hist.* **17**:169–214.

Osborn, H. F., and Matthew, W. D., 1909, Cenozoic mammal horizons of western North America, with faunal lists of the Tertiary Mammalia of the West, *Bull. U. S. Geol. Surv.* **361**:1–138.

Ostrander, G., 1983, New early Oligocene (Chadronian) mammals from the Raben Ranch local fauna, northwest Nebraska, *J. Paleontol.* **57**:128–139.

Ostrander, G., 1985, Correlation of the early Oligocene (Chadronian) in northwestern Nebraska, *Dakoterra* **2**:205–232.

Peterson, O. A., 1905, Description of new rodents and discussion of the origin of *Daemonelix*, Mem. Carnegie Mus. **2**:139–200.
Pomel, A., 1853, *Catalogue Méthodique et Descriptif des Vertébrés Fossiles*, J.-B. Bailliére, Libraire de l'Académe Impériale de Médecine.
Pratt, A. E., and Morgan, G. S., 1989, New Sciuridae (Mammalia: Rodentia) from the early Miocene Thomas Farm Local Fauna, Florida, *J. Vertebr. Paleontol.* **9**:89–100.
Qi, T., 1987, The middle Eocene Arshanto fauna (Mammalia) of Inner Mongolia, *Ann. Carnegie Mus.* **56**:1–73.
Reeder, W. G., 1956, A review of the Tertiary rodents of the family Heteromyidae, unpublished Ph.D. dissertation, University of Michigan.
Reeder, W. G., 1960, Two new rodent genera from the Oligocene White River Formation, *Fieldiana Geol.* **10**:511–523.
Reig, O. A., 1978, Reodores cricetidos del Pliocene superior del la Provincia de Buenos Aires (Argentina), *Publ. Mus. Mun. Cien. Nat. Mar Del Plata "Lorenzo Scaglia"* **2**:164–190.
Rensberger, J. M., 1971, Entoptychine pocket gophers (Mammalia, Geomyoidea) of the early Miocene John Day Formation, Oregon, *Univ. Calif. Publ. Geol. Sci.* **90**:1–163.
Rensberger, J. M., 1973a, *Sanctimus* (Mammalia, Rodentia) and the phyletic relationships of the large Arikareean geomyoids, *J. Paleontol.* **47**:835–853.
Rensberger, J. M., 1973b, Pleurolicine rodents (Geomyoidea) of the John Day Formation, Oregon, *Univ. Calif. Publ. Geol. Sci.* **102**:1–95.
Rensberger, J. M., 1975, *Haplomys* and its bearing on the origin of the aplodontoid rodents, *J. Mammal.* **56**:1–14.
Rensberger, J. M., 1979, *Promylagaulus*, progressive aplodontoid rodents of the early Miocene, *Contrib. Sci. Nat. Hist. Mus. Los Angeles Co.* **312**:1–16.
Rensberger, J. M., 1980, A primitive promylagauline rodent from the Sharps Formation, South Dakota, *J. Paleontol.* **54**:1267–1277.
Rensberger, J. M., 1981, Evolution in the late Oligocene–early Miocene succession of Meniscomyine rodents in the Deep River Formation, Montana, *J. Vertebr. Paleontol.* **1**:185–209.
Rensberger, J. M., 1983, Successions of meniscomyine and allomyine rodents (Aplodontidae) in the Oligo-Miocene John Day Formation, Oregon, *Univ. Calif. Publ. Geol. Sci.* **124**:1–157.
Rensberger, J. M., and Li, C. K., 1986, A new prosciurine rodent from Shantung Province, China, *J. Paleontol.* **60**:763–771.
Repenning, C. A., 1962, The giant ground squirrel *Paenemarmota*, *J. Paleontol.* **36**:540–556.
Repenning, C. A., 1968, Mandibular musculature and the origin of the subfamily Arvicolinae (Rodentia), *Acta Zool. Cracov.* **13**:29–72.
Repenning, C. A., 1980, Faunal exchanges between Siberia and North America, *Can. J. Anthropol.* **1**:37–44.
Repenning, C. A., 1984, Quaternary rodent biochronology and its correlation with climatic and magnetic stratigraphies, in: *Quaternary Chronologies* (W. C. Mahaney Norwich, ed.), Geoabstracts Ltd.
Repenning, C. A., 1987, Biochronology of the microtine rodents of the United States, in: *Cenozoic Mammals of North America: Geochronology and Biostratigraphy* (M. O. Woodburne, ed.) pp. 236–268, University of California Press, Berkeley.
Repenning, C. A., 1992, *Allophaiomys* and the age of the Olyor Suite, Krestovka sections, Yakutia, *Bull. U. S. Geol. Surv.* **2037**:1–98.
Repenning, C. A., and Grady, F., 1988, The microtine rodents of the Cheetah Room fauna, Hamilton Cave, West Virginia, and the spontaneous origin of *Synaptomys*, *Bull. U. S. Geol. Surv.* **1853**:1–32.
Repenning, C. A., Fejfar, O., and Heinrich, W.-D., 1990, Arvicolid rodent biochronology of the northern hemisphere, in: *International Symposium—Evolution, Phylogeny and Biostratigraphy of Arvicolids (Rodentia, Mammalia)*, Rohanov, Czechoslovakia (O. Fejfar and W.-D. Heinrich, eds.), pp. 385–418, Geological Survey, Prague.
Roger, O., 1885, Nachträgliche Bemerkung zu den *Diontherium*-Resten, *Ber. Naturwiss. Ver. Schwaben Beuburg* **28**:1–77.

Rovereto, C., 1914, Los stratos araucanos y sus fósiles, *An. Mus. Nac. Buenos Aires* **25**:1–247.
Russell, L. S., 1972, Tertiary mammals of Saskatchewan. Part II: the Oligocene fauna, nonungulate orders, *Life Sci. Contrib. R. Ontario Mus.* **84**:1–63.
Russell, R. J., 1968, Evolution and classification of the pocket gophers of the subfamily Geomyinae, *Univ. Kans. Publ. Mus. Nat. Hist.* **16**:473–579.
Sahni, A., 1985, Enamel structure of early mammals and its role in evaluating relationships among rodents, in: *Evolutionary Relationships among Rodents: A Multidisciplinary Analysis* (W. P. Luckett and J.-L. Hartenberger, eds.), pp. 133–150, Plenum Press, New York.
Savage, D. E., and Russell, D. E., 1983, *Mammalian Paleofaunas of the World*, Addison–Wesley, Reading, Mass.
Schaub, S., 1930, Fossile Sicistinae, *Ecol. Geol. Helv.* **23**:616–637.
Schaub, S., 1938, Tertiäre und quartäre Murinae, *Abh. Schweiz. Pal. Ges.* **61**:1–39.
Schaub, S., 1953, Remarks on the distribution and classification of the "Hystricomorpha," *Verh. Naturforsch. Ges. Basel* **64**:389–400.
Schaub, S., 1958, Simplicidentata (=Rodentia), in: *Traite de Paleo.* **6**:659–818.
Schlosser, M., 1884, Die Nager des europäischen Tertiärs nebst Betrachtungen über die Organisation und die geschichtliche Entwicklung der Nager überhaupt, *Palaeontographica* **31**:1–184.
Schlosser, M., 1924, Tertiary vertebrates from Mongolia, *Palaeontol. Sin.* **1**:1–119.
Schmidt-Kittler, N., and Vianey-Liaud, M., 1979, Evolution des Aplodontidae Oligocénes européens, *Palaeovertebrata* **9**:33–82.
Scott, W. B., 1895, *Protoptychus hatcheri*, a new rodent from the Uinta Eocene, *Proc. Acad. Nat. Sci. Philadelphia* **1895**:269–286.
Shevyreva, N. S., 1966, On the evolution of rodents from Middle Oligocene of Kazakhstan, *Bjull. Mosk. Obshch. Ispyt. Prirody, Otd. Geol.* **41**:143 (in Russian).
Shevyreva, N. S., 1971, The first find of Eocene rodents in the U.S.S.R., *Bull. Acad. Sci. Georgian SSR Tbilissi* **61**:745–747 (in Russian).
Shevyreva, N. S., 1972, New rodents in the Paleogene of Mongolia and Kazakhstan, *Paleontol. J. Moscow* **3**:134–145 (in Russian).
Shevyreva, N. S., 1984, New early Eocene rodents from the Zaysanski Depression, in: *Flora and Fauna of Zaysanski Depression*, Acad. Sci., Georgia S.S.R., pp. 77–113 (in Russian).
Shevyreva, N. S., 1989, New rodents (Ctenodactyloidea, Rodentia, Mammalia) from the Lower Eocene of Mongolia, *Acad. Sci. Paleontol. J.* **3**:60–72 (in Russian).
Shoshani, J., Goodman, M., Czelusniak, J., and Braunitzer, G., 1985, A phylogeny of Rodentia and other eutherian orders: Parsimony analysis utilizing amino acid sequences of alpha and beta hemoglobin chains, in: *Evolutionary Relationships among Rodents: A Multidisciplinary Analysis* (W. P. Luckett and J.-L. Hartenberger, eds.), pp. 191–210, Plenum Press, New York.
Shotwell, J. A., 1956, Hemphillian mammalian assemblage from northeastern Oregon, *Bull. Geol. Soc. Am.* **67**:717–738.
Shotwell, J. A., 1958, Evolution and biogeography of the aplodontid and mylagaulid rodents, *Evolution* **12**:451–484.
Shotwell, J. A., 1963, Mammalian fauna of the Drewsey Formation, Bartlett Mountain, Drinkwater and Otis Basin local faunas, *Trans. Am. Philos. Soc.* **53**:70–77.
Shotwell, J. A., 1967a, Late Tertiary geomyoid rodents of Oregon, *Bull. Univ. Oreg. Mus. Nat. Hist.* **9**:1–51.
Shotwell, J. A., 1967b, *Peromyscus* of the late Tertiary in Oregon, *Bull. Univ. Oreg. Mus. Nat. Hist.* **5**:1–35.
Shotwell, J. A., 1970, Pliocene mammals of southeast Oregon and adjacent Idaho, *Bull. Mus. Nat. Hist. Univ. Oreg.* **17**:1–103.
Shotwell, J. A., and Russell, D. E., 1963, Mammalian fauna of the Upper Juntura Formation, the Black Butte local fauna, *Trans. Am. Philos. Soc.* **53**:42–69.
Simpson, G. G., 1928, Pleistocene mammals from a cave in Citrus County, Florida, *Am. Mus. Novit.* **328**:1–16.
Simpson, G. G., 1945, The principles of classification and a classification of mammals, *Bull. Am. Mus. Nat. Hist.* **85**:1–350.

Sinclair, W. S., 1903, *Mylagaulodon*, a new rodent from the Upper John Day of Oregon, *Am. J. Sci.* **15**:143–144.

Skinner, M. F., and Taylor, B. E., 1967, A revision of the geology and paleontology of the Bijou Hills, South Dakota, *Am. Mus. Novit.* **2300**:1–53.

Skwara, T., 1986, A new "flying squirrel" (Rodentia: Sciuridae) from the early Miocene of southwestern Saskatchewan, *J. Vertebr. Paleontol.* **6**:290–294.

Skwara, T., 1988, Mammals of the Topham local fauna: Early Miocene (Hemingfordian), Cypress Hills Formation, Saskatchewan, *Nat. Hist. Contrib. Sask. Mus. Nat. Hist.* **9**:1–169.

Slaughter, B. H., 1981, A new genus of geomyoid rodent from the Miocene of Texas and Panama, *J. Vertebr. Paleontol.* **1**:111–115.

Slaughter, B. H., and James, G. T., 1979, *Saidomys natrunensis*, an arvicoline rodent from the Pliocene of Egypt, *J. Mammal.* **60**:421–425.

Slaughter, B. H., and Ubelaker, J. E., 1984, Relationship of South American cricetines to rodents of North America and the Old World, *J. Vertebr. Paleontol.* **4**:255–264.

Stehlin, H. G., and Schaub, S., 1951, Die Trigonodontie der simplicidentaten Nager, *Schweiz. Palaeontol. Abh.* **67**:1–385.

Stirton, R. A., 1934, A new species of *Amblycastor* from the *Platybelodon* beds, Tung Gur Formation, of Mongolia, *Am. Mus. Novit.* **694**:1–4.

Stirton, R. A., 1935, A review of Tertiary beavers, *Univ. Calif. Publ. Geol. Sci.* **23**:391–458.

Stock, C., 1935, A new genus of rodent from the Sespe Eocene, *Bull. Geol. Soc. Am.* **46**:61–68.

Storer, J. E., 1970, New rodents and lagomorphs from the Upper Miocene Wood Mountain Formation of southern Saskatchewan, *Can. J. Earth Sci.* **7**:1125–1129.

Storer, J. E., 1973, The entoptychine geomyid *Lignimus* (Mammalia: Rodentia) from Kansas and Nebraska, *Can. J. Earth Sci.* **10**:72–83.

Storer, J. E., 1975, Tertiary mammals of Saskatchewan. Part III: The Miocene fauna, *Life Sci. Contrib. R. Ont. Mus.* **103**:1–134.

Storer, J. E., 1978, Rodents of the Calf Creek Local Fauna (Cypress Hills Formation, Oligocene, Chadronian) Saskatchewan, *Nat. Hist. Contrib. Sask. Mus. Nat. Hist.* **1**:1–54.

Storer, J. E., 1984, Mammals of the Swift Current Creek Local Fauna (Eocene: Uintan), Saskatchewan, *Nat. Hist. Contrib. Sask. Mus. Nat. Hist.* **7**:1–158.

Storer, J. E., 1987, Dental evolution and radiation of Eocene and early Oligocene Eomyidae (Mammalia: Rodentia) of North America, with new material from the Duchesnean of Saskatchewan, *Dakoterra* **3**:108–117.

Storer, J. E., 1988, The rodents of the Lac Pelletier lower fauna, late Eocene (Duchesnean) of Saskatchewan, *J. Vertebr. Paleontol.* **8**:84–101.

Stout, T. M., and Stone, W. J., 1971, Fossil beavers in Tertiary caprocks in North Dakota and Montana, *Geol. Soc. Am. Abstr.* **3**:281–282.

Stucky, R. K., 1992, Mammalian faunas in North America of Bridgerian to early Arikareean "ages" (Eocene and Oligocene), in: *Eocene–Oligocene Climatic and Biotic Evolution* (D. R. Prothero and W. A. Berggren, eds.), pp. 464–493, Princeton University Press, Princeton, N.J.

Suchov, V. P., 1976, Remains of lemmings in the Bashkirian Pliocene deposits, *Acad. Sci. USSR Proc. Zool. Inst.* **66**:117–121 (in Russian).

Sulimski, A., 1962, Two new rodents from Weze 1 (Poland), *Acta Paleontol. Pol.* **7**:503–512.

Sulimski, A., 1964, Pliocene Lagomorpha and Rodentia from Weze 1 (Poland), *Acta Paleontol. Pol.* **9**:149–261.

Sutton, J. F., and Black, C. C., 1975, Paleontology of the earliest Oligocene deposits in Jackson Hole, Wyoming. Part 1. Rodents exclusive of the family Eomyidae, *Ann. Carnegie Mus.* **45**:299–315.

Swisher, C. C., III, and Prothero, D. R., 1990, Single-crystal $^{40}Ar/^{39}Ar$ dating of the Eocene–Oligocene transition in North America, *Science* **249**:760–762.

Sych, L., 1971, Mixodontia, a new order of mammals from the Paleocene of Mongolia. Results of the Polish–Mongolian Paleontologial Expeditions, Part III, *Palaeontol. Pol.* **25**:147–158.

Szalay, F. S., 1969, Mixodectidae, Microsyopidae, and the insectivore–primate transition, *Bull. Am. Mus. Nat. Hist.* **140**:193–330.

Szalay, F. S., 1977, Phylogenetic relationships and a classification of the eutherian Mammalia, in: *Major Patterns in Vertebrate Evolution* (M. K. Hecht, P. C. Goody, and B. M. Hecht, eds.), pp. 317–374, Plenum Press, New York.

Szalay, F. S., 1985, Rodent and lagomorph morphotype adaptations, origins, and relationships: Some postcranial attributes analyzed, in: *Evolutionary Relationships among Rodents: A Multidisciplinary Analysis* (W. P. Luckett and J.-L. Hartenberger, eds.), pp. 83–132, Plenum Press, New York.

Szalay, F. S., and McKenna, M. C., 1971, Beginning of the age of mammals in Asia: The late Paleocene Gashato fauna, Mongolia, *Bull. Am. Mus. Nat. Hist.* **144**:269–318.

Tedford, R. H., Skinner, M. F., Fields, R. W., Rensberger, J. M., Whistler, D. P., Galusha, T., Taylor, B. E., Macdonald, J. R., and Webb, S. D., 1987, Faunal succession and biochronology of the Arikareean through Hemphillian interval (late Oligocene through earliest Pliocene rocks) in North America, in: *Cenozoic Mammals of North America: Geochronology and Biostratigraphy* (M. O. Woodburne, ed.), pp. 153–210, University of California Press, Berkeley.

Teilhard de Chardin, P., and Young, C. C., 1931, Fossil mammals from the late Cenozoic of North China, *Palaeontol. Sin. Ser. C* **9**:1–66.

Thaler, L., 1966, Le rongeurs fossiles du Bas-Languedoc dans leurs rapports avec l'histoire des faunes et la stratigraphie du Tertiarie d' Europe, *Mem. Mus. Natl. Hist. Nat. Paris Ser. C* **17**:1–296.

Tomes, J., 1850, On the structure of the dental tissues of the order Rodentia, *Philos. Trans. R. Soc. London* **1850**:529–567.

Tomida, Y., 1987, Small mammal fossils and correlation of continental deposits, Safford and Duncan Basins, Arizona, USA, *Natl. Sci. Mus., Tokyo.*

Tong, Y.-S., 1992, *Pappocricetodon*, a pre-Oligocene cricetid genus (Rodentia) from central China, *Vertebr. PalAsiat.* **30**:1–16.

Topachevsky, V. A., Skorik, A. F., and Rekovets, L. I., 1978, The most ancient voles of the tribe Microtini (Rodentia, Microtidae) from southern Ukraine, *Vestn. Zool.* **2**:35–41 (in Russian)

Troxell, E. L., 1923a, *Pauromys perditus*, a small rodent, *Am. J. Sci.* **5**:155–156.

Troxell, E. L., 1923b, *Diplolophus*, a new genus of rodents, *Am. J. Sci.* **5**:157–159.

Tullberg, T., 1899, Ueber das System der Nagetiere: eine phlogenetische Studie, *Akad. Buchdr. Uppsala* **18**:1–514.

Turnbull, W. D., 1991, *Protoptychus hatcheri* Scott, 1895. The mammalian faunas of the Washakie Formation, Eocene age, of southern Wyoming. Part II. The Adobetown Member, middle division (=Washakie B), Twka/2 (in part), *Fieldiana Geol.* **21**:1–33.

Unay, E., 1981, Middle and Upper Miocene rodents from the Bayraktepe section (Canakkale, Turkey), *Proc. K. Ned. Akad. Wet. Ser. B Palaeontol. Geol. Phys. Chem.* **84**:217–238.

Van de Weerd, A., and Daams, R., 1979, A review of the Neogene rodent succession in Spain, 7th Internat. Congr. Medit. Neogene, *Ann. Geol. Pays Hell.* **1979**:1263–1273.

Van Valen, L., 1966, Deltatheridia, a new order of mammals, *Bull. Am. Mus. Nat. Hist.* **132**:1–126.

Vianey-Liaud, M., 1974, *Palaeosciurus goti* nov. sp., ecureuil terrestre de l'Oligocene moyen du Quercy; Donnees nouvelles sur l'apparition de sciurides en Europe, *Ann. Paleontol. Vertebr.* **60**:103–122.

Vianey-Liaud, M., 1979, Evolution des rongeurs á l'Oligocéne en Europe Occidentale, *Palaeontographica* **166**:136–236.

Viret, J., 1926, Nouvelles observations relatives á la faune de Rongeurs de Saint-Gérand-le-Puy, *C. R. Acad. Sci.* **183**:71–72.

von Koenigswald, W., 1980, Schmelzstruktur und Morphologie in den Molaren der Arvicolidae (Rodentia), *Abh. Senckenb. Naturforsch. Ges.* **539**:1–129.

von Koenigswald, W., 1985, Evolutionary trends in the enamel of rodent incisors, in: *Evolutionary Relationships among Rodents: A Multidisciplinary Analysis* (W. P. Luckett and J.-L. Hartenberger, eds.), pp. 403–422, Plenum Press, New York.

von Koenigswald, W., 1992, Microstructure of the molar enamel, a key character for myomorph rodents, (abstract), *J. Vertebr. Paleontol. (Suppl.)* **12**:37A.

Voorhies, M. R., 1975, A new genus and species of fossil kangaroo rat and its burrow, *J. Mammal.* **56:**160–176.

Voorhies, M. R., 1981, A fossil record of the porcupine (*Erethizon dorsatum*) from the Great Plains, *J. Mammal.* **62:**835–837.

Voorhies, M. R., 1984, "*Citellus kimballensis*" Kent and "*Propliophenacomys uptegrovensis*" Martin, supposed Miocene rodents, are Recent intrusives, *J. Paleontol.* **58:**254–258.

Voorhies, M. R., 1990a, Vertebrate paleontology of the proposed Norden Reservoir area Brown, Cherry, and Keya Paha Counties, Nebraska, *Tech. Rep. 82-09, Div. Archeol. Res. Univ. Nebraska–Lincoln.*

Voorhies, M. R., 1990b, Vertebrate biostratigraphy of the Ogallala Group in Nebraska, in: *Geologic Framework and Regional Hydrology: Upper Cenozoic Blackwater Draw and Ogallala Formation, Great Plains* (T. C. Gustavson, ed.), pp. 115–151, Bureau of Economic Geology, University of Texas, Austin.

Wahlert, J. H., 1968, Variability of rodent incisor enamel as viewed in thin section, and the microstructure of the enamel in fossil and recent rodent groups, *Breviora* **309:**1–18.

Wahlert, J. H., 1972, The cranial foramina of protrogomorphous and sciuromorphous rodents; an anatomical and phylogenetic study, unpublished Ph.D. dissertation, Harvard University.

Wahlert, J. H., 1973, *Protoptychus*, a hystricomorphous rodent from the late Eocene of North America, *Breviora* **419:**1–14.

Wahlert, J. H., 1974, The cranial foramina of protrogomorphous rodents; an anatomical and phylogenetic study, *Bull. Mus. Comp. Zool.* **146:**363–410.

Wahlert, J. H., 1976, *Jimomys labaughi*, a new geomyoid rodent from the early Barstovian of North America, *Am. Mus. Novit.* **2591:**1–6.

Wahlert, J. H., 1977, Cranial foramina and relationships of *Eutypomys* (Rodentia, Eutypomyidae), *Am. Mus. Novit.* **2626:**1–8.

Wahlert, J. H., 1978, Cranial foramina and relationships of the Eomyoidea (Rodentia, Geomorpha). Skull and upper teeth of *Kansasimys*, *Am. Mus. Novit.* **2645:**1–16.

Wahlert, J. H., 1983, Relationships of the Florentiamyidae (Rodentia, Geomyoidea) based on cranial and dental morphology, *Am. Mus. Novit.* **2769:**1–23.

Wahlert, J. H., 1984a, Hystricomorphs, the oldest branch of the Rodentia, *Ann. N.Y. Acad. Sci.* **435:**356–357.

Wahlert, J. H., 1984b, *Kirkomys*, a new florentiamyid (Rodentia, Geomyoidea) from the Whitneyan of Sioux County, Nebraska, *Am. Mus. Novit.* **2793:**1–8.

Wahlert, J. H., 1985, Skull morphology and relationships of geomyoid rodents, *Am. Mus. Novit.* **2819:**1–20.

Wahlert, J. H., 1989, The three types of incisor enamel in rodents, in: *Papers on Fossil Rodents in Honor of Albert Elmer Wood* (C. C. Black and M. R. Dawson, eds.), *Sci. Ser. Nat. Hist. Mus. Los Angeles Co.* **33:**7–16.

Wahlert, J. H., 1991, The Harrymyinae, a new heteromyid subfamily (Rodentia, Geomorpha) based on cranial and dental morphology of *Harrymys* Munthe, 1988, *Am. Mus. Novit.* **3013:**1–23.

Wahlert, J. H., 1993, The fossil record, in: *Biology of the Heteromyidae* (H. H. Genoways and J. H. Brown, eds.), *Am. Soc. Mammal. Spec. Publ.* 10, pp. 1–37.

Wahlert, J. H., and Sousa, R. A., 1988, Skull morphology of *Gregorymys* and relationships of the Entoptychinae (Rodentia, Geomyidae), *Am. Mus. Novit.* **2922:**1–13.

Wahlert, J. H., and von Koenigswald, W., 1985, Specialized enamel in incisors of eomyid rodents, *Am. Mus. Novit.* **2832:**1–11.

Wang, B. Y., 1986, On the systematic position of *Prosciurus lohiculus*, *Vertebr. PalAsiat.* **24:**285–294.

Wang, B. Y., 1987, Discovery of Aplodontidae (Rodentia, Mammalia) from middle Oligocene of Nei Mongol, China, *Vertebr. PalAsiat.* **25:**32–45.

Wang, B. Y., and Emry, R. J., 1991, Eomyidae (Rodentia: Mammalia) from the Oligocene of Nei Mongol, China, *J. Vertebr. Paleontol.* **11:**370–377.

Wang, B. Y., and Heissig, K., 1984, *Ephemeromys* nov. gen., a primitive prosciurine rodent from the Oligocene of southern Germany, *Mitt. Bayer. Staatssamml. Palaeontol. Hist. Geol.* **24:**105–199.

Wang, B. Y., and Li, C. T., 1990, First Paleogene mammalian fauna from northeast China, *Vertebr. PalAsiat.* **28:**165–205.

White, J. A., 1970, Late Cenozoic porcupines (Mammalia, Erethizontidae) of North America, *Am. Mus. Novit.* **2421:**1–15.

Whistler, D. P., 1984, An early Hemingfordian (early Miocene) fossil vertebrate fauna from Boron, western Mojave Desert, California, *Contrib. Sci. Nat. Hist. Mus. Los Angeles Co.* **355:**1–36.

Wilson, J. A., and Runkel, A. C., 1991, *Prolapsus*, a large sciuravid rodent and new eomyids from the late Eocene of Trans-Pecos, Texas, *Pearce-Sellards Ser. Texas. Mem. Mus.* **48:**1–30.

Wilson, J. A., and Westgate, J. W., 1991, A lophodont rodent from the middle Eocene of the Gulf Coastal Plain, Texas, *J. Vertebr. Paleontol.* **11:**257–260.

Wilson, R. W., 1934, Two rodents and a lagomorph from the Sespe of the Las Posas Hills, California, *Carnegie Inst. Washington Publ.* **453:**11–17.

Wilson, R. W., 1935a, Cricetine-like rodents from the Sespe Eocene of California, *Proc. Natl. Acad. Sci. USA* **21:**26–32.

Wilson, R. W., 1935b, *Simimys*, a name to replace *Eumyops* Wilson, preoccupied.—A correction, *Proc. Natl. Acad. Sci. USA* **21:**179–180.

Wilson, R. W., 1935c, A new species of porcupine from the later Cenozoic of Idaho, *J. Mammal.* **16:**220–221.

Wilson, R. W., 1936, A Pliocene rodent fauna from Smith Valley, Nevada, *Carnegie Inst. Washington Publ.* **473:**17–34.

Wilson, R. W., 1937a, New Middle Pliocene rodent and lagomorph faunas from Oregon and California, *Carnegie Inst. Washington Publ.* **487:**1–19.

Wilson, R. W., 1937b, Two new Eocene rodents from the Green River basin, Wyoming, *Am. J. Sci.* **34:**447–456.

Wilson, R. W., 1938a, Review of some rodent genera from the Bridger Eocene, *Am. J. Sci.* **35:**123–137.

Wilson, R. W., 1938b, Review of some rodent genera from the Bridger Eocene. Part III, *Am. J. Sci.* **35:**297–304.

Wilson, R. W., 1938c, Review of some rodent genera from the Bridger Eocene. Part II, *Am. J. Sci.* **35:**207–222.

Wilson, R. W., 1940, Two new Eocene rodents from California, *Carnegie Inst. Washington Publ.* **514:**85–95.

Wilson, R. W., 1949a, Early Tertiary rodents of North America, *Carnegie Inst. Washington Publ.* **584:**67–164.

Wilson, R. W., 1949b, On some White River fossil rodents, *Carnegie Inst. Washington Publ.* **584:**27–50.

Wilson, R. W., 1949c, Rodents and lagomorphs of the Upper Sespe, *Carnegie Inst. Washington Publ.* **584:**51–65.

Wilson, R. W., 1949d, Additional Eocene rodent material from southern California, *Carnegie Inst. Washington Publ.* **584:**1–25.

Wilson, R. W., 1960, Early Miocene rodents and insectivores from northeastern Colorado, *Univ. Kans. Paleontol. Contrib. Vertebr.* **7:**1–92.

Wilson, R. W., 1986, The Paleogene record of rodents: Fact and interpretation, in: *Vertebrates, Phylogeny, and Philosophy* (K. M. Flanagan and J. A. Lillegraven, eds.), pp. 163–176, *Contrib. Geol. Univ. Wyo. Spec. Pap.* 3.

Winge, H., 1924, *Pattedyr-Slaegter*, Vol. 2, *Rodentia, Carnivora, Primates*, H. Hagerups Forlag, Copenhagen.

Wood, A. E., 1931, Phylogeny of the heteromyid rodents, *Am. Mus. Novit.* **501:**1–19.

Wood, A. E., 1935a, Evolution and relationships of the heteromyid rodents with new forms from the Tertiary of western North America, *Ann. Carnegie Mus.* **24:**73–262.

Wood, A. E., 1935b, Two new genera of cricetid rodents from the Miocene of western United States, *Am. Mus. Novit.* **789:**1–8.

Wood, A. E., 1936a, A new rodent from the Pliocene of Kansas, *J. Paleontol.* **10**:392–394.
Wood, A. E., 1936b, Geomyid rodents from the middle Tertiary, *Am. Mus. Novit.* **866**:1–31.
Wood, A. E., 1936c, A new subfamily of heteromyid rodents from the Miocene of western United States, *Am. J. Sci.* **31**:41–49.
Wood, A. E., 1936d, Two new rodents from the Miocene of Mongolia, *Am. Mus. Novit.* **865**:1–7.
Wood, A. E., 1936e, The cricetid rodents described by Leidy and Cope from the Tertiary of North America, *Am. Mus. Novit.* **822**:1–8.
Wood, A. E., 1937, The mammalian fauna of the White River Oligocene. Part II. Rodentia, *Trans. Am. Philos. Soc.* **28**:155–269.
Wood, A. E., 1939, Additional specimens of the heteromyid rodent *Heliscomys* from the Oligocene of Nebraska, *Am. J. Sci.* **237**:550–561.
Wood, A. E., 1942, Notes on the Paleocene lagomorph, *Eurymylus*, *Am. Mus. Novit.* **1162**:1–7.
Wood, A. E., 1949, A new Oligocene rodent genus from Patagonia, *Am. Mus. Novit.* **1435**:1–54.
Wood, A. E., 1950, A new geomyid rodent from the Miocene of Montana, *Ann. Carnegie Mus.* **31**:335–338.
Wood, A. E., 1955a, A revised classification of the rodents, *J. Mammal.* **36**:165–187.
Wood, A. E., 1955b, Rodents from the Lower Oligocene Yoder Formation of Wyoming, *J. Paleontol.* **29**:519–524.
Wood, A. E., 1958, Are there rodent suborders? *Syst. Zool.* **7**:169–173.
Wood, A. E., 1959, Eocene radiation and phylogeny of the rodents, *Evolution* **13**:354–361.
Wood, A. E., 1962, The early Tertiary rodents of the family Paramyidae, *Trans. Am. Philos. Soc.* **52**:1–261.
Wood, A. E., 1965, Grades and clades among rodents, *Evolution* **19**:115–130.
Wood, A. E., 1968, Early Cenozoic mammalian fauna Fayum Province, Egypt. Part II. The African Oligocene Rodentia, *Bull. Peabody Mus. Nat. Hist. Yale Univ.* **28**:23–105.
Wood, A. E., 1970, The European Eocene paramyid rodent, *Plesiarctomys*, *Verh. Naturforsch. Ges. Basel* **80**:237–278.
Wood, A. E., 1972, An Eocene hystricognathous rodent from Texas: Its significance in interpretations of continental drift, *Science* **175**:1250–1251.
Wood, A. E., 1973, Eocene rodents, Pruett Formation, southwest Texas; their pertinence to the origin of the South American Caviomorpha, *Pearce-Sellards Ser. Tex. Mem. Mus.* **20**:1–40.
Wood, A. E., 1974, Early Tertiary vertebrate faunas, Vieja Group, Trans-Pecos Texas: Rodentia, *Bull. Tex. Mem. Mus.* **21**:1–112.
Wood, A. E., 1975, The problem of hystricognathous rodents, *Univ. Mich. Pap. Paleontol.* **12**:75–80.
Wood, A. E., 1976a, The paramyid rodent *Ailuravus* from the middle and late Eocene of Europe, and its relationships. *Palaeovertebrata* **7**:117–149.
Wood, A. E., 1976b, The Oligocene rodents *Ischyromys* and *Titanotheriomys* and the content of the family Ischyromyidae, in: *Athlon: Essays on Paleontology in Honor of L. S. Russell* (C. S. Churcher, ed.), pp. 244–277, University of Toronto Press, Toronto.
Wood, A. E., 1977, The evolution of the rodent family Ctenodactylidae, *J. Palaeontol. Soc. India* **20**:120–137.
Wood, A. E., 1980, The Oligocene rodents of North America, *Trans. Am. Philos. Soc.* **70**:1–68.
Wood, A. E., 1981, The origin of the caviomorph rodents from a source in Middle America: A clue to the area of origin of the platyrhine primates, in: *Evolutionary Biology of the New World Monkeys and Continental Drift* (R. L. Ciochon and A. B. Chiarelli, eds.), pp. 79–91, Plenum Press, New York.
Wood, A. E., 1984, Hystricognathy in the North American Oligocene rodent *Cylindrodon* and the origin of the Caviomorpha, in: *Papers in Vertebrate Paleontology Honoring Robert Warren Wilson* (R. M. Mengel, ed.), pp. 151–160, *Carnegie Mus. Nat. Hist. Spec. Publ.* 9.
Wood, A. E., 1985, The relationships, origins and dispersal of the hystricognathous rodents, in: *Evolutionary Relationships among Rodents: A Multidisciplinary Analysis* (W. P. Luckett and J.-L. Hartenberger, eds.), pp. 475–513, Plenum Press, New York.

Wood, A. E., and Patterson, B., 1959, The rodents of the Deseadan Oligocene of Patagonia and the beginnings of South American rodent evolution, *Bull. Mus. Comp. Zool.* **120**:281–428.

Wood, H. E., II, 1945, Late Miocene beaver from southeastern Montana, *Am. Mus. Novit.* **1299**:1–6.

Wood, H. E., II, Chaney, R. W., Clark, J., Colbert, E. H., Jepsen, G. L., Reeside, J. B., Jr., and Stock, C., 1941, Nomenclature and correlation of the North American continental Tertiary, *Bull. Geol. Soc. Am.* **52**:1–48.

Woodburne, M. O., 1987, *Cenozoic Mammals of North America, Geochronology and Biostratigraphy*, University of California Press, Berkeley.

Zakrzewski, R. J., 1969, The rodents from the Hagerman local fauna, Upper Pliocene of Idaho, *Contrib. Mus. Paleontol. Univ. Mich.* **23**:1–36.

Zakrzewski, R. J., 1984, New arvicolines (Mammalia: Rodentia) from the Blancan of Kansas and Nebraska, in: *Contributions in Quaternary Vertebrate Paleontology: A Volume in Memorial to John E. Guilday* (H. H. Genoways and M. R. Dawson, eds.), pp. 200–217, Carnegie Mus. Nat. Hist. Spec. Publ. 8.

Zhai, R.-J., 1978, More fossil evidences favoring an early Eocene connection between Asia and Nearctic, *Mem. Inst. Vertebr. Paleontol. Paleoanthropol.* **13**:107–115.

Zheng, S., 1982, Middle Pliocene micromammals from the second locality in Songshan Commune, Tianzhu, Gansu Province, *Gu Jizhui Dongwe yu Gu Renlei* **20**:138–147.

Zittel, K. A. von, 1893, *Handbuch der Palaeontologie*, Sect. I, Palaeozoologie, Vol. IV, Vertebrata (Mammalia), R. Oldenbourg, Munich.

Taxonomic Index

Abelmoschomys, 228, 236
acares, Eutypomys, 133
accessor, Castor, 147
Acritoparamys, 43, 44, 46, 50, 53
adamsi, Geomys, 211
Adjidaumidae, 159
Adjidaumo, 157–159, 161
adspectans, Gidleumys, 256
aewoodi, Marfilomys, 262, 263
agnewi, Florentiamys, 196, 197
Agnotocastor, 137, 141, 142, 144, 145, 148
agrarius, Diprionomys, 178, 182, 183, 187
Aguafriamys, 162
Ailuravinae, 43, 44, 53, 94
Ailuravus, 94
Akmaiomys, 163, 167, 170, 171
Alagomyidae, 25
alberti, Paradjidaumo, 161
alecer, Griphomys, 258
alexandrae
 Liodontia, 98
 Mojavemys, 212
alicae, Leidymys, 235
Allomyinae, 89, 91, 93, 97
Allomys, 91, 97
Allophaiomys, 246
alticuspis, Paronychomys, 236
altidens, Sciuravus, 65
altifluminus, Mookomys, 183, 187
altilophus, Protadjidaumo, 161
Alwoodia, 97
ambiguus, Scottimus, 235
Amblycastor, 127, 129, 131, 132, 133, 144, 146
ambos, Franimys, 52
Ameniscomys, 91
americanus, Parallomys, 97
amherstensis, Franimys, 42, 52

Ammospermophilus, 116, 123
amnicolus, Schizodontomys, 187
Anagalida, 25, 27–29
Anchitheriomys, 127, 129, 131–133, 147
andersontau, Hitonkala, 197
angulatus, Mylagaulodon, 95, 98
angusticeps, Protospermophilus, 124
annae, Parapliosaccomys, 211
annectens, Campestrallomys, 97
Anomaluridae, 32, 283, 284
Anomalurus, 115
Anomaluromorpha, 32
Anomeomys, 70, 75, 76
Anonymus, 53
antiquus
 Peromyscus, 235
 Pliolemmus, 246
anzensis, Geomys, 211
Apatosciuravus, 50, 51, 53, 61, 62, 65
Apletotomeus, 163, 164, 167, 170, 171
Aplodontia, 85, 87, 89, 91, 94, 99, 278
Aplodontidae, 32, 33, 48, 85–98, 108, 122, 265, 272, 278, 281, 282
Aplodontinae, 98
Aplodontoidea, 33, 108
aquilonius, Baiomys, 237
arctios, Notoparamys, 52
Arctomyoides, 123
arctomyoides, Arctomyoides, 123
Ardynomys, 70, 76
argonatus, Spermophilus, 123
Arikareeomys, 162
arizonae
 Calomys, 236
 Marmota, 124
arizonensis, Repomys, 236
Armintomyidae, 33, 220, 252–253

Armintomys, 63, 219, 220, 252, 253, 272
Arvicolidae, 232, 243
Arvicolinae, 243, 244, 246
Arvicolini, 246
asiatica, Pseudaplodon, 91
atavus, Acritoparamys, 51, 53
ateles, Eutamias, 123
Atlantoxerus, 119
Atopamys, 244
atwateri, Acritoparamys, 53
Atypognathes, 31
Aulolithomys, 149, 151, 162
auogoleus, Litoyoderimys, 162
avawatzensis, Cupidinimus, 186

Baiomys, 237
ballensis, Mesogaulus, 109
ballovianus, Miosciurus, 123
barbarae, Cupressimus, 162
barbouri
    Paenemarmota, 124
    Pseudopalaeocastor, 147
baroni, Anonymus, 53
barstowensis, Copemys, 235
Barunlestes, 25
basilaris
    Entoptychus, 211
    Tenudomys, 204, 212
baskini, Calomys, 237
Bathyergidae, 284
Bathyergomorpha, 31, 32
Bathyergomorphi, 31, 32
baumgartneri, Peromyscus, 235
beckensis, Peromyscus, 236
bensoni
    Onychomys, 236
    Pappogeomys, 212
    Spermophilus, 123
Bensonomys, 236
bidahochiensis, Cupidinimus, 186
bifax, Apatosciuravus, 50, 53, 65
bisulcatus, Pipestoneomys, 265
blacki
    Ischyromys, 54
    Leidymys, 235
    Niglarodon, 97
bodei,
Mookomys (= Tenudomys), 183, 209
    Tenudomys, 212
boothi, Spermophilus, 123
borealis, Peridiomys, 187
boronensis, Cupidinimus, 186
boskeyi, Pareumys, 74, 76
bounites, Aulolithomys, 162

brachyceps, Fossorcastor, 148
brachygnathus, Baiomys, 237
brachyodus, Eumys, 234
breugerorum, Euhapsis, 147
brevidens, Perognathus, 186
bridgerensis, Leptotomus, 52
bridgeri, Sciuravus, 65
browni, Hystricops, 146, 148
brulanus, Pelycomys, 96
bryanti, Miospermophilus, 124
buccatus, Franimys, 52
buisi, Pliogeomys, 211
bumpi, Yoderimys, 162
burkei
    Mytonomys, 53
    Zemiodontomys, 162
butleri, Barunlestes, 25

californicus
    Castor, 147
    Pseudotomus, 52
    Uriscus, 48, 53
Calomys, 236
Campestrallomys, 97
campestre, Sciurion, 124
candelariae, Subsumus, 264
cankpoepi, Capitanka, 147
Capacikala, 145, 147
Capacikalini, 147
Capitanka, 145, 147
carpenteri, Perognathus, 186
carranzai, Pliogeomys, 211
cartomylos, Schaubeumys, 211
caryophilus, Leptotomus, 51
Castor, 7, 140, 142, 144, 145, 147, 276–278
Castoridae, 31, 33, 131, 132, 135–148, 265, 273, 277, 278, 281, 283
Castorinae, 144, 147
Castoroidea, 33, 48, 49, 145
Castoroidinae, 144, 145, 148
Castoroides, 135, 137, 141, 144, 145, 148, 277, 278
Castorimorpha, 31, 145
cavatus, Allomys, 97
cavifrons, Entoptychus, 210
Caviidae, 285
Caviomorpha, 31, 32, 48, 82, 277, 284
Cedromurinae, 111, 117, 124
Cedromus, 95, 119, 122, 124
cedrus, Orelladjidaumo, 162
celtiservator, Eodipodomys, 188
Centimanomys, 161
centralis, Prodipodomys, 187
Cephalomyidae, 284
cerasus, Leidymys, 235

Index

Ceratogaulus, 107–109
chadwicki, Downsimus, 97
cheradius, Microparamys, 53
chihuahuensis, Prosigmodon, 237
Chinchillidae, 284
citofluminis, Pseudocylindrodon, 76
clasoni, Fanimus, 196, 197
clivosus, Plesiosminthus, 219, 221
Cocomyidae, 24
coelumensis, Leptotomus, 52
coffeyi, Calomys, 237
Coloradoeumys, 234
coloradensis
    Agnotocastor, 148
    Mytonomys, 53
collinus, Cylindrodon, 76
Comancheomys, 159, 160
complexus
    Elymys, 222
    Monosaulax, 148
compressidens, Paramys, 51
condoni, Protosciurus, 123
copei, Paramys, 51
Copemys, 228, 231, 232, 233, 235, 243, 276
coquorum, Perognathus, 186
corrugatus, Thisbemys, 52
Cosomys, 244, 246
costilloi, Notoparamys, 52
cragini
    Peromyscus, 235
    Spermophilus, 123
craigi, Adjidaumo, 161
crassus, Apletotomeus, 171
Cratogeomys, 212
cremneus, Knightomys, 65
Cricetidae, 33, 96, 233–247, 251, 264, 273–275, 277, 281, 284, 285
cricetodontoides, Eumys, 234
Crucimys, 107, 109
Cryptopterus, 124
Cseria, 245
Ctenodactylidae, 32, 284
Ctenodactylomorpha, 33
ctenodactylops, Reithroparamys, 46, 53
Cupidinimus, 178, 179, 182, 186
Cupressimus, 162
curtus
    Gregorymys, 210
    Monosaulax, 148
cuspidatus
    Taxymys (= Knightomys), 64
    Knightomys, 65
cuyamensis
    Cupidinimus, 186

cuyamensis (cont.)
    Pseudotheridomys, 161
Cylindrodon, 68, 70, 74, 76
Cylindrodontidae, 18, 33, 49, 67–76, 95, 271, 273, 282
Cylindrodontinae, 70, 71, 74, 76, 83
Cynomys, 117, 124, 277

Daemonelix, 145
dakotensis
    Pleurolicus, 211
    Tenudomys, 212
Dasyproctidae, 284
dawsonae
    Campestrallomys, 97
    Janimus, 133
Dawsonomys, 49, 71, 73, 74, 76
debequensis
    Reithroparamys, 53
    Lophiparamys, 53
delicatior, Paramys, 51
delicatissimus, Reithroparamys, 46, 52, 93
delicatus, Paramys, 51
dentalis, Copemys, 235
depressus, Knightomys, 64, 65
Dicrostonychinae, 240, 243
Dikkomys, 184, 188, 204, 207
Dinomyidae, 285
Diplolophus, 220, 251, 256, 258, 264, 274
Dipodidae, 82, 219, 283
Dipodoidea, 18, 33, 219, 220, 251, 253
Dipodomyinae, 175, 179, 182, 185, 187
Dipodomys, 177, 187, 277, 278
Dipoides, 142, 144, 145, 148
Diprionomys, 178, 183, 187
disjunctus, Goniodontomys, 246
dividerus, Eucastor, 148
dixonensis, Pliophenacomys, 246
douglassi
    Gregorymys, 210
    Ischyromys, 54
    Leptodontomys, 161
    Mylagaulus, 109
    Oligospermophilus, 124
Downsimus, 97
dubius
    Kansasimys, 159–161
    Microparamys, 53
    Pliosaccomys, 211
dunklei, Perognathus, 186
Duplicidentata, 27–29

Ecclesimus, 171, 189, 191, 192, 193, 195, 197
Echimyidae, 284

editus, Meniscomys, 97
edwardsi, Yatkolamys, 235
elachistos, Thisbemys, 52
elachys, Calomys, 237
elassos, Mylagaulus, 109
elegans, Eumys, 3, 234
eliasi, Calomys, 236
ellicottae, Euhapsis, 147
Ellobiini, 245
Elymys, 63, 216, 217, 219, 220, 222, 272
Entoptychinae, 184, 204, 207, 208, 210
Entoptychus, 203, 204, 207, 208, 210
Eocardiidae, 284, 285
Eodipodomys, 177, 187
Eoeumys, 231, 234
Eogliravus, 122
Eohaplomys, 43, 46, 53, 94
Eomyidae, 32, 34, 63, 149–162, 272–274, 276, 281, 282
Eomyinae, 161
Eomyini, 154, 158, 161
Eomyoidea, 34
Eomys, 155, 157–159
Eozapus, 217, 219
Epeiromys, 95, 96, 98
Ephemeromys, 91, 93
Epigaulus, 104, 107–109
Epitheria, 27
Erethizon, 146
Erethizontidae, 32, 278, 284
Erethizontomorpha, 32
Ernotheria, 28
esmeraldensis, Copemys, 235
Eucastor, 142, 144–148
Eucricetodontinae, 231, 234
eucristadens, Sciuravus, 65
eugenei, Pseudotomus, 52
Euhapsini, 147
Euhapsis, 145, 147
Eumyinae, 231, 234, 274
Eumys, 3, 225, 231, 234
eurekensis, Cupidinimus, 186
Eurymylidae, 18, 24, 28
Eurymyloidea, 24, 25
Eurymylus, 24
Eutamias, 119, 123
Eutypomyidae, 31, 33, 48, 49, 125–134, 145, 271, 272, 274, 275
Eutypomys, 124, 129, 131–133
exallos, Pauromys, 58, 66
excavatus, Paramys, 51
exiguus, Scottimus, 235

falkenbachi, Sanctimus, 197

Fanimus, 193, 195–197
fantasmus, Namatomys, 162
fayae, Hibbardomys, 246
fedti, Proheteromys, 187
fiber, Castor, 142
fieldsi, Entoptychus, 211
finlayensis, Spermophilus, 123
finneyi, Pliophenacomys, 246
Florentiamys, 189, 191–193, 195–197
Florentiamyidae, 33, 63, 164, 184, 189–197, 275
Florentiamyinae, 196
Floresomys, 58, 59, 62, 65
floridanus, Proheteromys, 187
fluminis, Amblycastor, 129, 132, 133
fontis, Cylindrodon, 76
formicorum, Mookomys (= *Proheteromys*), 183
Proheteromys, 187
formosus, Gregorymys, 210
fossilis
    Ammospermophilus, 116, 123
    Neotoma, 236
    Onychomys, 236
fossor, Palaeocastor, 147
Fossorcastor, 147
Fouisseurs, 29
francesi, Acritoparamys, 53
Franimorpha, 32, 33, 48, 260
Franimys, 42, 44, 52
fraternus, Mysops, 76
fricki
    Mimomys, 246
    Rapamys, 52
    Spermophilus, 123
fugitivus, Namatomys, 162
furlongi
    Liodontia, 98
    Perognathus, 186

gaitania, Mytonomys, 53
galbreathi
    Centimanomys, 161
    Coloradoeumys, 234
galushai
    Agnotocastor, 148
    Reithrodontomys, 236
Galushamys, 236
garbanii, Geomys, 211
Gaudeamus, 267
gemacollis, Macrognathomys, 221
Geomorpha, 32, 33, 63, 220
Geomyidae, 32, 33, 63, 96, 164, 182, 184, 185, 196, 199–212, 256, 258, 266, 274, 277

Index

Geomyinae, 184, 185, 203, 204, 207, 211, 277
Geomyoidea, 31, 33, 167, 170, 220, 258, 262, 266
Geomys, 203, 207, 208, 211
Geringia, 235
geringensis (= clivosus), Plesiosminthus, 221
germannorum, Entoptychus, 211
Gidleumys, 256
gidleyi
    Calomys, 237
    Dipodomys, 187
    Onychomys, 236
    Perognathus, 186
    Spermophilus, 123
    Thomomys, 212
gigans, Quadratomus, 52
gladiofex, Megasminthus, 222
Glires, 25, 27–29
Gliridae, 32, 48, 49, 122, 220, 279, 284
gloveri, Geringia, 235
Goniodontomys, 241, 244, 246
gradatus, Capacikala, 147
grandiplanus, Entoptychus, 211
grandis, Quadratomus, 52
grangeri
    Pareumys, 76
    Plesiosminthus (= Schaubeumys), 221
    Schaubeumys, 221
greeni
    Fossorcastor, 148
    Schizodontomys, 187
gregoryi, Heliscomys, 171
Gregorymys, 204, 207–210
gremmelsi, Proheteromys, 187
griggsorum, Prodipodomys, 187
Grimpeurs, 29
Griphomys, 161, 168, 179, 256, 257, 258, 272, 273
grossus, Quadratomus, 52
gryci, Phenacomys, 246
guanajuatoensis, Floresomys, 65
Guanajuatomys, 83, 258, 260, 262, 263, 273
guensburgi, Pareumys, 76
guildayi, Leptotomus, 52
Guildayomys, 244, 246
gustelyi, Repomys, 236

hagermanensis, Peromyscus, 235
halli, Cupidinimus, 186
hansonorum, Paradjidaumo, 161
Haplomys, 97
harkseni
    Alwoodia, 97
    Schizodontomys, 187

Harrymyinae, 175, 182, 184, 185, 188
Harrymys, 182, 184, 185, 188
hatcheri
    Epigaulus, 109
    Heliscomys, 171
    Protoptychus, 83, 260
Heliscomyidae, 33, 63, 163–171, 179, 196, 273–275
Heliscomys, 163, 164, 167, 168, 170, 171, 192, 193, 195, 207
hellenicus, Ruscinomys, 234
hemingfordensis, Pleurolicus, 211
hendryi, Metadjidaumo, 162
henryredfieldi, Perognathus, 186
Heomys, 24, 28, 46
hesperus
    Monosaulax, 148
    Pseudotheridomys, 161
Heteromyidae, 32, 33, 63, 160, 164, 170, 173–188, 196, 208, 266, 274, 275, 278
Heteromyinae, 175, 182, 185, 187
Heteromys, 182, 203
hians, Pseudotomus, 52
hibbardi
    Cynomys, 124
    Dipodomys, 187
    Guanajuatomys, 260, 263
    Pliophenacomys, 246
    Microtoscoptes, 246
    Parapliosaccomys, 211
    Sigmodon, 237
Hibbardomys, 246
higginsensis, Pliosaccomys, 211
hippodus, Meniscomys, 97
Hitonkala, 189, 191, 193, 195–197
hollisteri, Onychomys, 236
Horatiomys, 96, 97
horribilis, Pseudotomus, 52
howelli, Spermophilus, 123
huerfanensis
    Knightomys, 58, 65
    Reithroparamys, 49, 53, 91, 119
hugeni, Ziamys, 211
hulberti, Nototamias, 123
Hulgana, 70, 71
hunterae, Microparamys, 53
Hydrochoeridae, 278
hypsodus, Paradjidaumo, 161
Hystricidae, 146, 281, 283, 284
Hystricognathi, 31, 32, 48
Hystricomorpha, 30, 31, 32
Hystricomorphi, 31
Hystricops, 146, 148
Hystrix, 146

idahoensis
   Ondatra, 247
   Procastoroides, 148
   Prodipodomys, 187
incohatus, *Akaiomys*, 170, 171
individens, *Entoptychus*, 211
inexpectatus, *Eutypomys*, 133
insolens, *Diplolophus*, 256
insolitus, *Paciculus*, 234
intermedius
   Adjidaumo, 161
   Dipoides, 148
   Phenacomys, 245
ironcloudi, *Proheteromys*, 187
irvini, *Harrymys*, 188
Ischymomys, 243
Ischyromyidae, 18, 33, 37–54, 82, 83, 91, 94, 119, 271–274, 279, 281
Ischyromyinae, 39, 43, 44, 48, 54, 74, 273
Ischyromyoidea, 33
Ischyromys, 3, 40, 44, 49, 50, 54, 274
Ivanantonia, 252

jacobi, *Geomys*, 211
jacobsi, *Apatosciuravus*, 50, 53
jamesi, *Petauristodon*, 124
Janimus, 49, 129, 133
Jaywilsonomyinae, 70, 71, 74, 76
Jaywilsonomys, 74, 76
jeffersoni
   Protosciurus, 111, 113, 115, 119, 121–123
   Sciurus (= *Protosciurus*), 95, 122
Jimomys, 260–262, 266, 276
johanniculi, *Pseudotomus*, 52
johnsoni, *Spermophilus*, 123
junctionis, *Prolapsus*, 65
junctus, *Ischyromys*, 54
junturensis, *Ammospermophilus*, 123

kalicola, *Mattimys*, 129, 133
kansasensis, *Peromyscus*, 235
Kansasimys, 149, 154, 157, 159, 161
kansensis, *Prodipodomys*, 187
karenae, *Microeutypomys*, 133
kayi, *Gregorymys*, 209, 210
kellamorum, *Scottimus*, 235
kelloggi, *Protospermophilus*, 124
kennethi, *Florentiamys*, 197
kinseyi
   Florentiamys, 197
   Mylagaulus, 109
Kirkomys, 189, 191, 195, 197
kleinfelderi, *Cupidinimus*, 186
Knightomys, 50, 55, 58, 59, 61–65

koerneri, *Niglarodon*, 97
kolbi, *Baiomys*, 237
Kubwaxerus, 119

laevis, *Mylagaulus*, 109
labaughi, *Jimomys*, 260–262
lacus, *Namatomys*, 162
Lagomorpha, 25, 27–29
landeri, *Simimys*, 252
landesi, *Mictomys*, 245
Laredomyidae, 34, 253, 254
Laredomys, 253, 254
larrabeei, *Onychomys*, 236
larsoni, *Gregorymys*, 210
lateriviae, *Pseudocylindrodon*, 76
lecontei, *Eucastor*, 148
Leidymys, 228, 232, 235
lemhiensis, *Trilaccogaulus*, 109
Lemminae, 240, 243, 245
lemredfieldi, *Paronychomys*, 236
Leptictidae, 24, 28
Leptictimorpha, 27
Leptodontomys, 153, 157, 161
leptodus, *Leptotomus*, 51
Leptotomus, 51
leucopetrica, *Neotoma*, 236
lewisi
   Anomeomys, 75, 76
   Pareumys (= *Anomeomys*), 75, 76
Lignimus, 184, 188, 209
lindsayi, *Cupidinimus*, 186
Liodontia, 98
liolophus, *Haplomys*, 97
Liomys, 182, 278
Litoyoderimys, 162
littoralis, *Pseudotomus*, 52
lloydi, *Namatomys*, 162
lockingtonianus, *Leidymys*, 235
lohiculus, *Prosciurus*, 91
loneyi, *Niglarodon*, 97
longidens, *Copemys*, 235
longiquus, *Scottimus*, 235
loomisi, *Florentiamys*, 191, 197
lophatus
   Mojavemys, 212
   Presbymys, 83
   Scottimus, 235
Lophiparamys, 53
loxodon, *Copemys*, 235
lucarius, *Taxymys*, 65
lulli
   Florentiamys (= *Jimomys*), 262
   Jimomys, 260
lustrorum, *Litoyoderimys*, 162

macdonaldi, Tenudomys, 212
macdonaldtau, Hitonkala, 197
Macrognathomys, 216, 219, 221
Macroscelida, 28
Macroscelidea, 27, 29
magheei, Spermophilus, 123
magilli, Mimomys, 246
magna, Alwoodia, 97
magniscopuli, Viejadjidaumo, 162
magnumarcus, Mojavemys, 212
magnus
    Anchitheriomys, 132, 133
    Capitanka, 147
    Eutypomys, 131, 132
    Oregonomys, 187
    Pliosaccomys, 211
    Proheteromys, 187
    Prosciurus, 96
major, Centimanomys, 161
maldei, Perognathus, 186
malheurensis
    Eucastor, 148
    Protospermophilus, 124
Manitsha, 42, 44, 52
Manitshini, 42, 44, 52
Marcheurs, 29
Marfilomys, 262, 263, 273
Marmota, 117, 124
Marmotini, 121, 123
marthae, Hibbardomys, 246
martini, Onychomys, 236
mascallensis, Prodipodomys, 185–187
matachicensis, Spermophilus, 123
matthewi
    Dikkomys, 188
    Proheteromys, 187
    Spermophilus, 123
matthewsi, Petauristodon, 124
Mattimys, 49, 126, 129, 131, 132
matutinus, Eohaplomys, 53
maximus
    Adjidaumo, 161
    Proheteromys, 187
maxumi, Repomys, 236
mcgregori, Geringia, 235
mcgrewi
    Heliscomys, 167
    Nebraskomys, 246
    Rudiomys, 97
mckayensis, Spermophilus, 123
mcknighti, Mimomys, 246
meadensis
    Calomys, 236
    Mimomys, 246

meadensis (cont.)
    Pliopotamys, 247
    Spermophilus, 123
medius
    Pseudocylindrodon, 76
    Sigmodon, 237
    Thisbemys, 52
Megasminthus, 215, 219, 222
Meliakrouniomys, 150, 160–162, 168, 258
meltoni, Spermophilus, 123
mengi, Protosciurus, 113, 123
Meniscomyinae, 91, 93, 97
Meniscomys, 97, 107
Mesogaulus, 95, 107–109
Metadjidaumo, 150, 160, 162
Microeutypomys, 133
Microparamyinae, 42, 49
Microparamyini, 42, 53
Microparamys, 43, 44, 49, 53, 62, 122, 129, 279
Microsyopidae, 28
Microtidae, 33, 232, 239, 243
Microtoscoptes, 241, 242, 244, 245
Microtus, 240, 278
Mictomys, 244, 245
milleri
    Crucimys, 109
    Kirkomys, 197
    Pareumys, 76
mimomiformis, Plioctomys, 245
Mimomys, 244, 245, 246
Mimotonida, 28
Mimotonidae, 24, 25, 28, 29
mimus, Promimomys, 246
minimus
    Adjidaumo, 161
    Baiomys, 237
    Diprionomys, 187
    Mysops, 76
    Petauristodon, 124
minor
    Dipodomys, 187
    Entoptychus, 210
    Epigaulus, 109
    Geomys, 211
    Knightomys, 65
    Marmota, 124
    Pliopotamys, 247
minutus
    Adjidaumo, 161
    Microparamys, 53
    Neotoma, 236
    Perognathus, 186
Miosciurus, 123
Miospermophilus, 124

mirus, Janimus, 133
Mixodectidae, 23
Mixodontia, 24, 28, 29
Mojavemys, 185, 186, 203, 204, 207, 208, 212
monahani, Mimomys, 246
monodon, Mylagaulus, 109
Monosaulax, 144–147
Montanamus, 153, 157
montanensis
  Entoptychus, 211
  Eutypomys, 133
  Trilaccogaulus, 109
montanus
  Horatiomys, 96, 97
  Paciculus, 234
montis, Lignimus, 188
Mookomys, 183, 187
Morosomys, 70
Multituberculata, 23
Muridae, 232, 243
murinus
  Lophiparamys, 53
  (= *simplex) Simimys,* 251
Muroidea, 33, 220, 251, 264
Mylagaulidae, 33, 70, 93, 94, 99–110, 131, 275, 276
Mylagaulinae, 109
Mylagaulodon, 95, 97
Mylagaulus, 101, 104, 105, 107, 109
Myocastoridae, 285
Myodonta, 33, 220
Myomorpha, 30–33, 49, 158, 170, 220, 251
Myomorphi, 31
Myophiomyidae, 252
Mysops, 71, 72, 74, 76
mytonensis, Mytonomys, 53
Mytonolagus, 76
Mytonomys, 43, 46, 53

Namatomyini, 154, 155, 158, 162
Namatomys, 63, 155, 157, 158, 162
nanus, Macrognathomys, 221
Napaeozapus, 214, 278
nebracensis
  Palaeocastor, 147
  Stenofiber (= *Palaeocastor),* 3
nebraskensis
  Cupidinimus, 186
  Cylindrodon, 76
  Paciculus, 234
  Proheteromys, 178, 187, 274
  Zetamys, 267
Nebraskomys, 244, 246
neglectus, Pseudocylindrodon, 76

nematodon, Leidymys, 235
Neotoma, 236
Neotomini, 236
Neotomodon, 234
Nerterogeomys, 211
nevadans, Tardontia, 98
nevadensis, Paenemarmota, 124
nexodens, Pseudallomys, 97
Niglarodon, 95–97
nimius, Microparamys, 53
nini, Paramys, 51
niobrarensis, Copemys, 235
nitidens, Allomys, 97
nitidus, Sciuravus, 65
Nonomyinae, 264
Nonomys, 220, 251, 256, 263, 264, 273, 274
nosher, Peromyscus, 236
Notoparmys, 52
Nototamias, 123
Nototrogomorpha, 31
novellus, Mesogaulus, 108, 109

obliquidens, Eutypomys, 133
occidentale, Tardontia, 98
occidentalis, Ardynomys, 76
Octodontidae, 284
Ogmodontomys, 244, 246
ohioensis, Castoroides, 148
ojinagaensis, Jaywilsonomys, 76
Oligospermophilus, 124
Omegodus, 158, 159
Ondatra, 7, 244, 247
Ondatrinae, 244, 247
Ondatrini, 247
Onychomys, 236
Ophiomys, 244, 246
oregonensis
  Leptodontomys, 161
  Parapliosaccomys, 209, 211
  Peridiomys, 187
  Protospermophilus, 124
Oregonomys, 183, 186
Orelladjidaumo, 162
Oropyctis, 97, 107
oroscoi, Prosigmodon, 237
Oryzomys, 237
osborni, Pliophenacomys, 246
ostranderi, Heliscomys, 167, 168, 170, 171
Otospermophilus, 116, 123
ovatus, Trilaccogaulus, 109

Paciculus, 234
paenebursarius, Geomys, 211
Paenemarmota, 113, 124

Index

pagei
  Copemys, 235
  Pseudotheridomys, 161
Palaeocastor, 145, 147
Palaeocastorinae, 140, 145, 147
Palaeocastorini, 147
Palaeomys, 144, 145
Palaeoryctidae, 28
panacaensis, *Repomys*, 236
paniensis, *Mesogaulus*, 109
pansus
  Monosaulax, 146, 148
  Stenofiber (= Monosaulax), 146
Pantolestidae, 28
Pappocricetodon, 225, 228
Pappogeomys, 212
Paracricetodon, 225
Paradjidaumo, 151, 153, 157, 158, 161
Paralactaga, 219
Parallomys, 97
Paramyidae, 43, 49
Paramyinae, 41, 42, 51
Paramyini, 51
Paramys, 42, 51, 58
Paranamatomys, 151, 160, 162
Paraneotoma, 236
Parapliosaccomys, 203, 207, 208, 209, 211
Parasminthus, 217, 219, 220
Pareumys, 70, 71, 74–76
parkeri
  Cseria (= Propliophenacomys), 245
  Propliophenacomys, 245, 246
Paronychomys, 236
parvidens
  Eumys, 234
  Ischyromys, 54
  Tillomys, 65
parvus
  Diplolophus (= Proheteromys), 256
  Diprionomys, 183, 187
  Eutypomys, 133
  Leidymys, 235
  Leptotomus, 51
  Mimomys, 246
  Mysops, 76
  Pliogeomys, 211
  Proheteromys, 187
  Prosciurus, 96
pattersoni
  Acritoparamys, 53
  Petauristodon, 124
  Pipestoneomys, 265
  Spermophilus, 123
pauli, *Protadjidaumo*, 161

Pauromys, 58, 63, 66, 219, 231
pearlettensis, *Perognathus*, 186
pebblespringsensis, *Oregonomys*, 187
Pedetidae, 32, 283, 284
pediasius, *Oropyctis*, 97
pedroensis, *Onychomys*, 236
Pelycomys, 96
peninsulatus, *Palaeocastor*, 147
Pentalophodonta, 31
perditus
  Pauromys, 66
  Thisbemys, 52
perfossus, *Microparamys*, 44, 53
Peridiomys, 185–187
Perognathinae, 175, 177, 179, 182, 186, 275
Perognathus, 183, 186, 277, 278
Peromyscini, 235
Peromyscus, 7, 227, 228, 232, 233, 235, 278
Petauristinae, 113, 117, 121, 124
Petauristodon, 124
petersonensis, *Niglarodon*, 97
petersoni, *Pseudotomus*, 52
Phenacomys, 245, 247
Phiomorpha, 32
Phiomyidae, 283
pintoensis, *Jaywilsonomys*, 76
Pipestoneomys, 264, 265, 273
pisinnus, *Copemys*, 235
placidus, *Pelycomys*, 96
Plagiodontia, 267
planidens, *Wilsoneumys*, 234
planifrons, *Entoptychus*, 210
planus, *Eucastor*, 148
Plattypitamys, 263
platyceps, *Euhapsis*, 147
Plesiadapis, 23, 24
Plesiarctomys, 39
Plesiosminthus, 216, 217, 219, 220, 221
Pleurolicinae, 208, 209
Pleurolicus, 204, 207–209, 211
pliacus, *Ischyromys*, 54
plicatus, *Thisbemys*, 52
pliocaenicus
  Allophaiomys, 246
  Oryzomys, 237
pliocenicus, *Peromyscus*, 232–235
Plioctomys, 244, 245
Pliogeomys, 203, 208, 210
Pliolemmus, 244, 246
Pliophenacomyini, 246
Pliophenacomys, 244–246
Pliopotamys, 244, 247
Pliosaccomys, 203, 207–209, 211
Pliotomodon, 234, 237

Pliozapus, 219, 222
Poamys, 233, 235
poaphgus, *Mimomys*, 246
popi, *Sciuravus*, 65
praecursor, *Mesogaulus*, 109
praetereadens, *Agnotocastor*, 148
pratincola, *Reithrodontomys*, 236
predontia, *Sewelleladon*, 97
Presbymys, 77, 80–83
primaevus, *Pliophenacomys*, 246
Primates, 23
primitivus
   Pliotomodon, 237
   Spermophilus, 123
primus, *Mimomys*, 246
pristinus
   Eumys, 234
   Mesogaulus, 109
Procastoroides, 141, 145, 148
Prodipodomys, 177, 185–187, 277
Progeomys, 208, 210
progressus
   Monosaulax, 148
   Niglarodon, 97
   Taxymys, 63, 65
Proheteromys, 171, 178, 179, 182–184, 187, 204
Prolapsus, 55, 56, 58, 59, 62, 64, 65
Promeniscomys, 93, 94
Prometheomyinae, 241, 244, 245
Promimomys, 241, 243, 246
Promylagaulinae, 108
Promylagaulus, 94, 101, 104, 105, 107, 108
Proneofiber, 244
Propalaeocastor, 141
Propliophenacomys, 241, 244, 244, 246
Prosciurinae, 49, 75, 89, 91, 93, 96
Prosciurus, 75, 85, 89, 91, 94, 96, 119, 122
Prosigmodon, 237
Protadjidaumo, 157, 158, 161
Protalactaga, 219, 282
Protoptychus, 77, 80–83, 260
Protoptychidae, 34, 77–84, 260, 272, 273
Protosciurus, 111, 113, 115, 119, 121–123
Protospermophilus, 124
Protrogomorpha, 30–32
proximus, *Mesogaulus*, 109
Pseudallomys, 97
Pseudaplodon, 91
Pseudictopidae, 25
Pseudocylindrodon, 71, 74, 76
Pseudopalaeocastor, 147
Pseudoparamyinae, 41, 44
Pseudoparamyini, 42, 52

Pseudotheridomys, 153, 155, 157, 158, 161
Pseudotomus, 42, 52
pycnus, *Paramys*, 51

Quadratomus, 52
quadratus, *Nototamias*, 123
quadriplicata, *Neotoma*, 236
quartus, *Cupidinimus*, 186
quartzi, *Leptodontomys*, 162
quatalensis, *Protospermophilus*, 124
quinni, *Geomys*, 211

rachelae, *Protosciurus*, 123
raineyi, *Aguafriamys*, 162
Rapamys, 52
rarus, *Sciuravus*, 62, 66
readingi, *Agnotocastor*, 148
redingtonensis, *Galushamys*, 236
reginensis, *Knightomys*, 65
Reithrodontomys, 236
Reithroparamyidae, 46, 93
Reithroparamyinae, 41, 42, 46, 48–50, 52, 93, 122, 129
Reithroparamyini, 42, 52
Reithroparamys, 37, 40, 43, 46, 48, 49, 52, 89, 91, 93, 119
relictus, *Prosciurus*, 96
Repomys, 233, 236
reticulatus, *Allomys*, 97
rexroadensis
   Dipoides, 148
   Nebraskomys, 246
   Perognathus, 186
   Reithrodontomys, 236
   Spermophilus, 123
   Zapus, 222
rexroadi, *Baiomys*, 237
reynoldsi, *Paradjidaumo*, 161
rhinocerus, *Ceratogaulus*, 109
rhinophilus, *Janimus*, 133
Rhizomyidae, 232, 283
Rhodanomys, 267
Rhombomylidae, 28
riggsi
   Gregorymys, 210
   Promylagaulus, 108
rinkeri
   Plioctomys, 245
   Zapus, 222
riograndensis
   Gregorymys, 210
   Laredomys, 254
ritchiei, *Texomys*, 266
rivicola, *Poamys*, 233, 235

robustus
  Mytonomys, 53
  Pseudotomus, 52
Rodentia, 27–29, 33, 251, 256
rogersi, Comancheomys, 160
Ronquillomys, 159
Rudiomys, 97
rufa, Aplodontia, 91, 278
rugosus, Pelycomys, 96
rupinimenthae, Trogomys, 186
Ruscinomys, 234
russelli
  Copemys, 235
  Leptodontomys, 161
rusticus, Peridiomys, 187

sabrae, Schaubeumys, 221
sambucus, Microparamys, 53
Sanctimus, 189, 191, 193, 195, 197
sandersi, Zapus, 222
sansimonensis, Pappogeomys, 212
sargenti, Oregonomys, 187
saskatchewanensis, Cupidinimus, 186
sawrockensis
  Baiomys, 237
  Mimomys, 246
  Neotoma, 236
  Paenemarmota, 124
saskatchewaensis, Ardynomys, 76
Schaubeumys, 216, 219–221
schaubi, Pauromys, 66
Schizodontomys, 185, 187, 208
schlaikjeri, Kirkomys, 197
Sciuravidae, 34, 55–66, 271, 272, 273
Sciuravus, 55, 59, 62–66, 142
Sciuridae, 33, 48, 94, 111–124, 273, 275, 277, 281, 284
Sciurinae, 115, 116, 121, 122
Sciurini, 113, 116, 117, 121, 123, 277
Sciurion, 124
sciuroides, Reithroparamys, 53
Sciuroidea, 33
Sciurognathi, 31, 32
Sciuromorpha, 30–33, 145, 146
Sciuromorphi, 31
Sciurus, 7, 95, 116, 122
scopaiodon, Microparamys, 62, 66
scotti, Spurimus, 96
Scottimus, 235
selbyi, Spurimus, 96
selenoides, Ameniscomys, 91
sellardsi, Pleurolicus, 211
senex
  Heliscomys, 171

senex (cont.)
  Tillomys, 65
senior, Knightomys, 65
senrudi, Monosaulax, 146, 148
serus, Eohaplomys, 53
Sespemys, 75, 94, 95
sesquipedalis, Mylagaulus, 109
Sewelleladon, 97
sheppardi, Entoptychus, 211
shotwelli
  Spermophilus, 116, 123
  Tregomys, 235
sibilatoris, Prolapsus, 65
Sicista, 216, 217, 219
Sigmodon, 237
Sigmodontinae, 235
Sigmodontini, 236
Simiacritomys, 216, 221
Simimyidae, 33, 220, 249–252, 272
Simimys, 219, 220, 249, 251, 264, 272
simonisi, Sanctimus, 197
simplex
  Pauromys, 66
  Simimys, 251, 252
simplicidens
  Allomys, 97
  Nanomys (= *Nonomys*), 264
  Nonomys, 264
  Palaeocastor, 147
  Symmetrodontomys, 237
Simplicidentata, 27–29
simpsoni
  Abelmoschomys, 236
  Paramys, 51
sioxensis, Campestrallomys, 97
skinneri
  Arikareeomys, 162
  Hibbardomys, 246
  Meliakrouniomys, 160, 162
Sminthozapus, 219
smithi
  Dipoides, 148
  Geomys, 211
  Protoptychus, 82, 83
solidus, Microparamys, 53
solus, Pliozapus, 222
Spalacidae, 232
spanios, Epeiromys, 95, 96, 98
Spermophilus, 116, 117, 123
Spurimus, 89, 96
Stenofiber, 3, 142, 144–146
stevei, Perognathus, 186
stewarti
  Texomys, 266

stewarti (cont.)
    Yoderimys, 162
stirtoni
    Allomys, 91, 97
    Calomys, 236
    Dipoides, 148
    Leptodontomys, 162
storeri, Paranamatomys, 162
Stratimus, 187
strobeli, Stratimus, 187
stuartae, Sanctimus, 197
subtilis
    Heliscomys, 171
    Mookomys (= *Heliscomys*), 183
Subsumus, 264
sulcatus, Progeomys, 211
sulcidens, Schizodontomys, 187
sulcifrons, Pleurolicus, 211
sulculus, Proheteromys, 187
sundelli, Quadratomus, 52
sweeti, Procastoroides, 148
Symmetrodontomys, 237
Synaptomyini, 244, 245
Synaptomys, 240, 244

Taeniodontidae, 28
Tamias, 7, 113
Tamiini, 123
Tamquammyidae, 24
tanka, Manitsha, 42, 44, 52
tapensis, Tapomys, 52
Tapomys, 48, 52
Tardontia, 98
taurus, Paramys, 51
Taxymys, 58, 59, 62–65, 179
taylori
    Mimomys, 246
    Neotoma, 236
tecuyensis, Protosciurus, 123
tedfordi, Ziamys, 211
Tenudomys, 183, 195, 204, 207–209, 212, 274
tenuiceps, Ecclesimus, 171, 197
tenuis, Copemys, 235
tephrus, Spermophilus, 123
tertius, Cupidinimus, 186
tessellatus, Allomys, 97
texanus, Pseudocylindrodon, 76
texensis, Pauromys, 66
Texomys, 262, 265, 266, 276
Theridomyidae, 18, 31, 44
Theridomyoimorpha, 31
Thisbemys, 52
Thomomys, 207–209, 212
thompsoni, Eutypomys, 131, 133

Thryonomyidae, 283, 284
thurstoni, Sespemys, 75
tiheni
    Megasminthus, 222
    Prodipodomys, 187
tilliei, Microeutypomys, 133
Tillomys, 58, 59, 62, 64, 65
timmys, Pseudotomus, 52
tiptoni, Florentiamys, 197
Titanotheriomys, 49, 50
tobeyi, Pseudocylindrodon, 76
tobinensis, Geomys, 211
toltecus, Griphomys, 258
tortus, Eucastor, 146–148
tradux, Eohaplomys, 53
transitionalis, Mimomys, 246
transitorius, Entoptychus, 211
transversus, Lignimus, 188
Tregomys, 235, 276
tricus, Microparamys, 53
Trilaccogaulus, 104, 107, 108
trilophus, Paradjidaumo, 161
Trogomys, 186
Trogontherium, 144, 145
trojectioansrum, Perognathus, 186
troxelli, Pareumys, 76
Tsaganomyidae, 75, 282
Tsaganomyinae, 70, 75
tuitus, Spermophilus, 123
tullbergi, Armintomys, 220, 253
tungurensis, Anchitheriomys, 129, 132
tuttlei, Paronychomys, 236
typicus, Monosaulax, 148
typus
    Ischyromys, 3, 54
    Protadjidaumo, 161

uhtoffi, Meniscomys, 97
uintensis, Thisbemys, 52
ultimus, Fanimus, 197
uphami, Petauristodon, 124
uptegrovensis, Propliophenacomys, 245
Uriscus, 48, 53, 119

valensis, Copemys, 233, 235
validus, Paradjidaumo, 161
vallicula, Dipoides, 148
vasquezi, Copemys, 233, 235
venustus
    Hystricops, 148
    Hystrix (= *Hystricops*), 146
veterior, Ischyromys, 54
vetus
    Cynomys, 124

vetus (cont.)
  Eoeumys, 234
  Heliscomys, 167, 168, 171
  Marmota, 124
  Mictomys, 245
  (= *simplex*) *Simimys*, 251
vetustus, Prosciurus, 96
viduus, Scottimus, 235
Viejadjidaumo, 162, 168
voorhiesi, Hibbardomys, 246
vortmani, Protospermophilus, 124

wardi, Cedromus, 95, 124
webbi, Cryptopterus, 124
wetmorei, Reithrodontomys, 236
wheelerensis, Entoptychus, 211
whistleri, Simiacritomys, 222
whitlocki, Cupidinimus, 186
wiedemanni, Anchitheriomys, 129
williamsi, Dipoides, 148
Wilsoneumys, 234
wilsoni
  Cedromus, 124
  Dipoides, 148
  Kansasimys, 161
  Meliakrouniomys, 160, 162
  Pliophenacomys, 246
  Pliosaccomys, 211
  Rapamys, 52
  Ronquillomys (= *Kansasimys*), 159, 160
  Sciuravus, 58, 65

wilsoni (cont.)
  Spermophilus, 123
Woodi
  Dawsonomys, 71, 76
  Harrymys, 188
  Heliscomys, 167, 171
  Lophiparamys, 53
  Paciculus, 234
wortamani, Mytonomys, 53
wyomingensis
  Acritoparamys, 53
  Miospermophilus, 124

xylodes, Orelladjidaumo, 162

yarmeri, Yoderimys, 162
Yatkolamys, 235
yazhi, Calomys, 236
yeariani, Niglarodon, 97
Yoderimyinae, 151, 154, 157, 162
Yoderimys, 162, 265
Yuomys, 83

Zalambdalestidae, 25
Zapodidae, 33, 213–222, 251, 272, 277, 278, 282
Zapodinae, 222
Zapus, 214, 216, 222, 227, 278
Zelomys, 59, 155, 282
Zemiodontomys, 153, 162
Zetamys, 266, 267, 275
Ziamys, 211